Fire Otherwise

Ethnobiology of Burning for a Changing World

Edited by
Cynthia T. Fowler and James R. Welch

THE UNIVERSITY OF UTAH PRESS
Salt Lake City

Copyright © 2018 by The University of Utah Press. All rights reserved.

 The Defiance House Man colophon is a registered trademark of the University of Utah Press. It is based on a four-foot-tall Ancient Puebloan pictograph (late PIII) near Glen Canyon, Utah.

Library of Congress Cataloging-in-Publication Data

Names: Fowler, Cynthia, 1970- editor. | Welch, James R., editor.
Title: Fire otherwise : ethnobiology of burning for a changing world / edited by Cynthia T. Fowler and James R. Welch.
Description: Salt Lake City : The University of Utah Press, [2018] | Includes bibliographical references and index. |
Identifiers: LCCN 2017056379 (print) | LCCN 2017058903 (ebook) | ISBN 9781607816157 () | ISBN 9781607816140 (pbk.)
Subjects: LCSH: Fire ecology. | Ethnobiology. | Prescribed burning.
Classification: LCC QH545.F5 (ebook) | LCC QH545.F5 F576 2018 (print) | DDC 577.2/4--dc23
LC record available at https://lccn.loc.gov/2017056379

Printed and bound in the United States of America.

Fire Otherwise

*The organizers dedicate this book to Nina Etkin and William Balée
for their steadfast support and excellent mentorship in
ecological anthropology and ethnobiology.*

Contents

List of Figures   ix

List of Tables   xi

List of Abbreviations   xiii

Acknowledgments   xv

1. Lifeways Enhancing Fire Ecology: An Introduction
   Cynthia T. Fowler and James R. Welch   1

2. Anthropogenic Fire History, Ecology, and Management in Fire-Prone Landscapes: An Intercontinental Review
   James R. Welch, Joyce K. LeCompte, Ramona J. Butz, Angela May Steward, and Jeremy Russell-Smith   22

3. Fire in the African Savanna: Identifying Challenges to Traditional Burning Practices in Tanzania and Malawi
   Ramona J. Butz   63

4. Fire Management in Brazilian Savanna Wetlands: New Insights from Traditional Swidden Cultivation Systems in the Jalapão Region (Tocantins)
   Ludivine Eloy, Silvia Laine Borges, Isabel B. Schmidt, and Ana Carolina Sena Barradas   82

5. Fire Use among Swidden Farmers in Central Amazonia: Reflections on Practice and Conservation Policies
   Angela May Steward   104

6. Restoration, Risk, and the (Non)Reintroduction of Coast Salish Fire Ecologies in Washington State
   Joyce K. LeCompte   128

7. The Critical Role of Firefighters' Place-Based Environmental Knowledge in Responding to Novel Fire Regimes in Hawai'i
   Lisa Gollin and Clay Trauernicht   151

8. Burning Lands: Fire and Livelihoods in the Navosa Hill Region, Fiji Islands
   Trevor King   183

CONTENTS

9. Assessing Causes and Effects of Survival Emissions from Global to Local Scales: Agropastoral Communities in the North Kodi Subdistrict of Sumba Island, Indonesia
    Cynthia T. Fowler   213

Contributors   232

Index   233

Figures

3.1. Map of northern Tanzania showing the location of Engikareti. 66

3.2. Map of Malawi showing the location of Nyika National Park and other protected areas. 68

3.3. Bracken fern on the Nyika Plateau. 71

3.4. Contraction of Miombo woodland in Nyika National Park. 72

3.5. Adjacent landscapes in Engikareti that were burned one and eight years prior. 73

4.1. Map of the Jalapão region of Brazil showing the location of the Serra Geral do Tocantins Ecological Station, other protected areas, and areas of deforestation. 86

4.2. Representation of landscape diversity in the Rio Novo Valley, Serra Geral do Tocantins Ecological Station. 90

4.3. Burning of a swidden plot in peat swamp forest (*vereda*). 91

4.4. The main ditch of a drained peat swamp swidden field (*roça de esgoto*) after burning. 91

4.5. The common crop-fallow rotation cycle in a peat swamp. 95

4.6. Map of two large fallows from peat swamp swidden cultivation in the Riachão Valley of the Serra Geral do Tocantins Ecological Station. 96

5.1. Map of the Mamirauá and Amanã Sustainable Development Reserves, Amazonas State, Brazil. 106

5.2. Study communities within the Mamirauá and Amanã Sustainable Development Reserves, Amazonas State, Brazil. 111

5.3. People practicing *coivara* in the community of Baré within the Amanã Sustainable Development Reserve, Brazil. 116

6.1. Big huckleberry (*Swədaʔχ*). 129

6.2. Study area in northwest Washington State. 133

7.1. Trends in the spatial extent of wildfire occurrence and agricultural production in Hawai'i over the past century. 153

7.2. An experimental fire in guinea grass illustrating high fire intensity. 155

7.3. Environmental cues used by firefighters in Hawai'i to determine fire danger and guide decision-making. 160

7.4. Illustration of the Maui wind system. 161

7.5. Wildfire burning through *uluhe* fern in native forest on O'ahu. 168

7.6. Fountain grass colonizing lava substrate. 170

7.7. Changes in Hawaii's environment identified by project participants as having an effect on fire occurrence and suppression 172

7.8. White ash rings from burnt *mamane* and *naio* trees during the 2010 Mauna Kea fire. 173

Figures

8.1. Fresh regrowth from a fire-resistant clump of invasive mission grass. 188

8.2. A senior research participant scoring reasons for burning using cards and heaps of *vesi* seeds. 191

8.3. A boy helping an older interviewee use a fire wheel to estimate the percentage of land affected by uncontrolled burning. 192

8.4. The reasons identified by Navosans for burning land and their relative importance. 193

8.5. A study participant using cards to identify so-called instigators and custodians in his community. 203

9.1. Emissions of methane and nitrous oxide (CO_2eq) from burning savannas and woody savannas in Indonesia, 1997–2014. 216

9.2. Emissions of CO_2eq from burning grasslands on organic soils worldwide, 1997–2014. 217

9.3. Percentages of emissions of CO_2eq in Indonesia by land cover, 1997–2014. 217

9.4. Emissions of CO_2eq from agricultural activities by region, 1997–2014. 217

9.5. Emissions of CO_2eq from all agricultural activities in Indonesia, 1997–2014. 218

9.6. CO_2eq emissions from all agricultural activities ranked by country, 1997–2014. 218

9.7. Land area planted in 2014 with the five most widely planted crops in Southwest Sumba Regency. 225

9.8. Emissions of CO_2eq from burning crop residues in Indonesia, 1997–2014. 225

9.9. Emissions of CO_2eq from burning crop residues in world regions, 1997–2014. 226

Tables

1.1. Changes in fire environments discussed in this volume.   6

1.2. Inventory of positive and negative impacts of anthropogenic fires discussed in this volume.   9

4.1. Cultivated plants associated with different field types.   93

4.2. Vegetation cover in drained peat swamp swidden field fallows, Serra Geral do Tocantins Ecological Station.   95

4.3. Locations of drained peat swamp swidden fields in relation to the mouths and sources of rivers.   96

5.1. Production calendar showing *roça* management activities in the *terra firme*, Amazonas State, Brazil.   113

5.2. Production calendar showing *roça* management activities in the *várzea*, Mamirauá Sustainable Development Reserve, Amazonas State, Brazil.   114

6.1. Northwest Forest Plan land allocations.   137

7.1. Novel fuel types and corresponding fire behavior and characteristics described by project participants.   169

9.1. Regional mean carbon emissions from 1997 to 2013 for savanna and grassland fires ranked from most to least carbon emissions.   215

9.2. Detrimental and beneficial effects of North Kodi's anthropogenic fire regimes.   226

9.3. Implications of climate change for human health in Indonesia.   227

Abbreviations

APP: Permanent Preservation Areas (Áreas de Preservação Permanente), Brazil

BioREC: Biodiversity Conservation and Sustainable Use in Conservation Areas (Conservação e Uso Sustentável da Biodiversidade em Unidades de Conservação), Mamirauá, Brazil

CIFOR: Center for International Forestry Research

DOFAW: Division of Forestry and Wildlife, Department of Land and Natural Resources, Honolulu

DOI: U.S. Department of the Interior

EMBRAPA: Brazilian Agricultural Research Corporation (Empresa Brasileira de Pesquisa Agropecuária)

ENGOV: Environmental Governance in Latin America (Gobernanza Ambiental en América Latina y el Caribe)

ENSO: El Niño Southern Oscillation

EPA: Environmental Protection Agency, United States

FAO: Food and Agricultural Organization of the United Nations

FHA: fire hazard area, Fiji

FRCC: Fire Regime/Condition Classes

GFED: Global Fire Emissions Database

GIZ: German Corporation for International Cooperation (Gesellschaft für Internationale Zusammenarbeit)

IBAMA: Brazilian Institute of Environment and Renewable Natural Resources (Instituto Brasileiro do Meio Ambiente e dos Recursos Naturais Renováveis)

IBGE: Brazilian Institute of Geography and Statistics (Instituto Brasileiro de Geografia e Estatística)

ICMBio: Chico Mendes Institute for Biodiversity Conservation (Instituto Chico Mendes de Conservação da Biodiversidade), Brazil

IDSM: Mamirauá Institute for Sustainable Development (Instituto de Desenvolvimento Sustentável Mamirauá), Brazil

IPCC: Intergovernmental Panel on Climate Change

KBDI: Keetch-Byram Drought Index

LSR: late successional reserve or managed late successional areas, USDA Forest Service

MBSNF: Mount Baker-Snoqualmie National Forest, Washington State

MDA: Ministry of Agrarian Development (Ministério do Desenvolvimento Agrário), Brazil

Abbreviations

MFARD: Ministry of Fijian Affairs and Rural Development

MMA: Ministry of the Environment (Ministério do Meio Ambiente), Brazil

NASF: National Association of State Foresters, United States

NEPA: National Environmental Policy Act, United States

NFPA: National Fire Protection Association

NGO: nongovernmental organization

NVT: Nyika Vwaza Trust, Malawi

PCCSP: Pacific Climate Change Science Program

PWD: Public Works Department, Fiji

REDD: a program aimed at reducing emissions from deforestation and forest degradation

REDD+: a program aimed at reducing emissions from deforestation and forest degradation in developing countries, and also examining the role of conservation, sustainable management of forests, and enhancement of forest carbon stocks in developing countries

SGTES: Serra Geral do Tocantins Ecological Station, Brazil

UNDHA: United Nations Department of Humanitarian Affairs

UNDP: United Nations Development Program

UNESCO: United Nations Educational, Scientific, and Cultural Organization

USDA: US Department of Agriculture

USFS: USDA Forest Service

WUI: wildland–urban interface

Acknowledgments

The editors and contributors extend our heartfelt gratitude to the communities who graciously welcomed us into their lives to conduct research and whom we came to think of as friends and family. We hope our readers will sense the immense respect we have for these communities: the Xavante people of Pimentel Barbosa and Etênhiritipá; Maasai pastoralists and the Phoka people in the savannas of Tanzania and Malawi; *quilombolas* in Jalapão (Tocantins); *ribeirinhos* in the Mamirauá and Amanã Sustainable Development Reserves; Muckleshoot, Tulalip, Nisqually, Umatilla, and Warm Springs tribal members in the US Northwest; firefighters in Hawai'i; Fijian smallholders in Navosa; and Kodi agropastoralists on Sumba Island.

The editors wish to thank our dear friends and colleagues in the Society of Ethnobiology, one of our main professional homes. We especially appreciate Steve Wolverton for hosting the Society of Ethnobiology's thirty-sixth annual meeting at the University of North Texas, where we organized the "Fire Ecology and Ethnobiology" session that is the historical root of this book. We acknowledge the remarkable patience of Reba Rauch, who attended that 2013 session and encouraged us to produce this edited volume for the University of Utah Press, where she is the acquiring editor for anthropology and archaeology. We also thank our copyeditor, Alexis Mills, and the production team at the University of Utah Press for their excellent work. Finally, we are grateful for the sage advice from Gene Anderson and Carol Colfer during the preparation of this volume.

CHAPTER 1

Lifeways Enhancing Fire Ecology

An Introduction

CYNTHIA T. FOWLER AND JAMES R. WELCH

AN ETHNOBIOLOGY OF PEOPLE AND FIRE

Fire Otherwise: Ethnobiology of Burning for a Changing World presents new evaluations of knowledge about human interactions with fires in contexts of changing environmental conditions. Human-fire relationships change through time and vary across space. Their dynamism emerges materially as transforming fire environments and reactively as shifting fire regimes. "Fire is a reaction" (Pyne 2016) in the chemical sense of involving the combustion of fuels and oxidants, and also in the evolutionary sense of involving the emergence of traits in response to species' relationships within biotic and abiotic milieu. Humans, fires, and ecological communities coevolve by adjusting their behaviors in response to one another and also in response to changing environmental stimuli. The contributors to this volume examine the interplay of socioecological changes with fire ecologies. They bring to light new connections between human activities related to fire environments (i.e., fuel properties) and variations in socioecological space-time.

Humans and fire are ecosystem engineers—independently and in tandem. In the socioecological communities where Indigenous, tribal, Aboriginal, traditional, and rural peoples live, they are keepers of knowledge about the histories of fire in their homelands, and they are recorders of the coevolution of humans and fire. In those places where traditional fire ecologies continue to function, scientific and traditional fire ecologies provide complementary insights, often based on different premises and different questions. In other locations, where the links between traditional ecological knowledge and Indigenous and other traditional land managers have degraded or been impeded, the science of fire ecology is our main source of knowledge about how fire shaped ecosystems in the past and continues to shape them in the present. In chapter 2, Welch and his colleagues review case studies from different continents to illustrate the value of involving diverse sci-

The title of this chapter paraphrases Benyus's (2014) statement about biomimicry being both art and science, and recalls her evocation of the "life enhancing lifeways" of Earth's 30 million nonhuman species to inspire more cooperative human lifeways.

entific and Indigenous approaches and voices in efforts to understand complex human-fire interrelationships in the past and present. Likewise, the authors of chapter 3 through chapter 9 illustrate how the combination of scientific and ethnobiological information enhances fire ecology studies of contemporary human-fire interactions. The authors of *Fire Otherwise* demonstrate the transdisciplinary character of fire ecology studies by integrating approaches from numerous branches of the biological and social sciences and by employing diverse techniques.

This book contributes new data and interpretations to the disciplines of fire ecology, ethnobiology, environmental change, and conservation. Original field data described and analyzed in the chapters that follow constitute a robust compendium of fire management techniques employed by people in diverse field sites worldwide. The following list summarizes the fire-related knowledge and management techniques collected by the authors from Maasai pastoralists, Phoka, and other users of Nyika National Forest (Butz, chapter 3); *quilombola* swiddeners (Brazilian descendants of African slaves) (Eloy et al., chapter 4); Amazonian smallholders (Steward, chapter 5); members of the Muckleshoot and Tulalip tribes and national forest personnel (LeCompte, chapter 6); incident commanders of wildland firefighter crews (Gollin and Trauernicht, chapter 7); rural Fijian smallholders (King, chapter 8); and Kodi agropastoralists (Fowler, chapter 9).

MODULATION OF FIRE FREQUENCY
- Maasai pastoralists burn plots every one to eights years.
- Nyika Vwaza Trust (NVT) managers aim to burn one-third of the plateau annually, which would amount to a three-year fire return interval, but other park users shorten the fire return intervals by burning more frequently.
- Swiddeners in Serra Geral do Tocantins Ecological Station (SGTES) cultivate well-managed peat swamp forest gardens continually for ten to twenty years after fallow, without the need for additional burning.
- SGTES swiddeners burn plots that are not in the peat swamps once a year in the early dry season.

CONTROL OF FIRE EXTENT
- Maasai burn off pasture grasses by firing them in smaller patches rather than larger spaces.
- Firefighters in Hawai'i douse flames with water using hoses and buckets hanging from helicopters.

EXCLUSION OR PROTECTION OF AREAS NOT APPROPRIATE FOR BURNING
- Maasai do not burn near settlements or in heavily forested areas.
- Kodi agropastoralists manage fuels and the fire environment to protect assets and to prevent fires from entering nontarget areas.

CLEARING OF FIREBREAKS AROUND BURN PLOTS
- *Quilombola* agropastoralists in Brazil's SGTES slash fuels around their peat swamp forest gardens several days prior to burning.
- Maasai pastoralists use riverbeds as natural firebreaks and dig trenches to make firebreaks.
- NVT managers build firebreaks around sensitive resources to protect them from controlled burns.
- NVT managers clear fuels from roads and tracks so that they function as firebreaks.
- Firefighters in Hawai'i build firebreaks and fire lines with bulldozers in 'a'ā lava fields and other land cover types.

Use of fire lines
- Maasai tamp down flames with green tree branches to mind fire lines and extinguish fires.
- Firefighters in Hawai'i create blacklines in lava fields to burn away fuels and to create fire lines.
- Firefighters in Hawai'i construct fire lines close to the edges of fires burning through *uluhe* thickets.

Use of backfires
- Maasai in Tanzania back-burn away from burn plot boundaries.
- *Ribeirinhos* (riverine people in Brazil) prefer backing fires.
- Some firefighters in Hawai'i witnessed backfires in sugarcane fields as children and use the technique in fire management operations.
- Firefighters in Hawai'i anchor backfires to fire lines.

Use of weather to achieve burn goals
- Maasai prefer to burn when they have lighter winds, cooler temperatures, and higher humidity.
- Ribeirinhos select for desirable wind conditions.
- Hawaiian fire managers monitor wind and relative humidity to plan prescribed burns.

Selection of burn season
- Maasai control the size and intensity of fires by burning in the early dry season.
- NVT managers burn in the early dry season (May–July).
- Kodi agropastoralists burn savannas and grasslands during the late dry season (August–September).
- Selection of best times to burn
- Firefighters in Hawai'i conduct prescribed burns at night.

Fuel management
- Maasai use their grazing livestock to maintain low fuel loads in pastures.

Organization of participants for controlled burn operations
- Ribeirinho communities in the Mamirauá and Amanã Sustainable Development Reserves mobilize small social groups to manage burns.
- Firefighters in Hawai'i determine where their agency can manage a fire on its own or whether they need to request aid from other agencies based on the fire's characteristics.

Protecting the safety of participants in burn operations
- Firefighters in Hawai'i send crew members ahead of bulldozers to spot dangerous lava tubes when they are working in *pāhoehoe* fields.
- Firefighters in Hawai'i sometimes use burned-over grasses as safety zones for firefighting operations, but they caution against using burned-over areas where the fuels have only partially burned.
- Firefighters in Hawai'i do not build firebreaks and fire lines with bulldozers in *pāhoehoe* lava fields.

The contributors to this volume apply original evidence collected from people who ignite, manage, and extinguish fires to the task of understanding complex factors shaping fire environments and fire regimes in locations where humans manage their landscapes with fire. Understanding local human–fire relationships is necessary for interpreting the current dynamics of specific ecosystems and the processes through which human-occupied landscapes developed. Ecosystem managers can more effectively incorporate fire into their conservation efforts through knowledge of the past and present roles of fire in their ecosystems. Robust and inclusive knowledge of the temporal and spatial character of fire within a particular setting enables land managers to design suitable fire regimes that also take into account local stakeholders' unique cultural understandings and value systems. Planning relevant fire regimes requires ecological managers to access the fullest story possible about how human–fire interactions vary through time and space, the implications of fire presence or absence, and how the outcomes of different frequencies, intensities, return intervals, and spatial patterns (the fire regime) coincide with local values regarding human well-being and ecosystem functioning. The authors of *Fire Otherwise* evaluate "human dimensions of fire regimes" (Roos et al. 2014) by drawing on social theories and ethnographic methods that are especially apropos for translating the fire-related ecological knowledge of Indigenous, traditional, and rural peoples. We amplify the voices of our interlocutors by disseminating knowledge in the forums where information about fire science and management circulate.

FIRE OTHERWISE

The term *otherwise* in the book's title references Restrepo and Escobar's (2005, 99–100) "other . . . otherwise" advocacy for a more inclusive anthropology where diverse epistemologies have equal standing with Euro-American ways of knowing. Faye Harrison (2016) explains that deploying the "otherwise" concept means "taking the subaltern, the indigenous, the indigen*ized* and the minorit*ized* seriously":

> To the extent that we truly recognize people's full humanity, that of course means we recognize their wisdom, their intelligence, their capacity to produce forms of knowledge that include potentially powerful interpretations and explanatory accounts of

the world, which give us the clues to then create strategies to change the world.

The authors of this book have written about Indigenous, local, and rural peoples' interactions with fire in ways that clearly call for decolonizing the discipline of fire ecology as powerfully as Restrepo and Escobar (2005) and others' directives for decolonizing anthropology (e.g., Harrison 2016; Manalansan 2016; Povinelli 2011, 2012; Weiss 2016). By attaching the term *fire* to the "other . . . otherwise" chisel, we bring this important anthropological movement to the discipline of fire ecology. The extraordinarily rich accounts of local fire regimes in this volume provide abundant evidence to support our argument that fire science and fire management ought to be inclusive, plural, dynamic knowledge systems. This concept of "fire otherwise" is a critical intervention that encourages the powerful networks of Western fire science and management institutions to open up into a "plural [intellectual] landscape" (Restrepo and Escobar 2005, 100).

The word *fire* in this book's title does additional work, too. In the course of referencing social theory, the title simultaneously evokes the National Fire Protection Association's impactful Firewise campaign in the United States. The NFPA is a historically and politically entrenched American trade institution, founded in 1896 by insurance underwriters, that aims to protect people and their property from harmful fire. NFPA (2016a) asserts it can achieve its aims by delivering "information and knowledge through more than 300 consensus codes and standards, research, training, education, outreach and advocacy." It implements these programs through partnerships with the US Department of Agriculture's National Forest Service (USFS), the Department of the Interior (DOI), and the National Association of State Foresters (NASF). The Firewise campaign promotes widespread adherence to NFPA's brand of ideas based on "scientific" information about the dangers of fire and formally recognizes communities that comply with its "codes and standards" (NFPA 2016b).

NFPA endorses a particular brand of "information and knowledge" born in American colonialism and developed through the twentieth and twenty-first centuries in the context of a situated fire history marked by the industrialization of natural resources, the suppression of fires in the nation's wildlands, and the subsequent "management" of fires (Fowler and Konopik 2007). In this sense, NFPA represents paradigmatic and overreaching fire agendas throughout the world that encourage adherence to sanctioned principles in the name of community participation. By failing to acknowledge local ecological knowledge and diverse citizens' skills related to fuel and fire management, such approaches discourage true inclusiveness and participation in developing ecologically effective and culturally informed fire solutions.

The 2017 fires in California, which destroyed thousands of buildings and killed dozens of people, are a poignant example that this omission has tangible consequences. In a comment prompted by these tragic events, Daniel M. Leavell (2017) wrote:

> It has always been easier and safer to suppress fires in responsibly managed forests, where ecosystem health, fuels reduction, wildlife habitat and overall diversity are the primary objectives. This is true today.... Managing fuels through responsible forest management reduces wildland fire risks, hazards, intensity and severity. It also improves overall forest health and wildlife habitat.

In this volume, we argue that traditional and local expertise is a valuable resource that should be brought to bear on managing fire landscapes in the United States and elsewhere in the world. We encourage NFPA, USFS, DOI, NASF, and similar science/management organizations around the globe to value alternative skills and knowledges under a big Fire Otherwise tent with room for plural knowledges in diverse landscapes (Welch et al. 2016).

This book promotes a shift toward the Fire Otherwise paradigm of pluralizing fire science and management by bringing to the table robust information about myriad anthropogenic fire regimes. We provide deeply descriptive accounts of local fire related knowledge and practices enacted by local communities with the goal of democratizing the paradigmatic version of fire science/management that arose through the globalization of fire science as exemplified by Firewise. Our project operates in two directions: infusing global fire science/management with nuance from local fire ecologies, and opening fire ecology to access and participation by local communities everywhere. We believe liberating and decolonizing fire ecology by bringing social theory to bear has enormous potential for promoting scientific fire management excellence and local human fire ecology integrity.

This book's authors rely on both local and scientific knowledge to construct analyses of links between fire ecologies and environmental changes. Their epistemologically pluralist approaches are similar to knowledge construction by Hawaiian firefighters that combines "native Hawaiian local ecological knowledge, ranching and plantation management practices, technical and analytical training and new developments in fire science" (Gollin and Trauernicht, chapter 7).

The authors of each chapter gathered information about local peoples' observations of and engagement with changes in fire regimes and fire environments with a variety of field research methods, including interviews, observations, vegetation surveys, and participation in development projects (Table

Table 1.1. Changes in fire environments discussed in this volume

Community or Location	Weather and Climate	Drought	Land Use, Land Cover	Fire Behavior	Fire Environment, Fuels, Fire Susceptibility	Development	Ecological Succession	Ignition Sources
Maasai land	√	√	√			√		√
Nyika National Park	√		√	√	√	√	√	√
SGTES, Tocantins, Brazil				√	√	√	√	
Mamirauá and Amanã reserves, Amazonas, Brazil		√			√	√	√	
MBSNF, Washington State, US					√			
Hawai'i, US			√	√	√	√	√	√
Navosa, Fiji	√	√	√	√	√		√	√
Kodi, Sumba, Indonesia	√	√	√	√	√	√	√	

1.1). Combining these approaches with review and analysis of literature from diverse academic traditions, we embrace multiple frames of reference and understandings of knowledge production. The local people whom authors interviewed and whose work they observed are firsthand, embedded eyewitnesses. They have very special "place-based understandings of local environmental conditions and fire behavior" (Gollin and Trauernicht, chapter 7) that offer many advantages to remotely gathered information about larger geographical scales, even when taken by highly sophisticated instruments (King, chapter 8; Fowler, chapter 9).

The contributors to this volume demonstrate that local and scientific knowledge are complementary by drawing on multiple epistemologies to more fully understand the connections between changing environmental conditions and changing fire ecologies. Our disciplinary backgrounds are diverse: agronomy, anthropology, archaeology, botany, ecology, ethnobiology, ethnoecology, fire ecology, geography, and public health. We share a theoretical and methodological commitment to deeply describing human-environment interactions through fieldwork in places where people interact with fire and shape fire environments. Considered as a whole, our multidisciplinary collection documents why and how human-influenced fire regimes vary around the world and through time.

Changing Fire Regimes

Fire regimes have always been changing and continue to change, as the chapters in this volume demonstrate. Changes in fire regimes are related to the social and ecological contexts within which fires are ignited and spread. In this book we explore how spatial and temporal variations in fire regimes are related to variations in land uses, ecosystem characteristics, fire environments, seasonal and other weather patterns, precipitation, population densities, individual life histories, political economies, symbolic ecologies, cultural politics, and additional social and ecological elements. Among the notable trends and repercussions of fire regime change are increases or decreases in fire frequency, severity, spread, behavior, and seasonality; fire causes and effects on vegetation cover and landscapes; emissions and climate change; fire locations and distributions; fuels and the fire environment; and fire-related policies and institutional management.

This book is an engaging compendium of variations in fire regimes across socioecological space and time. It constitutes a state-of-the-world report on changing fire regimes by focusing on several diverse biosocieties in Africa, Southeast Asia, Australia, Polynesia, South America, and North America. Readers will learn about how variations in fire regimes link to biosocial diversity and biosocial change.

All of the case studies presented in this volume are situated within the larger stories of the multiscale environmental changes occurring within broader planetary-scale contexts. Linking global-level changes to specific local-level fire ecologies is challenging (Fowler, chapter 9). One of the clearer linkages between scales is that local-level weather patterns and global-level climate are simultaneously changing. We see evidence of changing weather patterns in nearly all of the settings described in this book. All of the authors examine the effects environmental change has on local fire ecologies as well as how local fire regimes contribute to climate change. For example, Butz (chapter 3) links broad-scale climate change to local-level problems ranging from changing political economies, privatization and reallocation of resources, social conflict, water and food shortages, human well-being, and the health of ecosystems in African savannas. Fowler (chapter 9) attempts to assess the global warming impacts of anthropogenic fires by sizing up the frequent small-scale, fast burning grass fires in North Kodi's savannas and gardens. Stymied by the lack of local emissions data, Fowler reviews data on emissions generated from biomass burning at the national, regional, and global levels to argue that biomass burning is a significant source of carbon, carbon dioxide, methane, and other global warming gases. Lack of sufficient information, however, impedes the goal of evaluating the consequences of anthropogenic fire for environmental change and the effects of climate change on the livelihoods of subsistence-oriented communities. Moreover, predictions about how climate change will affect sea levels, temperatures, precipitation, seasonal phenomena, water and food availability, and human health are variable and inconsistent enough to hinder anticipating change, buffering its effects, and preparing to adapt to it.

The contributors to this volume explore how changes in fire regimes are linked to environmental changes, but they have unique approaches. Analytical tactics for exploring how fire regimes and environmental changes are linked range from examining decision-making (Gollin and Trauernicht, chapter 7) to discussing relational ecologies (LeCompte, chapter 6); from rewriting land management organizations' extension plans (Steward, chapter 5) to supporting decolonization efforts (LeCompte, chapter 6); from critiquing "reducing deforestation" rhetoric (Fowler, chapter 9; Steward, chapter 5) to voicing justice concerns (LeCompte, chapter 6); and from comparing epistemologies across societies to considering climate change predictions relative to local fire regimes (Fowler, chapter 9). Within this variety of scholarly approaches, the authors express a unified respect for place-based ecological knowledge and resource management practices grounded in local knowledge. The authors also share an interest in advocating for the integration of local and scientific knowledge into fire management and land management. As King argues in chapter 8, "a locally participatory, livelihood-nuanced, and climate-adaptive set of regulations" would be both just and smart.

The contributors to this volume thoughtfully weigh what they perceive to be the beneficial and detrimental effects of anthropogenic fire for both human well-being and ecosystems health. Their analyses demonstrate that fire can be beneficial in some circumstances and detrimental in others, and its effects can be mixed or neutral. Fire management is more sensible than unfettered fire use or control. Where it is an issue, we attempt to explain why and how anthropogenic fires sometimes become components of environmental mismanagement. Table 1.2 presents an inventory of the positively and negatively valued impacts of anthropogenic burning, and includes points that could support cases both for and against the argument that anthropogenic fires sustain human well-being and/or ecosystems health. One difficulty in making policies related to fire use is that what promotes human well-being does not always support ecosystems health. In other words, fires affect people and ecosystems differently. Additionally, stakeholders evaluate these outcomes as desirable or undesirable according to variable human values. The idea that we need to find new ways of integrating views about human well-being and ecosystems health points the way toward future research and publications that consider the possibility that what benefits people does not necessarily benefit nonhuman species.

Fire Ecology in Social and Evolutionary Theory

Fires have the potential to cause monumental changes in vegetative communities. The extent of vegetation changes caused by fires varies depending on fire type and landscape conditions. Illustrating this point, the landscapes of the Mount Baker–Snoqualmie National Forest that LeCompte discusses in chapter 6 are in Fire Regime/Condition Classes (FRCC) One and Two. FRCC is a classification system for natural fire regimes based on fire frequency and severity, especially in terms of the amount of overstory vegetation that is replaced in fires. Only the alpine zones of the Mount Baker glacier are classified as FRCC One, with frequency of lightning-ignited fires from zero to thirty-five years, and low to mixed severity with less than 75 percent of the vegetation replaced. The majority of the national forest is FRCC Two, defined by the occurrence of high-severity stand replacement fires every zero to thirty-five years, although fire return intervals can vary widely at smaller scales. Fire return intervals in upper elevation huckleberry meadows are 150–500 years. Stand replacement fires in huckleberry meadows are infrequent, but so highly severe that they kill most of the canopy layer. Thus, the fire regime in huckleberry meadows differs from the regimes in FRCC One and Two zones. According to Tulalip tribal member and retired Forest Service employee Russell Moses, fires "*completely* change the *whole* kind of microsite" (LeCompte, chapter 6, emphasis added).

In these Mount Baker–Snoqualmie landscapes, stand replacement fires occur in old growth forests as well as in shrublands and grasslands. They initiate succession, which includes fostering the growth of early succession species and fire-dependent or fire-resistant species. Stand replacement fires in the meadows where big huckleberry grows kill the tops of these shrubs, which then regrow from rhizomes or root crowns. Less severe fires may not kill the big huckleberry because their foliage has low flammability. A key role for fire in big huckleberry shrublands is to reduce competition from other plants, which is achievable by a variety of fire types. Those that constitute the Indigenous fire ecologies in the territory that is now Mount Baker–Snoqualmie National Forest range from lower severity fires to stand replacement fires when particular weather and moisture conditions facilitate the growth of vegeta-

TABLE 1.2. Inventory of positive and negative impacts of anthropogenic fires discussed in this volume

Location/Community	Positive Impacts	Negative Impacts
Savannas in Tanzania and Malawi	Historically, fire promoted ecosystem diversity, health, and function.	Decline in dominant range plant species; increase in invasive/alien plants; change in species composition due to changing fire regimes; changing fuel loads; decline of species diversity.
Peat swamp forests in SGTES, Brazil	Increases diversity; catalyzes succession; increases germination in some species that prefer nonflooded soils; increases spatial heterogeneity.	Changes biodiversity, vegetation, water regimes; more destructive wildfires in cerrado.
Mamirauá and Amanã Sustainable Development Reserves, Brazil	Efficiently clears agricultural sites; eliminates debris; enhances soil fertility; eases tuber harvesting; improves crop growth; improves manioc and banana crops; assists with weed control.	
Muckleshoot, Tulalip Nisqually, Umatilla, and Warm Springs tribal members, US	Encourages big huckleberry patches; prevents woody species invasions; cleans and invigorates; creates canopy gaps; releases nutrients; reduces disease and insect infestations; improves deer forage; eliminates old to make room for new; has metaphorical value; maintains reciprocal relationships with the environment; contributes to cultural identity, sense of self, and attachment to land/place; restores Indigenous fire ecologies, which can be considerd a form of social justice and decolonization.	
Mount Baker–Snoqualmie National Forest, Washington State, US	Contributes to long-term development and maintenance of huckleberry habitats; enhances human well-being via maintenance of huckleberry resources, which have ecological, social, and cultural benefits; prevents encroachment of woody species.	Endangers firefighters; can be so severe that everything, including huckleberry, is killed, making recovery difficult.
Hawai'i		Damages archaeological sites, religious worship sites, and economic assets; promotes invasive grass species; increases landscape flammability.
Navosa, Fiji	Clears gardens and tracks and pine stands; refreshes livestock forage; eases yam, turmeric, and guava fuelwood harvests; increases livestock containment capacity when confined to smaller patches; deters wild pigs; stimulates sprouting of *Solanum americanum*.	Promotes invasive grass species; damages gardens and fallows, timber, wild foods and medicines; eliminates materials used for roof thatch; reduces livestock containment capacity when burn areas are extensive; can force communities to deal with more extensive and destructive fires, such as pasture fires lit during droughts and intentional fires that escape control, especially during drought.

continued on next page

TABLE 1.2.—*continued*

Location/Community	Positive Impacts	Negative Impacts
Kodi, Sumba, Indonesia	Temporarily eliminates unwanted vegetation; enables crop production; promotes forage for livestock production; has cultural significance.	Opens growing spaces for non-native and/or invasive species; produces emissions of gases and particles; can cause injury and damage property, as with accidental and escaped fires; can increase vulnerability of an already marginalized Indigenous community.

tion. Lightning-ignited and anthropogenic fires have sculpted Mount Baker–Snoqualmie landscapes for thousands of years.

Studying the phenomenon of fire is one means for contributing to efforts in the field of evolutionary ecology to understand biological and anthropological drivers of environmental change and biosocial evolution. Fire embodies tensions between respecting the anthropogenic habitats of subsistence-oriented peoples and so-called natural ecosystems supposedly unaffected by human histories. Alternatively, we might see fire as caught in the crossfire between social scientists and natural scientists who argue over whether fire is socioecologically beneficial or detrimental. Land managers and policymakers similarly debate whether people ought to strategically use or explicitly prohibit landscape burning.

Reconstructing fire histories is an essential step in developing effective landscape management plans. Fire histories are used by restoration ecologists and conservationists to design future landscapes (Gonzalez 2005; Schoennagel et al. 2011; Vernon et al. 2008). Conventional fire ecology has reconstructed fire histories for specific locations with dendrochronological and paleoecological methods, among others. In contrast, Stephen Pyne (2009) has firmly established the value of applying historical social science methods to fire ecology. His approach exemplifies the value of using historical approaches to tell stories about the entanglements of human histories with fire histories (e.g., Pyne 1997) and with the history of planet earth (e.g., Pyne 2016).

Humans, according to Pyne's (2016) historical reconstructions, are "fire creatures" who function as a keystone species because we use fire to shape our habitats. Humans are not capable of producing "natural" fire[1] consistent with its historical precedence as an ecosystem engineer, but anthropogenic fires have shaped the course of environmental history since early humans discovered how to use it. Historically, anthropogenic fire regimes typically mimicked lightning's fire regimes or enhanced them by strategically increasing fire frequencies. Anthropogenic fire regimes—in particular, state-sponsored fire regimes—also have the effect of suppressing, managing, and/or controlling fires. As LeCompte (chapter 6) tells us, "Fire on Earth does go on with or without human assistance." Fires have been revolutionizing ecosystems and initiating succession longer than humans have and do not depend on humans to effect environmental change. Nevertheless, we humans sometimes impose our understandings of fire and ecologies on fire behavior in the process of constructing fire regimes.

Changing fire regimes are associated with ecological transitions that increase wildfire frequency, severity, and distribution. This trend has been documented in numerous sites around the world, including Malawi's Nyika National Park, where 90 percent of the area's grasslands burned in 2010; Hawaiʻi, where 8,000 ha burn every year; and on Viti Levu, where increased wildfires are associated with the El Niño Southern Oscillation (ENSO). In some settings, the proportion of human-ignited fires relative to lightning-ignited fires has been increasing. This trend is evident in Tanzanian and Malawian savannas (where lightning causes 30 percent of ignitions); in the Brazilian *cerrado*; in the Hawaiian Islands, though lightning fires occur at a very low rate of 1 percent of fires; and in the huckleberry habitats in Mount Baker–Snoqualmie National Forest, where lightning causes as much as 50 percent of ignitions.

As summarized in Table 1.2, the case studies in chapters 3 through 9 document how local-level fire environments are changing through the combination of human activities, weather, climate, and fire histories. Human activities influence a place's fire ecology by shaping the fuels composition, which includes species composition, distribution, and abundance (i.e., fuel loads). Humans alter vegetative communities in the landscapes where fires burn and therefore influence fire behavior. The ways people affect fire environments vary depending on the socioecological setting. Human actions and activities enable the conversion of nonforests into forests, as we see in some of this book's case studies. Burn bans in Tanzania's Longido Game Controlled Area and Malawi's Nyika National Park have led to "green desertification" (Butz, chapter 3) wherein grazing lands convert to shrublands and woodlands. Fire exclusion in Mount Baker–Snoqualmie National Forest has catalyzed the conversion of big huckleberry meadows into forests (LeCompte, chapter 6).

Some human actions and activities have led to the conversion of forests into nonforests, as illustrated in several case studies presented in this book. For example, agropastoralists in Brazil's Jalapão region converted peat swamp palm forests into gardens; swiddeners and loggers in the Middle Solimões deforested upland and floodplain forests; plantation workers and ranchers cleared Hawaiian forests in the late nineteenth and early twentieth centuries; alien grasses invaded burned forests in a similar way as non-native species took over former plantations, eucalyptus stands, and other ecotypes; and horticulturalists on Fiji have made gardens out of forests since the archipelago's early settlement. Non-native grasses and trees are changing fire regimes in Hawaiʻi by creating novel grass-dominated, flammable ecosystems where fire behavior can be unprecedented for the islands.

Human activities together with fire–grass cycles and weather and climate conditions cause some landscapes—such as North Kodi's coastal plain (Fowler, chapter 9)—to operate as dynamic grass climax communities, where ecological succession cycles repeatedly from cleared to grassy to cleared to grassy. The invasive pyrophytic grass *Pennisetum polystachyon* (L.) Schult. maintains a "pyric disclimax" (King, chapter 8) in Fiji's Navosa region, where native grasses have been excluded, and shrubs and trees have few opportunities to gain footholds. Fire–grass cycles in grasslands sometimes lead to the replacement of native species by non-native grasses (King, chapter 8). In some cases, "human activities have dramatically increased landscape 'flammability'" (Gollin and Trauernicht, chapter 7) by increasing the grassy components in vegetative communities. This has occurred in Hawaiʻi where abandoned plantations and ranches have converted to grasslands and have emerged as novel ecosystems with non-native species. These new post-plantation ecosystems in Hawaiʻi have transformed into wildlands and function like wicks because they form flammable corridors between the new wildlands and residential areas. The wildlands are encroaching upon urban and suburban residential areas rather than the other way around, which is more typical on the North American mainland, where homebuilders usually move out toward the flammable countryside. Both directions of demographic-ecological change impact fire environments and regimes.

Many of the fire regimes described in this book occur in novel ecosystems (Hobbs et al. 2013). Changes in the species composition of landscapes can both contribute to and result from fire regime shifts. Invasive plant–fire regime cycles (Brooks et al. 2004) lead to the development of novel ecosystems and novel fire regimes (Hobbs et al. 2013). The introduction and continued use of anthropogenic fire in ecosystems such as the native forests in Fiji's Sigatoka River valley and Hawaii's Honuaʻula and Hamakua areas change the fuels, "which can in turn affect fire behavior and, ultimately, alter fire regimes' characteristics such as fire frequency, intensity, extent, type, and seasonality of fire" (Brooks et al. 2004, 677).

The emergent pathways to novel ecosystems involve diverse land use practices that do not necessarily entail anthropogenic fire. Farming is one practice through which people deliberately or unintentionally make spaces for alien plants to move in, some of which are invasive. Others include animal husbandry and hunting, which can affect fuel properties by introducing grazers and reducing populations of wild or feral grazers. Converting nonpasture to pasture is a major cause of change in fuel properties (De Sy et al. 2015) with the potential to cause changes in fire regimes.

The contributors to this volume analyze recent changes in fire regimes with the understanding that they have historical precedents. Maasai pastoralists in Tanzania have been influencing rangelands for more than 150 years. In Brazil's Jalapão region, Indigenous hunter-gatherers and agroforesters have been shaping fire ecologies for 10,000 years, and Euro-Brazilian settlers began influencing fire ecologies indirectly via their impact on Indigenous peoples' lives, as well as directly through their own interactions with landscapes for the last 300 years. On the Salish Coast of North America, Native peoples have been changing fire regimes for 7,000 years via their interactions with berries and other plant and animal resources, and in the last 200 to 300 years Euro-Americans have been amplifying the human influence on fire ecologies. In Hawaiʻi, increased periodicity and distribution of wildfires, spread of grassy fuels, and growing frequency of ignitions have been changing fire regimes for nearly 200 years, since the beginning of the sugar plantation era.[2] The current transformation follows on the heels of changes that occurred after Europeans began settling the islands about 200 years ago, which followed the significant changes that occurred after Polynesians began settling the islands about 800 to a 1,000 years ago. In Fiji, the archaeological record shows early Lapita peoples converting forests to gardens beginning about 3000 BP.

Initial human settlement in different regions was accompanied by significant impact on fire regimes. Although disturbances are natural agents in ecological evolution, humans add distinctive registers to disturbance ecologies. The authors of chapters 3 through 9 address contemporary situations and emphasize significant recent shifts in disturbance ecologies. Nevertheless, their accounts reference other important historical moments involving major changes in disturbance ecologies. Three more recent starting points are evident in the account of Eloy et al. (chapter 4) about changing fire ecologies: establishment of burn ban policies associated with installation of the Serra Geral do Tocantins Ecological Station (SGTES) in 2001; intensification of logging, fires, and land conversion throughout the cerrado region in the 1980s; and quilombola migration into what is now the SGTES the 1960s and 1970s. Notable moments in Steward's story (chapter 5) about shifting fire regimes are the establishment of the Mamirauá and Amanã Sustainable Development Reserves; the founding of the Mamirauá Institute for Sustainable Development (IDSM) in 1999; the initiation of its agricultural ecosystems management program in 2010, and the establishment of the Amazon Fund; the development of the REDD+ Forest Conservation Allowance Program in 2007; and the beginning of Steward's employment with IDSM and the attendant (attempt toward) attitudinal change about swidden fires. In the setting for LeCompte's story (chapter 6), the timeline for changing fire ecologies has many notable moments: President Cleveland's establishment in 1893 of the 2.25 million acre Pacific Forest Reserve, where fire suppression policies were implemented; the 1974 Boldt Decision, which established precedent for state-tribal co-management of resources; the launching in 1994 of the Northwest Forest Plan, which reduced logging; and attention to the cultural value of big huckleberries at the 2007 Big Huckleberry Summit.

Indigenous peoples around the world face many challenges in maintaining the traditional fire regimes that enhance ecosystem health and in preventing shifts to alternative fire regimes that can be detrimental to ecosystem health and human well-being. Butz (chapter 3) mentions the following reasons for the detrimental shifts that are occurring in fire regimes in Tanzania and Malawi: fire suppression policies, food insecurity, unpredictable rainfall patterns and climate change, increasing population pressure, poaching, alternative agricultural practices, and an increasing number of catastrophic accidental fires.

Changes in fire regimes—especially increases in fire frequency—have been noticed by subsistence-oriented people elsewhere in the world. For example, the Crow people living in the temperate grasslands, savannas, and shrublands of their reservation in Montana's Little Bighorn River valley associate increased fire frequency with other environmental changes, including increased drought, decreased snowfall, lower streamflow, drier weather, warmer summers, and milder winters (Savo et al. 2016). The Crow perceive changes not only in fire and weather, but also in living species. They have noticed declining populations of freshwater mussels and frogs, relocating trout populations, and shifting plant phenologies. In another example, subsistence-oriented people in the Mediterranean forests, woodlands, and shrublands of Italy's Amalfi Coast are also noticing increases in fire frequencies at the same time as they observe warmer and drier weather, decreased precipitation with increased drought, changing seasonality, and shifts in plant locations and animal species' activities (Savo et al. 2016).

Managing fires in dynamic contexts such as these—with ever-changing conditions in weather, climate, and abiotic and biotic environments—is challenging for local residents and for the entities that purport to govern them. Savo et al. (2016, 462) suggest that "local observations can make an important contribution to understanding the pervasiveness of climate change on ecosystems and societies." Their assertion is similarly true for the complex relationships between local fire dynamics and climate change, as evidenced by the observations of changing fire regimes documented in the chapters 2 through 9.

Transformations in Fire-Related Knowledge and Management

Fire-related knowledge and management transform along with and in response to changes in socioecological contexts and fire regimes. Communities around the world are working proactively to adjust and adapt fire management systems amidst continually changing social and ecological contexts. These efforts are challenged by the diverse ecological, social, political, and economic forces impacting the functionality and sustainability of local fire-related practices. This book includes examples of ecological management practices deriving from traditional or locally produced knowledge that range from intact or well-functioning to impeded or dysfunctional. An example of the former situation is the Kodi community of Sumba in Eastern Indonesia (Fowler, chapter 9), where traditional fire ecologies are relatively adaptive, flexible, autonomous, and persistent. Examples of the latter are evident throughout the world, as illustrated in the regional reviews of traditional anthropogenic fire on four continents presented in chapter 2 (Welch et al.).

Many of the communities described in this book exemplify intermediate situations with local fire knowledge maintenance in the face of abundant socioecological challenges. The community living in Fiji's Navosa Valley (King, chapter 8) is an example. Most anthropogenic fires in Fiji's Navosa region are controlled burns, but the number of uncontrolled fires increases under climatic and political stresses. ENSO is one noticeable climatic stressor. When El Niño–induced droughts cause the pasture grasses and other fuels to become unusually dry, a different type of fire regime emerges. In normal years, Navosa's traditional resource management system functions well, but the Indigenous fire ecologies become somewhat problematic in abnormal years.

Another example of a community whose fire-related knowledge and resource management system occupy a middle ground is Hawaii's firefighter community (Gollin and Trauernicht, chapter 7). Combining elements from professional firefighting and from the place-based knowledge of its crew members and members of the communities where they work, firefighters in Hawai'i have innovated to protect human lives and property and to manage novel ecosystems. The traditional ecological knowledge and resource management system that existed in Hawai'i prior to colonialism crossed a threshold—perhaps several thresholds—at some point(s) in history to become what it is today. The system that we read about in chapter 7 is a novel one that emerged in communities of people with diverse heritages who collaboratively care for cultural and natural resources. Contemporary Hawaiian firefighting culture is adaptive and flexible, and includes some persistent elements.

A third middle ground example is the Brazilian descendants of African slaves (quilombolas) whose fire-related knowledge/management has evolved through the generations as members of the community have forcefully, coercively, and perhaps sometimes willingly migrated over long distances and into a series of enormously variable ecosystems. Eloy and her colleagues (chapter 4) explain that when the quilombolas migrated into the Jalapão region of Tocantins State, fire was one of the tools they used as they figured out how to make a living in an environment that, for them, was a novel territory. Social pressures forced the quilombolas to migrate multiple times over the course of hundreds of years—from Africa to the east coast of Brazil, and subsequently from eastern Brazil into the Jalapão region. They were able to survive along the way by applying a generalized swiddening toolkit in their new territories, which made it possible for them to produce their own subsistence and trade items. The origins of their landscape management system in the peatlands of SGTES and the larger cerrado biome are traced to Indigenous subsistence-oriented populations occupying the region since prehistoric times. Using diverse management techniques including crop diversification, distribution of gardens in multiple local ecosystems, and burning, quilombolas shape succession processes in ways that help them adapt to fluctuating social and environmental conditions. This quilombola swiddening system is similar in some respects to the agricultural traditions of ribeirinhos elsewhere in Brazil (Steward, chapter 5).

The quilombolas in the SGTES cultivate their peatland gardens within a difficult governance context. A multilayered nest of regulations usually prohibits burning in conservation areas such as SGTES in order to "protect" the cerrado against swiddeners, despite legislation guaranteeing quilombolas the right to preserve their traditional practices and customs. Within such a challenging political ecological setting, it is especially notable that quilombolas entrusted Eloy and her collaborators with such rich information about their swiddening practices, including burning techniques. Fortunately, some of the quilombola families with whom Eloy and colleagues conducted research benefit from "a pioneering agreement" with conservation area managers that allows some fire uses and research activities.

Collaborating to Deal with Changing Fire Ecologies

Strongly collaborative relationships between local peoples, land managers, and policymakers facilitate the exchange of knowledge and advance conservation goals. For example, Eloy and her colleagues (chapter 4) argue that knowledge exchanges could potentially serve two crucial goals in Brazil's Jalapão region: decolonizing fire science and fire management, and sustaining cerrado ecosystems, which range in their fire adaptations from fire-dependent grasslands to fire-sensitive peat swamps. These goals link together because answering questions about how people use fire, how ecosystems respond to fire, and how to manage ecosystems in relation to fire benefit from dialectical and iterative approaches.

Fire managers in Hawai'i are modeling what strongly collaborative relationships between local people and fire personnel could look like. There, a "burn boss" credits local fishermen for his ability to plan and execute a successful prescribed burn in a western region of Hawai'i Island (Gollin and Trauernicht, chapter 7). The fishermen shared with the burn boss what they knew about wind patterns based on their long-term experiences fishing and observing how winds affect fishing success on the Big Island coast. In turn, he trusted the fishermen's knowledge in the face of strong winds shortly before the prescribed start time. This persuasive story about the cross-fertilization of knowledge reveals how fishermen and other local people can inform fire management through horizontal collaborative and trustful interaction. Although the research literature about anthropogenic fire appears weighted toward issues of fire use among smallholder famers and, to a lesser extent, pastoralists, perhaps we ought to direct more attention to people who make their livings through different or diverse subsistence modalities. For example, people in Navosa (King, chapter 8) and Kodi (Fowler, chapter 9) make their living through combinations of farming, pastoralism, fishing, agroforestry, and other pursuits.

Gollin and Trauernicht (chapter 7) argue that firefighters know how to manage prescribed burns and wildland fires through intensive technical training in combination with personal experiences sensing, observing, and carefully thinking about environmental phenomena such as wind, humidity, smoke behavior, and topography in the islands' microclimates. At least some of the firefighters in Hawai'i also derive fire management knowledge and experience from genealogical ties within the areas they work. This case exemplifies how we might look for traditionalism in populations conventionally viewed as nontraditional.

Hawai'i is a unique place, as we learn from the incident commanders whose voices are heard in chapter 7. They tell us how their firefighting expertise differs from that of firefighters on the mainland and how the models and guides (e.g., FARSITE, BEHAVE, grassland curing guides) that dominate wildfire management efforts on the mainland do not work well in the Hawaiian Islands. For example, firefighters who participated in Gollin and Trauernicht's research point out the importance of spotting dangerous lava tubes in the *pāhoehoe* lava, which is rare on the US mainland. They explain how firefighter safety and effective fire control depend on knowing the locations of smoldering earth ovens (*imu*) and mesquite (*kiawe*) stumps. Also, they tell us of their obligations to protect ancient Hawaiian worship sites (*heiau*) as well as threatened and endangered species.

Although firefighting and fire ecology knowledge in Hawai'i may be institutionally and historically framed in terms of policies derived from mainland locations, it exhibits many similarities with the Pacific and Indo-Pacific island contexts also discussed in this volume. Hawai'i has no fire season per se, but hosts fires year-round, a typical pattern among tropical islands. Other islands, such as Viti Levu (King, chapter 8) and Sumba (Fowler, chapter 9) may have comparable weather patterns (e.g., in terms of wind, precipitation) and climate trends (e.g., links between ENSO and drought). Similar links may exist between fuels and windward/leeward polarities. Pacific and Indo-Pacific islands may also have comparable issues related to topography (e.g., mountains, wind tunnels in saddles) and geology (e.g., vol-

canoes, lava as land cover). Also, the social and ecological diversity of islands within the Polynesian region and across the Pacific and Indo-Pacific is extraordinarily high. In fact, the level of biodiversity may be too high for broader-scale models to capture. Modeling fire behavior at the Indo-Pacific regional level would be interesting, as would performing comparative analyses of fire ecologies that focus on the applicability of knowledges across modalities. These studies could seek answers to questions such as:

How do local communities integrate knowledge gained from engaging in fishing, pastoralism, horticulture, orcharding, forestry, and other pursuits? How do their fire management schemes benefit from interacting with grasslands, savannas, woodlands, coastal strands, forests, volcanic terrain, and other ecosystem types?

WHAT DO PEOPLE WHO ENGAGE IN SUCH DIVERSIFIED ACTIVITIES AS OCEAN FISHING, ANIMAL HUSBANDRY, FIELD CROPPING, AND AGROFORESTRY KNOW ABOUT:

- ideal conditions for conducting prescribed burns
- conditions likely to foster wildfires
- relative proportions of lightning-ignited to human-ignited fires
- relationships between fire behavior and fire environments
- temporal and spatial variations in fuels, winds, humidity, precipitation, and terrain
- temporal and spatial variations in fire behavior across microecosystems, and
- changes in fire regimes, fire environments, and the conditions that affect these over time?

The authors of this volume's chapters about smallholder communities point out that we know very little about their immensely complex and suppressed traditional fire regimes. Similar compensatory statements in other chapters (e.g., Welch et al., chapter 2; Eloy et al., chapter 4; Steward, chapter 5; LeCompte, chapter 6; and Fowler, chapter 9) point to a pattern of insufficiently robust and inclusive research in neo-/postcolonial settings about the projects, programs, and prohibitions that affect fire use and management by local peoples. This volume seeks to help balance the scale with reports about four regional fire regimes (chapter 2) and nine local fire regimes (chapters 3 through 9), thereby contributing to the expanding body of literature on human-environment interactions involving fire.

Research on local peoples' fire ecologies can help liberate subjugated fire-related knowledges borne by the "otherwise" (Povinelli 2012). The Fire Otherwise project seeks to reform persistent, narrow understandings of how fire ought to occur and who ought to manage it. Foremost, scholars and practitioners working within the field of fire ecology can use the information in this book to more explicitly promote respect for local knowledge and practices, as well as advocate for Indigenous, local, and rural peoples. Indigenous fire ecologies may be incorporated or reintroduced in appropriate settings to restore highly valued ecosystems (Wilder et al. 2016) such as big huckleberry patches and to more holistically remedy "cultural disturbances wrought by colonialism" (LeCompte, chapter 6). As King (chapter 8) argues for Fiji, neo-/postcolonial fire management systems should be socially just and "involve and empower local community leadership as an important adjunct to top-down official authority . . . power should be divested to community leaders . . . [and] should be recognized with appropriate official salaries."

Fire researchers who work in rural communities know that some burn incidents are at least partly acts of opposition against neo-/postcolonial governance (Fowler 2013). Burning is an everyday form of resistance (Scott 1987) against colonial rule, and a weapon used in conflicts related to identity politics within and between neighboring Indigenous communities, or between communities of different ethnic groups in neocolonial nations (Fowler 2013; King, chapter 8). For example, hunters—or "poachers," as they are frequently called—use fire in attempts to undermine managers and law enforcers from interfering with their hunting practices in Malawi's Nyika National Park (Butz, chapter 3). When hunters ignite these fires for subsistence purposes, they burn landscapes to drive game and make them easier to capture. But when they "burned wooden bridges . . . to restrict the movement of game scouts and antipoaching patrols" (Butz, chapter 3), their fires transcend subsistence. These are resistance fires whose purposes—in addition to facilitating the hunters' ability to obtain food—include subverting dominant social groups who claim ownership of local resources. Another example is in Fiji, where forest fires increased during the 1987 political coup (King, chapter 8). These cases illustrate that even subsistence fires often involve identity politics.

Pluralistic Fire Ecology

Researchers who study subsistence-oriented peoples' fire ecologies serve the decolonizing agenda by amplifying subaltern critiques of fire science, fire management, and land management. Eloy and her colleagues (chapter 4) take political action by writing about the Jalapão quilombolas' disagreement with the criminalization of their prescribed burns. When they do burn, it is not because they are ignorant of the laws, nor are they simply lawbreakers. Rather, by burning Jalapão, quilombolas are actively critiquing and partially invalidating the overarching narrative of internal colonialism. In some cases, burning is just one component among many in a relational ecology through which a community expresses its opposition to postcolonialism. The Coast Salish in LeCompte's case study (chapter 6), for example, fervently continue harvesting berries and other resources from their ceded territories as part of their identity work. Those berries grow in habitats partly formed and sustained by fire.

The Fire Otherwise project has emerged from the challenges to common knowledge about the fire management practices of Indigenous, traditional, and rural peoples that are explored in this volume. Research on the fire ecologies of local communities is a powerful tool for examining sociocultural processes related to disconnections between environmental science and climate politics in popular Western culture. Armed with deeply descriptive details about Indigenous fire ecologies, fire researchers are well positioned to critique the politics of blame that we find in both science and popular culture, and to desubjugate those who are blamed. The authors of *Fire Otherwise* effectively demonstrate that diverse "other" knowledges construct "wise" fire ecologies.

Who places and who receives blame for global warming? Which social groups benefit and which are harmed by blame games? Fowler (chapter 9) discusses the state of science on emissions and agriculture in Indonesia, where climate politics is intensely contentious. The blame game over environmental degradation, however, goes on elsewhere in the world, such as in the Navosa region where, according to King (chapter 8), resident smallholders seek to identify culprits for the destructive wildfires that occur more frequently during El Niño–induced droughts. Throughout the world, government and law enforcement agencies, nongovernmental organizations, and research teams are trying to identify the causes and the causers of anthropogenic fires. Meanwhile, blame for air pollution is passed between representatives of agribusiness interests and smallholder farmers and pastoralists. Local, national, and international media politicize and sensationalize anthropogenic fires during the dry season and under El Niño conditions. Unfortunately, many attempts to investigate, publicize, and remediate wildfire disasters exacerbate or reinforce existing social inequalities that cause smallholders to be impoverished, dispossessed, and disempowered. The case studies in this volume encourage nuanced dialogue about complex socioecological dimensions of local ecological knowledge and practices related to fire.

By documenting social dimensions of local-level fire regimes, we hope to encourage conservation scientists and policymakers to consider new possibilities for integrating local stakeholders' fire-related knowledge into scientific discussions about land management systems. Social theory and ethnographic data have the potential to contribute to the democratization of fire ecology and management. In knowledge "democracies," local ways of producing knowledge are held in equal regard as scientific, bureaucratic, and Western knowledges without the expectation that they reach consensus. Fire science and fire management systems in neo-/postcolonial nations typically are not knowledge democracies. In the history of colonized nations, where the colonizers ignored or eliminated local peoples' fire ecologies, the power of local knowledges has been undermined by the power of state-sanctioned knowledge. Remedying the negative effects of colonial processes requires affirming the value of rural peoples' fire ecologies while also recognizing that not all traditional knowledge is effective or appropriate in contemporary transformed landscapes. The process of collaboration and compromise leading to joint management has begun to occur in some places, such as the Mount Baker–Snoqualmie National Forest, where efforts to comanage treaty lands are "measures" of decolonization (LeCompte, chapter 6). The contributors to this volume are hopeful that Indigenous, local, and rural peoples elsewhere gain the autonomy to manage their own fire ecologies and have access to collaborations with scientists, managers, and policymakers based in respectful dialogue.

We envision this collection as impetus for the construction of a global Fire Otherwise effort by expanding the boundaries of fire science through engagement with diverse epistemologies and social theories. We envision a fire ecology that is inclusive of "more international voices and perspectives" (Restrepo and Escobar 2005, 119), and in which "power equity . . . [and] social justice" are included among the guiding principles (Armstrong and McAlvay 2016). Neo-/postcolonialists are "builders of exclusive and impenetrable silos" (Nabhan 2016), which is anathema to the Fire Otherwise project. Inclusive scholarship, suggests ethnobiologist Nabhan (2016), builds "bridges between different ways of knowing . . . [and] participates in a century-old tradition of valuing both 'indigenous knowledge' about the natural world and 'western' scientific knowledge." Decolonizing fire ecology requires fire scientists and managers to seek new kinds of relationships with local people that facilitate the exchange of knowledge and sharing of skills. Moreover, fire scientists and managers must hear what local residents desire for their own fire ecologies.

Like the richly diverse biomes that fires create, a liberated fire ecology is dense with environmental information and teeming with technical skills. The participants in non-imperialistic fire ecology are as socioculturally diverse as species' responses to fires' effects. A pluralistic fire ecology adjusts to social, ecological, economic, and political conditions just as fire regimes do. Similar to the biological constituents of coevolved fire ecologies, a transformative fire ecology could emerge from equitable relationships among subsistence-oriented communities, rural peoples, scientists, and managers.

Notes

1. Lightning-ignited fires are often referred to as being natural; likewise, fire regimes are called natural if the fires are ignited by lightning and not by humans.
2. The 200-year figure is derived from a statement found on the website of the Grove Farm Sugar Plantation Museum: "The first recorded planting of sugar cane in Hawaii for the purpose of extracting sugar was in Mānoa Valley on Oahu in 1825. The plantation failed two years later. The first successful sugar cane plantation was started in 1835 by Ladd and Company at Koloa, Kauai. . . . Sugar cane plantings increased rapidly from the first twenty hectares (fifty acres) on Kauai in 1835, to 40,400 hectares (100,000 acres) in 1900 and 89,000 hectares (220,000 acres) in 1980."

References Cited

Armstrong, Chelsey, and Alex McAlvay. 2016. "Decolonizing Ethnobiology." Accessed June 5, 2016. https://ethnobiology.org/decolonizing-ethnobiology-resources.

Benyus, Janine. 2014. "Biomimicry: Emulating Life's Genius and Grace." *Bioneers* 2003 Annual Conference video, 21:22. https://www.youtube.com/watch?v=WPLDzro1kQo&index=14&list=PLcrF8lYZY1451c_DTJWnRBRoc35-XwRdb.

Brooks, Matthew L., C.M. D'Antonio, D.M. Richardson, J.B. Grace, J.E. Keeley, J.M. DiTomaso, R.J. Hobbs, M. Pellant, and D. Pyke 2004. "Effects of Invasive Plants on Fire Regimes." *Bioscience* 54 (7): 677–688.

De Sy, V., M. Harold, F. Achard, R. Beucle, J. G. P. W. Clevers, E. Lundquist, and L. Verchot. 2015. "Land Use Patterns and Related Carbon Losses Following Deforestation in South America." *Environmental Research Letters* 10 (12).

Fowler, Cynthia. 2013. *Ignition Stories: Indigenous Fire Ecologies in the Indo-Australian Monsoon Zone*. Durham: Carolina Academic Press.

Fowler, Cynthia, and Evelyn Konopik. 2007. "The History of Fire in the Southern United States." *Human Ecology Review* 14 (2): 165–176.

Gonzalez, Mauro E. 2005. "Fire History Data as Reference Information in Ecological Restoration." *Dendrochronologia* 22 (3): 149–154.

Grove Sugar Farm Plantation Museum. Accessed June 4, 2016. https://grovefarm.org/kauai-history/.

Harrison, Faye. 2016. "Decolonizing Anthropology: A Conversation with Faye V. Harrison, Part I." *Savage Minds: Notes and Queries in Anthropology* blog. Accessed March 14, 2017. http://savageminds.org/2016/05/02/decolonizing-anthropology-a-conversation-with-faye-v-harrison-part-i/.

Hobbs, R.J., Eric Higgs, and Carol M. Hall. 2013. *Novel Ecosystems: Intervening in the New Ecological World Order*. Oxford: John Wiley and Sons.

Leavell, Daniel M. 2017. "Fuels management can be a big help in dealing with wildfires." Herald and News. Accessed October 22, 2017. https://www.heraldandnews.com/members/forum/guest_commentary/fuels-management-can-be-a-big-help-in-dealing-with/article_161e4ca5-f923-5f79-b710-db0b657f1f84.html.

Manalansan, Martin. 2016. "Queer Anthropology: An Introduction." *Cultural Anthropology* 31 (4): 595–597.

Nabhan, Gary. 2016. "Ethnobiology as the Unifying Theory of All Things Biocultural." Accessed March 14, 2017. http://www.garynabhan.com/news/2016/03/ethnobiology-as-the-unifying-theory-of-all-things-biocultural/.

NFPA (National Fire Protection Association). 2016a. "About Firewise." Accessed May 30, 2016. http://www.firewise.org/about.aspx.

———. 2016b. "Firewise Communities/USA Recognition Program." Accessed May 30, 2016. http://www.firewise.org/usa-recognition-program.aspx?sso=0.

Povinelli, Elizabeth A. 2011. "Routes/Worlds." *e-flux* 27. Accessed March 14, 2017. http://www.e-flux.com/journal/27/67991/routes-worlds/.

———. 2012. "The Will to Be Otherwise / The Effort of Endurance." *South Atlantic Quarterly* 111 (3): 453–475.

Pyne, Stephen J. 1997. *World Fire: The Culture of Fire on Earth*. Seattle: University of Washington Press.

———. 2009. *Voice and Vision: A Guide to Writing History and Other Serious Nonfiction*. Cambridge, MA: Harvard University Press.

———. 2016. "Welcome to the Pyrocene." *Slate* (May 16). Accessed March 14, 2017. http://www.slate.com/articles/technology/future_tense/2016/05/the_fort_mcmurray_fire_climate_change_and_our_fossil_fuel_powered_society.html.

Restrepo, Eduardo, and Arturo Escobar. 2005. "'Other Anthropologies and Anthropology Otherwise': Steps to a World Anthropologies Framework." *Critique of Anthropology* 25 (2): 99–129.

Roos, Christopher, David M. J. S. Bowman, Jennifer K. Balch, Paulo Artaxo, William J. Bond, Mark Cochrane, Carla M. D'Antonio, Ruth DeFries, Michelle Mack, Fay H. Johnston, Meg A. Krawchuk, Christian A. Kull, Max A. Moritz, Stephen Pyne, Andrew C. Scott, Thomas W. Swetnam, and Robert Whittaker. 2014. "Pyrogeography, Historical Ecology and the Human Dimensions of Fire Regimes." *Journal of Biogeography* 41 (4): 833–836.

Savo, Valentina, Dana Lepofsky, Jordan P. Benner, Karen Kohfeld, Joseph Bailey, and Ken Lertzman. 2016. "Observations of Climate Change Among Subsistence-Oriented People

Around the World." *Nature Climate Change* 6: 462–474.

Schoennagel, Tania, Rosemary L. Sherriff, and Thomas T. Veblen. 2011. "Fire History and Tree Recruitment in the Colorado Front Range Upper Montane Zone: Implications for Forest Restoration." *Ecological Applications* 21(6): 2210–2222.

Scott, James. 1987. *Weapons of the Weak: Everyday Forms of Peasant Resistance*. New Haven: Yale University Press.

Vernon, C. Bleich, Heather E. Johnson, Stephen A. Holl, Lora Konde, Stephen G. Torres, and Paul R. Kraussman. 2008. "Fire History in a Chaparral Ecosystem: Implications for Conservation of a Native Ungulate." *Rangeland Ecology and Management* 61 (6): 571–579.

Weiss, Margot. 2016. "Always After: Desiring Queer Studies, Desiring Anthropology." *Cultural Anthropology* 31 (4): 627–638.

Welch, James R., John M. Marston, and Elizabeth A. Olson. 2016. "Plurality in Ethnobiology: A Look Towards 2017." *Ethnobiology Letters* 7(1): 106.

Wilder, Benjamin T., Carolyn O'Meara, Laurie Monti, and Gary Paul Nabhan. 2016. "The Importance of Indigenous Knowledge in Curbing the Loss of Language and Biodiversity." *Bioscience* 66 (6): 499–509.

CHAPTER 2

Anthropogenic Fire History, Ecology, and Management in Fire-Prone Landscapes

An Intercontinental Review

JAMES R. WELCH, JOYCE K. LECOMPTE, RAMONA J. BUTZ,
ANGELA MAY STEWARD, AND JEREMY RUSSELL-SMITH

INTERDISCIPLINARY INVESTIGATION OF COMPLEX INTERRELATIONSHIPS ACROSS SCALES

Conservation science and politics deal uncomfortably with fire as a human ecological challenge at the center of debates about global trends in land conversion, climate change, and human health. Although abundant evidence shows that wildfires and biomass burning are central to debates about the sustainability of global change, consensus regarding their specific causes, effects, and policy implications is elusive. Among the complexities presented by fire within conservation sciences are inherent difficulties in parsing natural and anthropogenic fire dynamics, reconciling multiple temporal and spatial scales, and ascertaining cause and effect interrelationships. Challenges also derive from multiple ideological frameworks bearing on how fire ecology questions and conservation priorities are framed, such as epistemological differences between knowledge production in the biological and social sciences. Further difficulties arise as scientists, policymakers, and local stakeholders assign values to human-influenced fire outcomes based on diverse notions of human well-being, healthy ecosystems, and appropriate governance systems. As a result, debates about potentially negative or positive outcomes of anthropogenic fire in contemporary landscapes are prone to talking past each other at cross-purposes (Park and Kleinschmit 2016; Perrot 2015; Roos et al. 2016; Scott et al. 2016).

A swell of recent literature has attempted to address these problems, among others, by promoting interdisciplinary approaches that attend to all temporal and geographical scales and contemplate diverse human interests based in different epistemological and ideological systems. Much of this literature is associated with two interrelated emergent research frameworks. Whereas pyrogeography, based in biogeography, proposes using a more robust set of tools and models to explain the full complexity of human ecological fire dynamics (Bowman et al. 2013; Krawchuk et al. 2009; O'Connor et al. 2011; Roos et al. 2014), recent calls for increased collaborative and integrative research about the Anthropocene propose a forward-thinking analytical framework that embraces multiple scientific and conservation points of view (Bai et al. 2016; Bowman 2014; Brondizio et al. 2016; Lövbrand et al. 2015). These approaches are united in calling for renewed attention to some of the core principles that emerged through the "new ecology" and its cross-fertilization with the social sciences several decades ago, including holistic approaches, non-equilibrium, historical transformation, the importance of human actors, and engagement with conservation politics (Odum 1964, 1977; Scoones 1999; Zimmerer 1994).

Consistent with the problems and solutions proposed by these recent scientific frameworks, the study of anthropogenic fire and, more specifically, burning regimes involving traditional ecological knowledge is remarkably diverse disciplinarily and epistemologically. As a field of knowledge production, it is not just multidisciplinary but also tolerant of many points of reference and systems of understanding, as evidenced by its association with such inclusive fields as anthropology, conservation biology, ethnobiology, historical ecology, political ecology, and ecological economics. For example, Fowler and Welch (chapter 1) make a persuasive argument for equipping fire ecology with social science perspectives in order to expand the kinds of knowledge considered legitimate and thereby democratize the scientific process informing anthropogenic fire politics at different scales.

Traditional ecological knowledge regarding the use of burning and management of fire is not well accommodated by traditional biological sciences or contemporary environmental politics in some countries and regions because it is not produced according to paradigmatic scientific conventions. Indigenous and other traditional peoples may attribute legitimacy to fire ecology knowledge derived from oral histories, mythology, imitation, and other culturally informed methods of information production and transmission. They may evaluate its merits based on value systems involving such "unscientific" foundations as cosmology, kinship, and personal experience. Scholarship based in different epistemological premises often reduces these bodies of knowledge to simplistic caricatures, whether positive or negative. For example, a common but uncritical characterization of contemporary traditional ecological knowledge asserts that it is the culmination (or vestige) of experience accumulated over millennia (for discussions of reductionist understandings of traditional ecological knowledge, see Berkes 2012; Gómez-Baggethun et al. 2013; Reyes-García 2015). This view fails to adequately consider anthropological understanding of people past and present as active subjects contributing to knowledge transformation under changing circumstances. It also fails to recognize that Indigenous or local knowledge systems are not homogenous, but also diverse in their character and distribution within local societies. Negative caricatures of burning by Indigenous peoples as self-interested, anti-conservationist, nonrational, and obsolete are contested by a large body of recent transdisciplinary research into the motivations and knowledge systems of Indigenous and non-Indigenous stakeholders (Eriksen 2007; Kronik and Verner 2010; Sletto and Rodriguez 2013; Smith 2001; Welch et al. 2013; Zander et al. 2013).

A growing body of conservation literature supports the conclusion that the supposed opposition between traditional and scientific knowledge, often construed in terms of one being inferior to the other, is fallacious because their practical compatibility depends on dialogue rather than comparing their legitimacies. As recently argued by Brondízio and his colleagues (2016), environmental governance and management efforts benefit from conversations between diverse stakeholders with distinct knowledge bases and value systems, even when these diverge. Conservation narratives that suppress the geographical, cultural, and individual forms of diversity inherent to complex human-ecological phenomena ignore what Chimamanda Ngozi Adichie (2009) calls "the danger of a single story." Thus, the value in embracing multiple traditional and scientific ways of knowing includes expanding how conservation questions are posed and approached (Smith 1999). This is particularly true when fire is at issue because its perceived threats and benefits are so closely linked to human values or, as Sletto (2008, 1939) argues, "imaginaries" of nature and landscapes, whether or not these values and ideals are supported by evidence conventionally understood as empirical.

Recent interdisciplinary treatments of anthropogenic fire, burning traditions, and fire ecology illustrate how basic and applied research questions are asked and answered by scholars working with different disciplinary or theoretical frameworks to address phenomena in unique ecological, cultural, and historical settings (Fowler and Welch 2015; Scott et al. 2016). Among the factors affecting this diversity are distinct regional, national, and continental scholarly traditions and governance frameworks. This chapter aims to explore these dynamics based on four case studies of anthropogenic fire history, ecology, and management in fire-prone landscapes on different continents (Australia, Africa, South America, and North America). We reflect upon how these distinct knowledge production trajectories interface with global narratives about traditional fire use in order to promote conversations about the overall direction of research in this field. Despite acceleration of globally informed research on all continents, differences between regions highlight important barriers to the advancement of transdisciplinary and inclusive scholarship with the potential to effectively address complex phenomena and inform policy at different scales.

The four case studies addressed in this chapter are the savannas and, to a lesser extent, deserts of Australia; the savannas of Sub-Saharan Africa; the tropical savannas of South America, specifically the *cerrado* in Brazil and the Gran Sabana in Venezuela; and the temperate rainforests of the central Northwest Coast of North America. Each case study includes brief reviews of key literature about fire regime reconstruction, human occupation histories, traditional burning practices, and conservation efforts involving Indigenous peoples. These examples draw on diverse historical, ecological, and ethnographic bodies of literature to discuss fire in very different cultural and environmental contexts. The traditional peoples addressed have unique historical experiences, sociocultural realities, and modes of engaging with the environment. Their local landscapes—which include deserts, tropical savannas, and temperate rainforests—exhibit very different ecological characteristics. Their burning practices are maintained, transformed, and interrupted to varying degrees. Also distinct are the scholarly traditions within which the bodies of information presented for each case study were produced, each with its own historical frameworks and questions.

Despite these differences, each of the regional reviews that follow illustrates that contemporary Aboriginal, tribal, and Indigenous burning traditions have deep histories and derive from extensive landscape knowledge and value systems based in local understandings of healthy and productive ecosystems. Each exemplifies the continued relevance of colonial power relationships and the importance of decolonizing scientific knowledge production (Fowler and Welch, chapter 1). Despite their heterogeneous ecological contexts, each case study shows that conservation impacts of traditional subsistence burning are dwarfed by those associated with broader pressures such as demographic growth, economic development, and globalization (see also Fowler, chapter 9). Although there are examples of collaborative fire conservation efforts involving native stakeholders in each region, the following case studies reveal that understanding of traditional burning and fire management approaches is imbalanced and insufficiently incorporated into conservation policy and management practices.

The Savannas and Deserts of Australia

Initial human occupation of Australia has, until very recently, generally been considered to have occurred between approximately 47,000 to 55,000 BP by people island hopping through the eastern Indonesian archipelago (O'Connell and Allen 2015; Roberts et al. 1990). However, reassessment of archaeological deposits in northern Australia now proposes that occupation of the continent occurred before 65,000 BP (Clarkson et al. 2017). Current archaeological evidence and demographic models suggest that occupation of the continent occurred rapidly, albeit selectively, in favorable well-watered, resource-rich habitats, and that relatively low population numbers (measured in tens of thousands of people) fluctuated throughout the late Pleistocene until the early Holocene (Williams 2013). By the time of European colonization in the late eighteenth century, it is estimated that rapid expansion of the Aboriginal (Indigenous) population in the late Holocene had resulted in a continental population on the order of one million people (Williams 2013) spread across all habitat types, including Australia's dominant semi-arid and arid core (Smith 2013).

While it is generally acknowledged that Aboriginal use of fire as a landscape management and hunting tool doubtless was practiced throughout the period of occupation, the extent and impact of prehistoric Aboriginal burning remain highly contentious. Continental-scale reconstructions of fire activity based on charcoal sedimentary records imply that over the past few hundred thousand years, fire activity has generally been associated with progressive amplification of glacial-interglacial cyclicity—with reduced activity in drier glacial periods and increased activity under relatively wetter interglacial conditions with increased fuel loads (Lynch et al. 2007; Mooney et al. 2011). Within the period of human occupancy, the admittedly sparse and geographically biased charcoal record "provides little support for large-scale impacts of human activity on Australasian fire regimes" (Mooney et al. 2012, 18). Rather, lack of correlation between archaeological and charcoal records points to a much stronger association with external climatic forcing (Williams, Mooney et al. 2015). However, interpreting landscape-scale changes from the charcoal record (including soot, microcharcoal, and macrocharcoal fractions) remains highly challenging (Lynch et al. 2007; Singh et al. 1981; Williams, Mooney et al. 2015).

Conversely, people's burning and associated hunting activities have been implicated in the demise of much of Australia's unique megafauna (reptiles, birds, and marsupials over 40 kilograms) (see, for example, Flannery 1994; Johnson et al. 2016; Miller, Mangan et al. 2005; Miller, Fogel et al. 2005; Rule et al. 2012; Saltré et al. 2016). Key assumptions advanced by proponents of this ecosystem collapse hypothesis are that megafauna extinctions occurred mostly in close association with human arrival (with a peak extinction probability at approximately 41,000 BP [Saltré et al. 2016]), with "little evidence for exceptional climate change around [that] time" (Johnson et al. 2016, 4). Conflicting evidence, however, indicates that: (a) substantial megafaunal (and smaller faunal) extinction associated with longer-term processes of aridification and habitat change had occurred prior to the arrival of people (Price et al. 2011; Wroe et al. 2013); (b) extensive landscape drying occurred across the central Australian region from about 50,000 BP (Cohen et al. 2015); and (c) "climate change overwhelmed any modifications to fire regimes by Aboriginal landscape burning or megafauna extinctions," affecting the continental-wide distribution of a fire-sensitive conifer species complex (Sakaguchi et al. 2013, 1). Nevertheless, against a background of precarious climate variability it is plausible that hunting and burning activities undertaken by small, localized human populations could have had landscape-scale impacts on surviving megafauna (with characteristically low reproductive rates [Johnson 2006]) in remnant habitat refugia.

An important corollary of the above discussion is that Aboriginal burning practices, and the scales over which these were applied, doubtless have changed markedly through time. Paleorecords attest that pre-European patterns of occupation, technological development, population size, and mobility have changed markedly even within the late Holocene (McGowan et al. 2012; Mulvaney and Kamminga 1999; Smith 2013; Williams, Ulm et al. 2015). It is therefore likely that the intensive Aboriginal patch-mosaic burning practices evident at the time of European settlement had been developed only over the past few thousand years, and especially the last 1,500 (Bliege Bird et al. 2008; Williams 2013; Williams, Ulm et al. 2015).

Much of our modern understanding of the cultural and ecological significance of Aboriginal fire regimes owes considerable debt to Rhys Jones's 1969 seminal paper, "Firestick Farming." Based on practical experience with Aboriginal landscape fire management in northern Australia, and being an archaeologist by profession, Jones queried the then prevalent view that the Australian landscape at the time of European settlement could be simplistically considered natural; instead, he posited that Aboriginal people had substantially modified (farmed) the ecological landscape, particularly through the agency of fire.

Today, a considerable body of knowledge concerning the Aboriginal traditional (customary) use of fire exists, including: (a) records and analyses of sparse historical records, (b) documentation of traditional practices, and (c) knowledge, which continues to be applied into the present.

Various early Australian settlers, explorers, and artists observed Aboriginal peoples' use of fire but, with few exceptions, did not translate those incidental observations into a broader appreciation of Aboriginal fire regimes and landscape management. A notable exception was the explorer Thomas Mitchell (1848, 412), who famously noted that "fire, grass, kangaroos, and human inhabitants, seem all dependent on each other for existence in Australia." While some made detailed and now invaluable keen observations of burning and burnt vegetation (for example, the explorer Ludwig Leichhardt [1847]), it was not until the latter half of the twentieth century that critical assessments of those records, typically regional in scope, have been undertaken: for example, southwestern Australia (Abbott 2003; Hallam 1975); central Australia (Kimber 1983); northern Australia (Braithwaite 1991; Preece 2002); southeastern Australia (Benson and Redpath 1997); northeastern Australia (Crowley and Garnett 2000; Fensham 1997); and northwestern Australia (Vigilante 2001). While these accounts typically include at least seasonal observations of burning, those of Hallam (1975) and especially Abbott (2003) also address fire frequency, extent, patchiness, and intensity. Drawing on a range of historical and artistic sources, Bill Gammage (*The Biggest Estate on Earth: How Aborigines Made Australia* [2011]) provides a compelling account of intensive and extensive landscape fire and resource management as practiced especially in southeastern Australia around the time of European settlement. His thesis, with broader relevance to Australia as a whole, finds strong support from the late Holocene paleorecord (Williams, Ulm et al. 2015).

Intriguingly, drawing on the same historical sources, especially for southeastern and southwestern Australia, in *Dark Emu*, Bruce Pascoe (2014) finds strong evidence in support of the notion that rather than uniformly being hunters and gatherers, in various locations Aboriginal communities evidently resided in large settlements (of many hundreds of people) and practiced sophisticated forms of aquaculture, agriculture, and associated food storage. The farreaching implications (including political) concerning Aboriginal societal organization, land management, and ownership practices stemming from both Gammage's (2011) and Pascoe's (2014) theses have yet to be fully appreciated.

A substantial ethnographic literature is available, especially for remote and relatively culturally intact northern and central Australia, describing the purposes for and elements of traditional fire regimes. For northern Australia, sources and respective details have been summarized by Russell-Smith et al. (2003). Various accounts indicate that substantial areas (for example, as much as half of a clan estate, the homeland of a land-owning patrilineal family group) might be burnt in any one year (Haynes 1985; Jones 1975, 1980; Thomson 1949; Yibarbuk 1998), but through application of many small fires (Jones 1975; Yibarbuk 1998). For central Australia, pertinent references include works by Gould (1971), Kimber (1983), Latz (1995), Bird et al. (2005), Burrows et al. (2006), and Bliege Bird et al. (2008, 2012).

The Western Desert study presented by Burrows and his colleagues (2006) is singularly important because it describes the size distribution of patches burnt under traditional fire management by Pintubi people in the 1950s, prior to contact with non-Aboriginal society. From the aerial photographic record for a 240,000 hectare region, they found that 22 percent of the study area had been burnt by people within five to six years of the 1953 photography, with 75 percent of burnt patches being less than 32 hectares and 50 percent more than 5 hectares in extent. The authors contrasted this fine-grained management with contemporary "boom and bust" fire patterns: in 1981 four unplanned fires burnt 87 percent of the study area. Another useful, if qualitative, account of precontact patch burning is given in *The Last of the Nomads* (Peasley 1983).

Characterization of spatial and temporal dimensions of Aboriginal patch-mosaic burning remains a core challenge for contemporary biodiversity conservation purposes (Bowman 1998; Bradstock, Williams, and Gill 2012) and land management applications generally. Hence, in recent decades there has been growing acceptance in the wider community, but especially in very sparsely settled northern savannas and central desert rangelands, of the need for strategic implementation of prescribed burning to reduce risks associated with Australia's notorious bushfire (wildfire) problems. Ground-borne fires, mostly anthropogenic in origin, can occur annually in areas of the northern savannas, promoted by reliable monsoonal rainfall (generally over 600 millimeters per year) and associated development of cured grassy fuel loads in the following dry season (Russell-Smith et al. 2007; Whitehead et al. 2014). In more semiarid and arid central Australia, extensive fires, often ignited by lightning, typically occur following occasional regional heavy rainfall events and subsequent fuel buildup (Allan and Southgate 2002). Implementing strategic prescribed management in these regions—which have limited firerestricting natural features (few rivers, generally flat topography), built infrastructure (roads, fence lines), and resources (people, aircraft)—is an ongoing contemporary challenge, reinforcing the value of the fine-scale fire mosaic previously imposed under Aboriginal custodianship (Bliege Bird et al. 2008; Bowman 1998; Burrows et al. 2006; Garde et al. 2009; Russell-Smith et al. 2003; Yibarbuk et al. 2001). The catastrophic and ongoing loss of Australia's small mammal fauna is attributable, at least in substantial part, to the replacement of fine-scale mosaic burning with contemporary boom-and-bust fire patterns, and associated impacts from introduced predators such as cats and foxes given loss of protective habitat (Burrows et al. 2006; Lawes et al. 2015; Woinarski et al. 2011; Ziembicki et al. 2014).

Vestiges of traditional fire management practice continue in parts of Australia, but especially in the Arnhem Land region of northern Australia and central Australia's Western Desert. Useful accounts exist, but often in the form of relatively inaccessible reports and film materials. The most detailed published description of traditional fire management practice is provided by Murray Garde and elderly Aboriginal coauthors for the western Arnhem Plateau region in their eloquent essay "The Language of Fire" (Garde et al. 2009). The essay provides a nuanced description drawing on direct translations of motivations for and patterns of burning over the seasonal cycle. Related regional descriptions are given by Haynes (1985), Russell-Smith et al. (1997), Yibarbuk (1998), Yibarbuk et al. (2001), Bowman et al. (2001, 2004), and Altman (2009). As described in some of these accounts, traditional Aboriginal landscape fire management was (and in places continues to be) systematically planned and highly organized. Today, however, such practices are typically highly modified, reflecting contemporary patterns of settlement, mobility, and adoption of new technologies (e.g., aerial ignition, satellite fire monitoring, GIS mapping), and variable policy settings (see, for example, Bird et al. 2005; Preece 2007; Russell-Smith et al. 2013).

Aboriginal fire management approaches and activities are increasingly recognized and supported throughout Australia, including the formal conservation estate comprising both national parks (e.g., Kakadu National Park Board of Management 2016) and the national system of Indigenous Protected Areas (Department of the Environment and Energy 2016a). From the 1990s, such activities have been supported, both on the conservation estate and other lands which Aboriginal people either own or have interests in, through government-funded Aboriginal ranger programs (e.g., Altman and Kerins 2012; Department of the Environment and Energy 2016b). Landscape-scale Aboriginal fire management programs are concentrated particularly in central and northern Australia given both the extent of Aboriginal land and the magnitude of fire management issues in those sparsely populated regions (Russell-Smith and Whitehead 2015). Despite this, there is rapidly developing momentum, and encouragement on the part of state authorities, for Aboriginal people to exercise their customary fire management responsibilities in more densely settled southern Australian jurisdictions (e.g., Hunt 2012).

A particular impetus for empowering Aboriginal fire management, especially across the fire-prone savannas of northern Australia, has been the development since the early 2000s of government-sanctioned, market-based savanna burning emissions abatement methodologies. The initial methodology, formally accepted in 2012, was based on traditional fire management practice and developed in partnership with western Arnhem Land Aboriginal land managers (Russell-Smith et al. 2013). At the time of writing, eighty-six landscape-scale savanna burning projects are operating, most extensively on Aboriginal lands (Clean Energy Regulator 2016). With the likely addition of complementary carbon sequestration methodologies in the next few years, savanna burning projects have the potential to transform current regional, pastorally focused economies across northern Australia (Russell-Smith and Whitehead 2015). Despite the success of the savanna burning program to date, substantial regulatory and legal hurdles remain concerning the rights of Aboriginal people to burn on, and derive economic benefits from, their traditional lands (Preece 2007; Dore et al. 2014). While fire management and carbon-based economic opportunities are not readily applicable in southern temperate Australia (Bradstock et al. 2012), savanna burning provides a powerful example of the future potential for developing market-based ecosystem services through enhanced fire management.

Savannas of Sub-Saharan African

It is perhaps most difficult to distinguish between the effects of human and nonhuman fire regimes in Africa, where hominins have lived for millions of years (White et al. 2009) and are thought to have affected its fire regimes for at least thousands of years (Archibald et al. 2012). Evidence of early Pleistocene fire in Africa is very limited. Some of the earliest fire evidence associated with human habitation comes from what are thought to be hearths from the Lake Turkana area in northern Kenya. Oldowan localities at East Turkana date to over 1.8 million years ago, but are outnumbered by those of the somewhat later Karari complex on the Karari Ridge (approximately 1.6 million years ago) (Gowlett and Wrangham 2013; Harris and Isaac 1997). Additional, less ambiguous early evidence of human control of fire dating from approximately one million years ago in Africa was identified from Wonderwerk Cave in the Northern Cape Province of South Africa (Berna et al. 2012). Fire incidence in the Sub-Saharan region increased about 400,000 BP during transition from interglacial to glacial periods, but it is likely human burning significantly influenced overall fire regimes only during the Holocene (Bird and Cali 1998), especially during the last 4,000 years (Archibald et al. 2012).

One of the earliest references to the use of fire by people was in the "Periplus," an account by Hanna the Carthaginian of a voyage along the west coast of Africa in 600 BCE (Busse 1908). Vasco da Gama called the Cape of Good Hope "Terra de fume" on account of the smoke pall discernible from the sea in 1497 (Sim 1907). Early threats of punishment for fire use were proposed in the late 1600s by the Dutch East India Company in what is now South Africa (Phillips 2012). Also, in contemporary South Africa, the Forest and Herbage Act of 1959 stated that fines, imprisonment, or a combination of the two should be imposed for fire use (Phillips 2012). Kayll (1974) noted that in the latter part of the nineteenth century, almost every African country visited and described by explorers showed evidence of deliberate burning by local peoples. Many colonial accounts describe human use of fire as a destructive and careless practice responsible for converting forested landscapes into what are often described as "degraded" savannas (see Fairhead and Leach 1996). An example of such sentiment from West Africa is illustrated by Green (1991, 20): "As forest people, the Kissi are not as careless as savanna people with regard to fire."

Landscape fires from both natural and human ignitions are key factors in shaping contemporary African ecosystems (Bond et al. 2005; Bowman et al. 2009; Scholes and Archer 1997). There has been a long-standing debate about the origins and spread of tropical savannas (Jones 1980), given the observation that fire is a recognized potent driver for maintaining grasslands and savannas as an alternate state to forested systems (e.g., Thomas and Palmer 2007; Staver et al. 2011a, 2011b). However, there is a vast literature describing the development of savanna systems in Africa since the Miocene, well before the appearance of modern humans (Cerling et al. 2011; Dupont et al. 2013; Linder 2014; Senut et al. 2009). Evidence from African savannas indicates that human ignitions are more frequent than lightning ignitions, accounting for nearly 70 percent of annual fires (Grégoire et al. 2013; van Wilgen et al. 1990). Lightning is largely absent during the dry season in African savannas (Archibald et al. 2012; Saamak 2001; Shea et al. 1996). The degree to which human-ignited fire regimes differ from historical, lightning-driven regimes remains largely unresolved (Frost 1999; van Wilgen et al. 2008). Similarly, the specific dynamics surrounding anthropogenic burning in Africa and the savanna landscape remain largely understudied, especially in regard to the linkages among human burning practices, fire regimes, and savanna vegetation (Laris 2002; Sheuyange et al. 2005).

Fire in African savannas is part of a complex system of landscape-level management by local people (Butz 2009; Gil-Romera et al. 2010; Homewood and Brockington 1999; Kamau and Medley 2014; Laris et al. 2015; Mbow et al. 2000; Shaffer 2010). Traditional management practices such as burning, coppicing, weeding, irrigating, tilling, and rotational grazing have been used for centuries to modify vegetation for the benefit of local communities, and the development of fire as a vegetation management tool enabled people to systematically alter the natural environment on a long-term basis and at a massive scale (Archibald et al., 2012).

Degradation or desertification narratives—which often draw on colonial stereotypes of destruction by Indigenous peoples due to overgrazing and other pastoral or agricultural activities—have facilitated the marginalization and oppression of Indigenous groups and often fail to recognize traditional ecological knowledge (Agrawal 1998). Despite common public views of fire as a factor of degradation in many ecological systems and the implementation of fire suppression policies in many countries, research on Indigenous uses of fire in wooded savannas in West Africa (Laris 2002; Mbow et al. 2000; Walters 2012, 2015), East Africa (Angassa and Oba 2008; Butz 2009; Gil-Romera et al. 2010; Kamau and Medley 2014), and southern Africa (Shaffer 2010; Sheuyange et al. 2005; Wigley et al. 2009) have suggested that seasonal burning not only prevents destructive late-season fires but also increases plant biodiversity.

In African savannas, hunter-gatherers often light fires to attract game to the subsequent green regrowth and to improve visibility and movement (Frost 1999). Pastoralists ignite fires to promote new, diverse, high-quality forage for livestock, prevent shrub encroachment, eliminate cover for predators of livestock, reduce tick loads, and promote habitat for plants of edible, medicinal, or construction value (Angassa and Oba 2008; Butz 2009). In contrast, agricultural communities light firebreaks to protect their crops and use fire to open new land for cropping and to burn agricultural residue (Laris 2002). In general, human ignitions are spatially clustered around settlements.

The level of maintenance of traditional fire management practices varies widely throughout the African savanna biome. Reasons given by inhabitants of a Maasai village in northeastern Tanzania for changing fire management practices included recent increases in the number of large accidental fires, inadequate and unpredictable rainfall, local and federal governmental policies restricting burning, and increasing population pressures that have reduced the amount of land available for burning (Butz 2009, 2013). There is a pressing need for further research focusing on the ecological significance of altering historical fire regimes, particularly in areas where traditional practices continue or have only recently begun to change. Substantial modifications to historical fire regimes could have cascading consequences for ecosystem health in savanna systems by increasing late-season fuel loads and the occurrence of large, catastrophic fires. These changes have multiple implications for the conservation and management of African savanna systems.

The savanna biome in Africa covers half of the continent's terrestrial landscape and is found broadly throughout Sub-Saharan Africa (Beerling and Osborne 2006; Kennedy and Potgieter 2003; Werner et al. 1991). African savannas support an unparalleled diversity of large wild and domestic herbivores and their predators, as well as some of the last remaining long-range migrations of these animals on Earth (Fynn et al. 2016; Harris et al. 2009; Homewood 2008). Their populations, however, are subject to intense human pressure by conversion to agricultural and grazing land (Bond and Parr 2010; Fynn et al. 2016). Because grass production in the wet season is followed by an extended dry season, a continuous cover of fuel creates the opportunity for high fire activity wherever human or natural ignition sources are available (Bond and Parr 2010; Higgins et al. 2000). The high biodiversity value of African savanna systems, the importance of fire in shaping these systems, and the major threats to restoration and maintenance of these systems highlight the importance of conservation.

Savannas have been extensively cleared for crops and plantation forestry or used as rangelands for livestock throughout Sub-Saharan Africa (Tiffen 2006). In southern Africa, afforestation is a major conservation threat to the savanna biome (Bond and Parr 2010; Neke and Du Plessis 2004). The loss of grassland and savanna habitat to encroaching woody species is a major concern of wildlife conservationists and nature protection organizations worldwide (Gil-Romera et al. 2010). Additionally, the African continent is responsible for approximately 70 percent of global burned area and 50 percent of fire-related carbon emissions (Giglio et al. 2013; van der Werf et al. 2010).

Challenging economic and environmental conditions, coupled with increasing populations, are leading to changes in the priorities and activities of rural Africa (Butz 2013). While savannas were used traditionally for extensive grazing and hunting by their seminomadic inhabitants, ongoing socioeconomic development and lack of institutional appreciation for traditional ways of living currently limit the mobility of livestock in many areas (Andela and van der Werf 2014; Turner 2011). Increasing population density, the introduction of land ownership, and increased agricultural activity are all leading to a general shift from nomadic pastoralism toward sedentary farming and grazing (Butz 2009; Tiffen 2006; Turner 2011). Conversion of land for agriculture and the development of road networks and more permanent settlements have broken up fuel continuity in remaining savannas, leading to a reduction in the overall area burned annually (Andela and van der Werf 2014; Shaffer 2010). Vegetation patterns and fire regimes are therefore likely to be affected by proximity to agricultural activity and settlements. Changes in climate, particularly those related to precipitation and seasonality, may also be driving changes in burned area in Africa (Andela and van der Werf 2014; Archibald et al. 2012).

A unique long-term experiment to test the effects of fire on vegetation was initiated in 1954 in Kruger National Park, South Africa (Van der Schijff 1958). Regular prescribed burning every three years was introduced in 1957, and these fixed-rotation burns continued until 1980, when a more flexible approach was put in place between 1981 and 1991 (van Wilgen 2009). A shift in fire policy in the park from 1992 to 2001 allowed all lightning-ignited fires to burn freely while attempts were made to prevent, suppress, or contain all other fires. From 2002 onward, point ignitions have been used to initiate prescribed burns, and wildfires are tolerated in areas where fire is deemed necessary (van Wilgen 2009; van Wilgen et al. 2008).

Acceptance of and support for traditional fire use in Sub-Saharan savannas is still quite limited, particularly compared with contemporary management approaches to fire in Australia. Fire regimes continue to change across Africa (Archibald et al. 2012), and the extent to which customary burning practices and knowledge have been retained varies widely (Butz 2009; Laris 2002; Laris and Wardell 2006; Mbow et al. 2000; Walters 2015). In many countries, colonial antifire policies became part of the centralized state bureaucracy that is still enforced today (Laris and Wardell 2006). The importance, however, of linking wildfire management programs with local communities in national parks and conservation areas throughout Sub-Saharan Africa is beginning to gain more traction (Jeffery et al. 2014; Parr et al. 2009; G. M. Walters 2015). A better understanding of historic and contemporary fire management practices can contribute to the more effective use of fire to manage savanna landscapes for livelihoods and maintain ecological and cultural diversity.

Tropical Savannas of South America

Tropical savannas are the second most extensive biome in South America and present unique conservation issues in a region where Amazonian rainforest ecology is paradigmatic, and consequently, fire, deforestation, and carbon emissions are often assumed to be an interrelated set of problems. For savanna areas where fire plays an adaptive role, the unique relationships between these factors are less understood and inadequately incorporated into conservation science and politics. This section reviews anthropogenic fire in South American savanna ecosystems, with emphasis on the cerrado of Brazil and the Gran Sabana of Venezuela.

The diverse savannas of South America—including the cerrado, Gran Sabana, Los Llanos, and Gran Chaco ecoregions, among others—share important characteristics. They exhibit great heterogeneity, mostly due to differences in underlying edaphic conditions such as soil fertility and water saturation percentages caused by differences in water table levels, as well as the degree of aluminum and iron soil toxicity (Furley 1999; Miranda et al. 2002; Ruggiero et al. 2002).

The cerrado is mainly located in Central Brazil but also reaches into Paraguay and occurs in discontinuous patches in Bolivia (Eiten 1972, 1994; Ratter et al. 1973; Ratter et al. 1997). Cerrado-like savanna vegetation known popularly as *lavrados* covers a large portion of northern Roraima State, Brazil, extending into the Guianas (Miranda and Absy 2000). The Gran Sabana refers to a mixed forest grasslands savanna located in a plateau region in the southern section of Bolívar State, Venezuela. It reaches the Brazilian border to the south and spreads to Guyana to the northeast. Most of the Gran Sabana occurs in the Caroní River basin, inside the boundaries of Canaima National Park (Kingsbury 2001; Sletto and Rodriguez 2013).

Fires of non-anthropogenic origin have been documented for the cerrado as early as 32,000 BP (Ledru 2002). The principal non-human ignition factor in the cerrado was lightning during the rainy and transitional seasons (Ramos-Neto and Pivello 2000), although very little is known about natural fire frequencies (Pivello 2006; Miranda et al. 2009). Ecological research has shown cerrado lightning fires to burn areas as much as 10,000 hectares or more (Ramos-Neto and Pivello 2000). Constant surface fires overwhelmingly consume the herbaceous layer (Miranda et al. 2002; Miranda et al. 2009; Pivello 2011; Simon et al. 2009). A predominantly (70 percent) subterranean vegetative biomass and numerous anatomical adaptations favor resistance to fire damage and vigorous postfire regrowth and flowering in the cerrado (Castro and Kauffman 1998; Coutinho 1990, 2002; Eiten 1972; Miranda et al. 2002; Rizzini and Heringer 1961; Simon et al. 2009; Warming 1892). Damage to crown foliage does not persist after the first fire-free year (Silvério et al. 2015). Furthermore, fires in the cerrado consume less biomass and produce less emissions and faster carbon uptake than those in Amazonian tropical evergreen forest, which has more than ten times the total aboveground biomass (Castro and Kauffman 1998).

In the Gran Sabana, fire is also one of the most important factors contributing to landscape dynamics. Fires are believed to be primarily of anthropogenic origin because most lightning strikes occur during the rainy season and pose little fire risk. Currently, 2,000 to 3,000 fires are recorded annually, burning 5,700 to 7,500 hectares (Bilbao et al. 2010).

The use of fire by Indigenous inhabitants of tropical savannas in South America appears to be widespread and has been well documented for some locations by naturalists and scientists from various disciplines. Ethnographic accounts for several Gê-speaking Indigenous groups—including the Apinajé, Kayapó, Krahô, and Xavante—demonstrate that these practices involve extensive traditional knowledge (Hecht 2009; Melo and Saito 2013; Mistry et al. 2005; Nimuendaju 1939; Pivello 2006; Posey 1985, 1986; Welch 2014, 2015). Traditional burning in cerrado landscapes is also practiced by *quilombolas* (descendants of escaped or displaced slaves) and *caboclos* (peoples of mixed Indigenous, European, and African backgrounds) (Eloy et al., chapter 4; Pivello 2006). Nineteenth-century accounts suggest Indigenous peoples of the cerrado and Gran Sabana have long used fire for hunting, agricultural clearing, and landscape management (França et al. 2004; Miranda et al. 2009; Welch 2014). The Pemon groups of the Gran Sabana had been using fire management for at least 400 years when they were first documented in southeastern Venezuela (Armellada 1960, cited in Sletto and Rodriguez 2013). Pemon traditional knowledge is crucial to subsistence practices involving fire, which is understood to have magical properties and its own sense of agency and power (Bilbao et al. 2010; Sletto 2008; Sletto and Rodriguez 2013).

Since the nineteenth century, when Danish botanist Eugene Warming first documented that burning by Indigenous peoples had positive effects on plant form and vegetation structure in the cerrado (Warming 1892), scholarship has shown that cerrado landscapes benefit from fire, including some anthropogenic burning regimes (Eiten 1972; Pivello 2011; Simon et al. 2009). Periodic fires result in less destructive (cooler, lower, and more fragmented) fires than occur in areas with histories of fire suppression; they also increase plant biodiversity and stimulate fire-dependent sexual reproduction and dispersal (Castro and Kauffman 1998; Coutinho 1990; Miranda et al. 2002; Pivello 2006; Pivello 2011; Ramos-Neto and Pivello 2000). Research has also shown cerrado fires, including those set by Indigenous peoples, increase the availability of foods (pollens, nectars, shoots, and fruits) for some animal taxa and are unlikely to have caused local animal extinctions (Cavalcanti and Alves 1997; Coutinho 1990; Frizzo et al. 2011; Miranda et al. 2002; Pivello 2011; Prada et al. 1995; Vitt and Caldwell 1993).

Most burning occurs in savanna areas of the Gran Sabana, although fires sometimes cross the savanna-forest boundary and can thus cause forest degradation. Some researchers defend the idea that the Gran Sabana is a degraded landscape and today's forest patches are remnants of a once more extensive and closed forest. According to this view, fires set annually by Pemon Indigenous peoples, historical residents of the region, have spurred a large-scale process of savanna expansion (Sletto 2008). Other researchers assert that "a mosaic of forest patches, grasslands and shrub lands has dominated the Gran Sabana since the last Ice Age" (Sletto 2008, 1943). This viewpoint draws on evidence that the decline of primary forests is not uniform across the Pemon territory. Soil studies also show that some complexes could have never supported forests, suggesting that widespread Pemon burning practices are not necessarily destructive to the Gran Sabana forests and do not account for deforestation and degradation (Sletto 2008; Sletto and Rodriguez 2013).

Because Indigenous burning traditions in South American savannas are poorly documented for most groups, it is difficult to characterize the degree to which they continue to be practiced. In the cerrado, the complex cultural knowledge involved in contemporary Indigenous burning traditions is best documented for the Kayapó, Krahô, and Xavante ethnic groups, among whom it is maintained through group decision making and participation by elders and younger individuals (Hecht 2009; Mistry et al. 2016; Welch 2015). Even among these groups, transfer of knowledge to younger generations is challenged by changing ecological and cultural circumstances, as well as prevalent negative public opinions (Welch 2014, 2015; Melo and Saito 2013). Unlike most other contemporary Indigenous ethnic groups in the cerrado, many Xavante communities continue to practice group hunting with fire, which is important for subsistence, ceremonial events such as weddings and rites of initiation, as well as cultural identity (Welch 2014, 2015; Melo and Saito 2013). In one Xavante Indigenous reserve, this cultural practice, which involves distinctly Xavante notions of environmental conservation, has been shown to produce no measurable negative environmental impacts and may in fact facilitate vegetation regrowth following non-Xavante agribusiness deforestation (Welch et al. 2013; Welch 2014). However, group hunting with fire and associated traditional knowledge may be modified or interrupted in other Xavante Indigenous reserves with higher population densities and more landscape degradation.

In a similar fashion, fire practices and knowledge are well documented for the Pemon of the Gran Sabana, particularly those of the Taurepan linguistic subgroup, among whom these traditional practices continue and their underpinning cosmologies are evident in daily life (Bilbao et al. 2010; Sletto 2008; Sletto and Rodriguez 2013). Practices and associated knowledge systems are more pronounced in isolated Pemon communities located farther from the Pan American Highway and urban settlements. Studies have shown that Pemon prescribed burning practices, which include frequent and widespread grassland burning, serve to create a mosaic landscape. Differences in plant physiognomies and fire histories across the landscape control against large wildfires, with recently burned areas serving as firebreaks (Sletto and Rodriguez 2013). Experimental studies show biomass accumulation levels permit burning every three to four years (Bilbao et al. 2010). For the Pemon, because fire is a powerful agent that humans cannot completely control, its use should be deliberate, careful, and controlled. From this perspective, the frequent small burns described above are said to prevent hotter and faster-spreading fires with more destructive potential (Sletto and Rodriguez 2013). Thus, the Pemon custom of burning grasslands to form firebreaks at the savanna-forest transition zone acts to minimize forest degradation (Sletto 2008; Sletto and Rodriguez 2013).

In the cerrado, large-scale ranching and intensive monoculture were largely responsible for converting 49 percent of the cerrado by 2010 (MMA/IBAMA 2011; Klink and Machado 2005; Miranda et al. 2009). Conversion of cerrado vegetation to crop/pasture mosaic has raised average daytime temperatures by 1.55°C (Loarie et al. 2011). More importantly, change in vegetation patterns due to human activity in conjunction with fire suppression has led to the dominance of African grasses (e.g., genera *Andropogon*, *Brachiaria*, and *Melinis*) and caused open cerrado forms to be replaced with taller vegetation cover, thus increasing the intensity of wildfires (Durigan and Ratter 2016; Klink and Machado 2005; Klink and Moreira 2002; Silva et al. 2006).

While land cover change is not as drastic in the Gran Sabana, over the past decades the region has undergone demographic changes related to development projects and highway building. The migration of the Pemon from isolated regions of the Gran Sabana to larger settlements closer to roads has resulted, for example, in less frequent small fires in the interior of Canaima National Park. This leads to buildup of combustible material across savanna areas and increases the risk of fast and intense wildfires (Sletto and Rodriguez 2013).

Cerrado conservation is a political priority in Brazil, but the region still has relatively few protected areas (Klink and Machado 2005; Carranza et al. 2014; Carvalho et al. 2009). Indigenous lands and strictly protected ecological areas are shown to be more effective in reducing conversion of cerrado vegetation than sustainable use conservation units such as sustainable development and extractive reserves (Carranza et al. 2014; Françoso et al. 2015). Despite scientific and policy recognition of the value of traditional and controlled burning, in practice the public administration of most cerrado conservation units presumes that all fire is destructive (Pivello 2006; Mistry et al. 2016; Falleiro 2011; Durigan and Ratter 2016; Pereira Júnior et al. 2014). Most governmental partnerships with Indigenous peoples of the Brazilian cerrado have focused on developing Indigenous fire brigades to combat wildfires on Indigenous lands, although the Ministry of the Environment has taken small steps toward developing a new policy paradigm based on integrated fire management and consultation with traditional knowledge holders, including some Xavante communities (Falleiro 2011; Falleiro et al. 2016; Mistry et al. 2016).

The Gran Sabana is also a significant conservation priority for Venezuela (Bilbao et al. 2010; Sletto and Rodriguez 2013). Conservation and management of the Gran Sabana has largely been handled by the parastatal company Electrificación del Caroní, which has a vested interest in protecting the region's power plants by conserving riparian forest to prevent erosion and sedimentation (Sletto and Rodriguez 2013). Despite efforts that demonstrate the ecological benefits of Pemon prescriptive burning, this agency largely views these practices as irrational and destructive to forests. They have thus prioritized demonstrably ineffective fire suppression initiatives including firefighting, surveillance, and education (Bilbao et al. 2010). Ongoing discussions, however, between agency representatives, the Pemon, and scientists are under way to discuss the strategic possibilities for intercultural fire management, inspired by patch-mosaic models used in Australia and Africa (Sletto and Rodriguez 2013). This case has potential to elucidate alternatives to fire suppression policies for tropical savannas, including the cerrado, where threats to

Temperate Rainforests of the Central Northwest Coast of North America

The central Northwest Coast extends from present-day Homalco on the coast of British Columbia to the Cowlitz River drainage in southwest Washington State. It is a narrow arc of land bordered by the Pacific Ocean to the west and the crests of the Cascade and British Columbia Coast Mountains to the east (Coupland 1998; Kroeber 1939). Humans have inhabited the Northwest Coast for more than 10,000 years (Moss 2011). Archaeological evidence indicates that Native peoples were utilizing ice-free areas along the outer coast even before the retreat of the last glacial advance (Waters et al. 2011; Wilson et al. 2009). Prior to Euro-American disease and colonization, Wakashan- and Coast Salish–speaking people occupied winter villages of one to several plank houses along the saltwater, rivers, and uplands of their territories and spoke at least twenty-one different languages (Thompson and Kinkade 1990). A conservative estimate of their peak precontact population is about 80,370 (Boyd 1990).

As with the entire Northwest Coast of North America, archaeologists have conventionally classified central Northwest Coast peoples as "complex hunter-gatherers" or, following Marshal Sahlins, "affluent foragers" (Ames and Maschner 1999, 24). These designations signal that Northwest Coast societies had hierarchical social structures, were partially to fully sedentary, lived in often rather densely populated winter villages with extended family households of 30 to 100 people or more, and produced and stored large amounts of surplus foods with specialized technologies. The Indigenous societies of the Northwest Coast are also often characterized as "paradoxical" due to their social complexity in the apparent absence of domestication and agriculture (Ames and Maschner 1999; Turner and Peacock 2005; Deur and Turner 2005).

The social complexity exhibited on the Northwest Coast is often attributed to the region's unusual natural abundance and species diversity (Ames and Maschner 1999; Haeberlin and Gunther 1930; Suttles 1987). However, while high species abundance and diversity may be the case for marine resources, it is not necessarily so for terrestrial resources, which were also a crucial part of the historical food system of central Northwest Coast peoples. With a mean annual temperature of about 10°C and a mean annual rainfall of approximately 100 centimeters, the region is characterized as a temperate rainforest. Thus, the Northwest Coast may be unusually productive in terms of overall biomass, but historically this was primarily banked in extensive mature forests. While producing the massive western red cedar (*Thuja plicata* Donn ex. D.Don)—a tree that was used for everything from plank houses to canoes, clothing, and baby diapers—these forests produced very little in the way of edible plants.

Regional paleoecology studies dating back to the late Pleistocene suggest that biomass combustion increase began about 5500 BP (Walsh et al. 2015). This period is generally understood to mark the end of the warmer, drier Hypsithermal of the early Holocene and the beginning of a cooler, wetter climate similar to the modern period. Possible explanations for the increase in fires when the climate was moister include increases in the variability and intensity of the El Niño Southern Oscillation cycle, increases in Native American burning practices, or a combination of both (Walsh et al. 2015; Gavin et al. 2013). Several archaeological and/or ethnographically documented sites in the region also suggest increases in biomass combustion during the mid- to late Holocene (Bach and Conca 2004; Burtchard 2009; Derr 2014; Lepofsky et al. 2005; Tweiten 2007; Weiser and Lepofsky 2009). An archaeobotanical synthesis of seventeen sites along the Green-Duwamish watershed in the Puget Sound basin also supports increased intensity of Indigenous burning beginning about 2800 BP (LeCompte-Mastenbrook 2014).

While scholarship regarding elevated resource cultivation on the Northwest Coast reflects an emergent paradigm (Boyd 1999; Deur and Turner 2005), the observation that the ancestors of contemporary central Northwest Coast peoples began to use the region's resources more intensively in the late Holocene is not new. Several hypotheses for resource intensification during this period have been proposed, including the desire to maintain open conditions that were characteristic of the more fire prone landscape in the Hypsithermal, and/or a combination of population increase and resource depression (Ames 2005; Burtchard 2009; Butler and Campbell 2004; Lepofsky et al. 2005). Yet there is increasing evidence that even though they had sufficient technologies and population sizes to intensify resource use, Indigenous peoples of the central Northwest Coast appear to have harvested ample resources without depleting them for perhaps thousands of years (Butler and Campbell 2004; Campbell and Butler 2010; Derr 2014). Although there was a general cooling trend in the mid- to late Holocene, there was also a great deal of deviation from that trend over the course of centuries and decades (Walsh et al. 2015). The most recent examples of long-term trends in variability throughout this period were the Medieval Climate Anomaly (circa 900–1300 CE) and the Little Ice Age (circa 1500–1900 CE). There is also some evidence of increasing variability and intensity of the El Niño Southern Oscillation cycle during this time. Given the possibility that there was greater interannual variability, burning and other forms of cultivation not only maintained open conditions but may also have strategically helped to minimize uncertainty and ameliorate risk in a dynamic environment (Cashdan 1985; Goland 1991; LeCompte-Mastenbrook 2014). Understanding whether and how this was accomplished will require more nuanced understandings of the historical uses of fire on the central Northwest Coast. It will also require adoption of new tools and methods to investigate human-scale ignitions and their effects at the species, habitat, and landscape levels since the conventional wisdom among forest and fire ecologists is that each subsequent high-severity fire destroys evidence of the one that came before (Agee 1993; Franklin and Dyrness 1988).

The vast majority of fire history research conducted in the region has been driven by economic concerns—namely, timber production. Much of it has also been conducted within a "wilderness" paradigm, whereby Indigenous peoples' influence on fire extent and frequency is assumed to be minimal. Thus, much fire history research on the central Northwest Coast tends to include only a cursory acknowledgement of the role of anthropogenic fire in shaping patterns and processes of forest disturbance and regeneration prior to Euro-American colonization (Agee 1993; Franklin and Dyrness 1988; Hemstrom and Franklin 1982). This large blind spot is exacerbated by a predominant focus on the role of marine resources in the archaeological record, and the accompanying predominant assumption on the part of anthropologists working in the region that the social complexity exhibited on the Northwest Coast can be attributed mostly to the natural abundance of Pacific salmon (*Oncorhynchus* spp. Suckley) and other key resources such as western red cedar (Ames and Maschner 1999; Hebda and Mathewes 1984).

Despite the popular impression that it always rains on the Northwest Coast, the region has considerable variability in precipitation and thus fire frequency and plant species composition. Stand-replacing fires are understood to be the norm in these biogeoclimatic zones. In the mild, wet Sitka spruce and western hemlock zones, a millennium or more may pass between fires (Agee 1993; Wong et al. 2003), while stand-replacing fires may occur as frequently as every 200 years in the coastal Douglas-fir zone (*Pseudotsuga menziesii* [Mirb.] Franco) (Wong et al. 2003). Despite this variability, these forests are considered in general to be so nonflammable that they are colloquially referred to as "asbestos" forests. These infrequent, high-severity fires stand in sharp contrast to the forests of the inland West, east of the Cascade Range, where the ubiquitous ponderosa pine forests are estimated to have missed eight or more low-severity fire events as a result of fire suppression (Agee 1993). Currently, substantial resources are expended to understand and restore historic fire regimes to eastside forests in an effort to prevent catastrophic wildfires (Ingalsbee 2010; USDA 2013).

Cultural burning practices are widely documented in the ethnographic literature in a variety of habitats from the Pacific Coast to the Cascade Mountains. Other forms of plant cultivation and management involving perennial plants include weeding, transplanting, pruning, and control of access to harvesting sites (Amoss 1978; Anderson 2009; Boyd 1999; Collins 1974; Gibbs 1978; Leopold and Boyd 1999; Norton 1979, 1985; Peter and Shebitz 2006; M. Smith 1940; A. Smith 2006; Storm 2002; Suttles 2005; Turner 1999; Turner and Peacock 2005; Welch 2013). The once extensive Garry oak (*Quercus garryana* Douglas ex Hook.) savannas of the Puget Sound lowlands and the Salish Sea islands were collectively tended through burning and are probably the best-known examples of Indigenous cultivation with fire to enhance plant abundance on the central Northwest Coast (Leopold and Boyd 1999; Pellatt et al. 2015; Storm 2002; Storm and Shebitz 2006; Weiser and Lepofsky 2009). Cranberry bogs and other wet prairies were regularly burned (Anderson 2009), and fire was also used to enhance berry production and ungulate browse from the lowlands to the mountains (Turner 1999; LeCompte-Mastenbrook 2015; Lepofsky et al. 2005; Smith 2006).

In addition to using fire to cultivate food plants, central Northwest Coast peoples also used it to maintain and enhance nonfood plant species that were integral to food harvesting. In their testimony before the Indian Claims Commission in the late 1940s (US Court of Claims 1947), Puget Sound Coast Salish tribal elders spoke of burning and weeding stinging nettle (*Urtica dioica* L.) patches on Whidbey Island for the purpose of producing quality material for crafting twine, and also burning underbrush in densely forested areas throughout their territories to promote the growth of quality cedar for plank houses and canoes. People from inland villages closer to the Cascade foothills also spoke of clearing and maintaining seasonal house sites and permanent village sites (US Court of Claims 1947). The areas typically selected were ones that were deemed productive not only in terms of aquatic resources, but also for the productivity—or potential productivity—of terrestrial resources. These sites tended to be areas with brushy vegetation and small trees, which would be cleared and burned. Large, potentially hazardous trees were also removed by cutting and burning, leaving a cleared area around the houses of about 1 to 2 hectares. Recent research on the Gulf Islands (Trant et al. 2016) and Hecate Island (Hoffman et al. 2016) in British Columbia indicates a strong association between settlement sites and frequent low-severity fires, suggesting that the inhabitants conducted controlled burns around villages and seasonal camps for perhaps millennia.

On the Olympic Peninsula, Peter and Shebitz (2006) have documented extensive anthropogenic beargrass savannas. Beargrass (*Xerophyllum tenax* [Pursh] Nutt.) is widely used for imbrication in basketry. There is considerable ethnographic and archival evidence suggesting these savannas were maintained through intentional burning, perhaps for millennia.

Over the course of the past two decades, the "affluent forager" model of Northwest Coast livelihood strategies has been increasingly challenged by anthropological and ethnobiological research throughout the broader Northwest Coast region. Two important edited volumes highlighting the widespread uses of fire and other forms of plant resource management by Indigenous peoples in the Pacific Northwest (Deur and Turner 2005; Boyd 1999) have fostered the beginnings of a major paradigmatic shift toward "low-level food production" (Smith 2005; Ames 2005). These volumes, along with broader discourse recognizing the influence of Indigenous resource cultivation throughout North America (e.g., Mann 2005), have attracted a wider readership among academics in other fields, as well as restoration ecologists, land managers, and First Nations people.

As indicated by LeCompte (chapter 6), there are numerous bureaucratic, economic, and epistemological barriers to the reintroduction of prescribed fire in anthropogenic ecosystems on the central Northwest Coast. Yet growing awareness of the role that Indigenous burning practices have played in shaping ecosystem structure and species composition has had some influence on recent ecological restoration practices. A case in point is the reintroduction of prescribed fire to restore and maintain Garry oak ecosystems, which, due to land conversion and fire suppression, are increasingly rare and fragmented (Hamman et al. 2011). It is estimated that more than a hundred plants and animals reliant on Garry oak habitat are endangered to some extent (Pellatt et al. 2015). Conservation of critically imperiled species is the primary motivator for funding and using prescribed fire in these habitats. Much less attention has been paid to the cultural significance of maintaining these prairies for Natives to harvest. During planning and implementation, communication and collaboration are rare between Native communities and the agencies and nongovernmental organizations that implement such projects. A handful of collaborative efforts committed to biocultural diversity conservation on public lands is emerging, however (Sarah T. Hamman, personal communication; Marlow G. Pellatt, personal communication), even though implementation is not without controversy (Dickson 2016). Perhaps the most promising emergence of such collaborations is in the portion of the central Northwest Coast within the United States where some federally recognized tribes, such as the Quinault Indian Nation, have initiated collaborative projects to reintroduce fire to anthropogenic landscapes on reservation lands (Preston 2005, 2016).

A Case for Inclusive Scholarship, Policy, and Management

More than sixty years have passed, and much has changed since Omer Stewart (2002) wrote in his groundbreaking historical analysis *The Effects of Burning of Grasslands and Forests by Aborigines the World Over*: "fire has been almost ignored as one of the important ecological variables determining vegetation. Particularly neglected have been man-made fires" (68). Once considered a nonfactor, anthropogenic fire is now inseparable from conservation narratives at different geographical scales worldwide. Unfortunately, the heterogeneous phenomenon of burning is also sometimes reduced to overly simplistic representations of environmental destruction, degradation, and loss (Kull 2000) consistent with persistent environmental orthodoxies about natural versus cultural processes (Forsyth 2003; Fairhead and Leach 1996). Historical ecological scholarship has begun to retell some of these stories based on ample and diverse evidence that many early landscapes once presumed to be "natural" were in fact anthropogenic, often modified through the human use of fire (Bird et al. 2016; Denevan 1992, 2001; Gammage 2011). Efforts to overcome "environmental mismanagement tropes" (Stephenson and Stephenson 2016, 304) that place undue blame on local peoples for systemic problems, often because of outdated understandings of human ecological relationships, are essential to addressing complex conservation challenges involving fire. For Indigenous and other traditional peoples, the relevance of this challenge transcends environmental concerns because their burning traditions are inexorably linked to human rights as expressions of cultural identity, means of physical sustenance, and strategies for territorial management. Our four regional case studies from Australia, Africa, South America, and North America demonstrate the importance of understanding these connections while illustrating the unevenness with which scientific and conservation narratives grapple with the complex interrelationships between Indigenous peoples, anthropogenic fire, and environmental conservation.

Every landscape has a complex history of human land use and natural disturbance (Aragón et al., 2003), and the distinction between natural and cultural landscapes is not always obvious (Eriksson et al. 2002). Evident in our case studies are examples of differences in past and present anthropogenic fire regime histories. Although there is evidence of fire associated with humans in Africa more than a million years ago, it did not become a widespread and broadly ecologically relevant factor on any continent before the mid- to late Holocene (Archibald et al. 2012; Bird and Cali 1998; Gouveia et al. 2002; Walsh et al. 2015; Williams, Ulm et al. 2015). Thus, as shown by evidence from Africa, Australia, and North America, the distinctive burning regimes observed in the early historic period through the present do not appear to have developed anywhere until relatively late in the history of modern humans (Bliege Bird et al. 2008; LeCompte-Mastenbrook 2014; Williams 2013; Williams, Mooney et al. 2015). However, lack of data contributes to difficulties distinguishing between anthropogenic and nonhuman fire regimes and assessing associations between burning and landscape-scale change.

Archaeological, historical, and ethnographic evidence of burning traditions in the modern era provide more nuanced accounts of the interrelationships between local environments, subsistence practices, and burning traditions. The degree to which burning by Indigenous peoples impacted environments in the past is a subject of debate. In some regions, degradation or desertification narratives predominate and serve to stigmatize Indigenous peoples. For example, subsistence burning in Australia has been implicated in megafauna extinctions (the ecosystem collapse hypothesis) about 40,000 BP, despite multiple lines of evidence suggesting more complex causality (Cohen et al. 2015; Price et al. 2011; Sakaguchi et al. 2013; Wroe et al. 2013). According to some scholarly perspectives, traditional fire use in Africa, Asia, and South America was responsible for transforming forests into savannas (Sankaran and Ratnam 2013; Jones 1980; Sletto 2008). In contrast, regional disinterest in the ecological impacts of traditional fire use in the

Northwest Coast of North America reflects a wilderness paradigm, according to which Indigenous peoples are assumed to have had minimal influence over fire patterns. In fact, ample evidence from the region points to extensive historical fire management practices in Garry oak savannas, cranberry bogs, and beargrass savannas, among others (Anderson 2009; Leopold and Boyd 1999; Pellatt et al. 2015; Peter and Shebitz 2006; Storm 2002; Storm and Shebitz 2006; Weiser and Lepofsky 2009).

Some of the earliest nineteenth-century observations of traditional burning practices in northern Australia and Central Brazil associate them with modification or management of landscape resources (e.g., Mitchell 1848; Warming 1892). More recent evidence from these regions suggests that landscape burning had relatively large overall footprints but was not evenly distributed (Jones 1975; Yibarbuk 1998; Welch et al. 2013). The idea that Indigenous peoples not only altered landscapes with fire but also managed them developed in conjunction with the notions of fire-stick farming in Australia (Jones 1969) and patch-mosaic burning in Africa (Parr and Brockett 1999). This new paradigm of cultural landscapes has begun to influence scholarship in savanna regions worldwide, even if it is not as widely accepted by ecologists and policymakers in some regions—such as South America, Asia, and Africa—as in others, such as Australia. Even in the temperate rainforests of the Northwest Coast of North America, landscape management with fire models has begun to replace the presumption that burning by Indigenous peoples was ecologically insignificant (Boyd 1999; Leopold and Boyd 1999; Lepofsky et al. 2005; Turner 1999; Weiser and Lepofsky 2009). Although the cultural and environmental importance of traditional burning has been recognized for decades in Australia, South America, and North America (Hallam 1975; Nimuendaju 1939; Norton 1979), it is only in the last twenty years that robust and ethnographically rigorous documentation has been produced on how burning for subsistence purposes by Aboriginal, Tribal, and Indigenous peoples was often planned, organized, and complex (e.g., Anderson 2009; Boyd 1999; Butz 2009; Frost 1999; Garde et al. 2009; Leopold and Boyd 1999; Mistry et al. 2005; Peter and Shebitz 2006; Russell-Smith et al. 1997; Turner 1999; Welch 2014, 2015).

Another recurrent theme across continents is the interruption or modification of burning traditions in the face of dramatic territorial, ecological, economic, and governance transformations in recent decades (Bird et al. 2005; Butz 2009; Preece 2007; Russell-Smith et al. 2013; Welch et al. 2013). In some regions, traditional practices are maintained in more isolated or ecologically conserved regions. In Australia, they most commonly continue in the Arnhem Land region in the northern part of the country (Garde et al. 2009) and the Western Desert (Kimber 1983). In Central Brazil and the Gran Chaco, recent and detailed accounts of landscape burning traditions are limited to relatively few ethnic groups, including but not limited to the Xavante and Pemon (Bilbao et al. 2010; Hecht 2009; Mistry et al. 2016; Sletto and Rodriguez 2013; Welch 2014). In the Pacific Northwest of North America, most firsthand accounts of traditional burning practices are from the historical literature (Boyd 1999).

Interrelationships between contemporary traditional burning practices and local ecosystems are intrinsically linked to environmental change and conservation processes at different temporal and geographical scales. Among the four case studies of fire-prone cultural landscapes discussed in this chapter, perhaps the savannas of Australia, Africa, and South America are most emblematic of fire-adapted flammable vegetation (Bond and Keeley 2005; Bond et al.

2005). Yet even the so-called "asbestos" forests of northwestern North America show evidence of fire since before the Holocene (Brown and Hebda 2003; Walsh et al. 2015). These deep fire histories condition how humans ignite and control fire to produce desirable ecological change, including within contemporary conservation frameworks. For example, whereas fire is paradigmatically associated with environmental destruction in many savanna landscapes throughout the world, wildfire risks on the central Northwest Coast of North America were not widely recognized until recently (Kaiser 2016). Conversely, academic evidence of positive relationships between periodic fires, biodiversity, and reduced destructiveness of wildfires is more abundant and therefore more available to inform conservation narratives about savanna settings (Bradstock et al. 2012; Laris 2002; Pivello 2011; Sletto and Rodriguez 2013).

Environmental orthodoxies about fire are not always adequately informed by empirical evidence and sometimes lead to important misconceptions by presuming the destructiveness of human alteration of landscapes in diverse environmental and cultural contexts. For example, contemporary swidden agroforestry is a dominant form of land use in Asia often characterized as a major contributor to deforestation, landscape degradation, and carbon emissions, even in the absence of scientific evidence (Alencar et al., 2005; Erni 2009; Ickowitz 2006; Lambin et al. 2001; Mertz 2009; Tacconi and Ruchiat 2006; Tacconi and Vayda 2006; van Vliet et al. 2013; Zhang et al. 2002; Ziegler et al. 2012). Misunderstandings about swiddeners in forested areas also lead to major oversights regarding the value of these practices for promoting biodiversity, preserving ecosystem services, and anticipating, mitigating, and adapting to the effects of climate change (Parrotta and Agnoletti 2012; van Vliet et al. 2012; van Vliet et al. 2013; Ziegler et al. 2012). Similar paradigmatic fallacies endure regarding forests in South America, Africa, and elsewhere (Colfer et al. 2015; Mistry et al. 2016; Steiner et al. 2004; Welch et al. 2013). Although the case studies of fire-prone landscapes in this chapter do not specifically address these widespread forest swidden systems, they are closely linked through overreaching conservation discourse and policies that wrongly demonize traditional and other smallholder cultivators in forest and nonforest settings.

For example, the geographical and scholarly juxtaposition of Amazonian rainforests and nearby tropical savannas in lowland South America illustrates how inappropriate overextension of conservation narratives results in consequential misconceptions about Indigenous and other traditional or local peoples' contributions to environmental mismanagement. Academic and policy discourses that borrow ideas about biodiversity losses and carbon emissions resulting from burning forests in Amazonia to tell similar stories about subsistence burning in fire-adapted cerrado and Gran Sabana landscapes have led to tropical savanna fire policies in Brazil and Venezuela that largely overlook the positive conservation value of controlled burning and collaboration with Indigenous fire managers (Mistry et al. 2016; Pivello 2006; Ramos-Neto and Pivello 2000; Sletto 2008; Welch et al. 2013). Analogous scenarios exist in several other regions discussed in this chapter and elsewhere (Ratnam et al. 2011; Bond and Parr 2010).

As explored in the final chapter of this book (Fowler, chapter 9), smallholder biomass burning for subsistence purposes is a very different environmental issue than industrial fossil fuel consumption, although their relative impacts can be difficult to quantify. As suggested, however, by the case studies presented in this and other chapters in this volume (Butz, chapter 3; Eloy et al., chapter

4; Steward, chapter 5; Fowler, chapter 9), traditional and smallholder subsistence burning is generally shown not to be a major cause of deforestation and decreased carbon stores; its environmental impacts are dwarfed in comparison to those resulting from agribusiness and fossil fuel combustion. Such subsistence burners are, in fact, victims rather than drivers of forest conversion and climate change processes (Colfer et al. 2015; Erni 2009; Li et al. 2014; Parrotta and Agnoletti 2012; Sorrensen 2009; Tacconi and Ruchiat 2006; van Vliet et al. 2012; van Vliet et al. 2013).

Mismatches between local socioenvironmental realities and fire conservation narratives may also be aggravated by non-inclusive viewpoints regarding appropriate governance systems, property systems, and human well-being, thereby amplifying short-term challenges to long-term solutions. An especially insidious illustration of this dynamic is ethnic shaming resulting from conservation discourse that inappropriately disparages traditional subsistence burning practices as incompatible with conservation values or environmentally destructive (Erni 2009; Kull 2002; Petty et al. 2015; Spoon et al. 2015; Whitehead et al. 2003). Several examples were mentioned in this chapter's case studies of traditional burning in savannas and deserts. In West Africa, the false assumption that Indigenous peoples degrade ecological systems by burning draws on colonial stereotypes and has facilitated their marginalization and oppression (Agrawal 1998). The Xavante in Brazil's cerrado region have been stigmatized by unsubstantiated claims that they do not deserve access to traditional lands because their ritualized cultural practice of group hunting with fire causes deforestation and local animal depletion (Welch et al. 2013; Welch 2014). In the Venezuelan Gran Chaco, government agencies reinforce misconceptions that Pemon fire use is destructive and backward, a viewpoint that has been internalized by some younger individuals and residents in more integrated settlements (Sletto and Rodriguez 2013). Such stigmatization perpetrated by the public, governments, academia, conservation organizations, and development interests negatively impacts local communities' resilience in the face of multidimensional marginality often involving ethnic minority, geographical isolation, poverty, lack of land tenure, and inadequate political representation (Erni 2009; Kull 2002; van Vliet et al. 2013).

The case studies in this chapter illustrate how insufficient cross-fertilization between Indigenous, academic, conservation, and governance systems of knowledge can impede implementation of proven conservation interventions as well as experimentation with innovative solutions. Each region exemplifies different degrees of tolerance for traditional burning and willingness to collaborate with Indigenous stakeholders. For example, Australia has relatively ample and enduring experience with collaborative management of conservation areas and policies favorable to incorporation of Indigenous people and their conservation knowledge. In contrast, efforts in the tropical savannas of Africa, Brazil, and Venezuela are incipient despite modest recent policy advances. In the central Northwest Coast of North America, old assumptions about the ecological nonrelevance of fire use by Native peoples have led to their exclusion from most policy formulation and fire management efforts in conservation areas.

Overcoming these many challenges requires new approaches. This chapter addresses an important step in this direction: addressing complex interrelationships across spatial and temporal scales in order to better understand the human ecological dynamics of anthropogenic fire. As indicated by recent calls to advance scholarship regarding the Anthropocene, conservation attention to anthropogenic fire will be improved by opening systems of knowledge production through collaboration and epistemological inclusion (Brondizio et al. 2016; Lövbrand et al. 2015; Bai et al. 2016). This involves crossing disciplines, reducing boundaries between pure and applied research, and embracing involvement by diverse stakeholders with distinct voices, understandings, and values (Wolverton 2013; Roos et al. 2016; Pyne 2007). Most importantly, it should draw on all of these human and intellectual resources in order to formulate forward-looking scientific and conservation approaches.

REFERENCES CITED

Abbott, Ian. 2003. "Aboriginal Fire Regimes in South-West Western Australia: Evidence from Historical Documents." In *Fire in Ecosystems of South-West Western Australia: Impacts and Management*, edited by Ian Abbott and Neil Burrows, 119–146. Leiden: Backhuys.

Adichie, Chimamanda Ngozi. 2011. The Danger of a Single Story. TED Talk, Accessed June 7, 2016. https://www.ted.com/talks/chimanda_adichie_the_danger_of_a_single_story.

Agee, James K. 1993. *Fire Ecology of Pacific Northwest Forests*. Washington, DC: Island Press.

Agrawal, Arun. 1998. "Indigenous and Scientific Knowledge: Some Critical Comments." *Antropologi Indonesia* 55. http://journal.ui.ac.id/index.php/jai/article/view/3331.

Alencar, Ane, Daniel Nepstad, and Paulo Moutinho. 2005. "Carbon Emissions Associated with Forest Fires in Brazil." In *Tropical Deforestation and Climate Change*, edited by Paulo Moutinho and Stephan Schwartzman, 23–33. Belém, PA: Instituto de Pesquisa Ambiental da Amazônia and Environmental Defense.

Allan, G. E., and R. I. Southgate. 2002. "Fire Regimes in the Spinifex Landscapes of Australia." In *Flammable Australia: The Fire Regimes and Biodiversity of a Continent*, edited by Ross A. Bradstock, Jann E. Williams, and Malcolm A. Gill, 145–176. Cambridge: Cambridge University Press.

Altman, Jon. 2009. "Manwurrk (Fire Drive) at Namilewohwo: A Land-Management, Hunting and Ceremonial Event in Western Arnhem Land." In *Culture, Ecology and Economy of Savanna Fire Management in Northern Australia: In the Tradition of Wurrk*, edited by Jeremy Russell-Smith, Peter Whitehead, and Peter Cooke, 165–180. Collingwood: CSIRO.

Altman, Jon C., and Seán Kerins. 2012. *People on Country: Vital Landscapes, Indigenous Futures*. Annandale, N.S.W.: The Federation Press.

Ames, Kenneth M. 2005. "Intensification of Food Production on the Northwest Coast and Elsewhere." In *Keeping It Living: Traditions of Plant Use and Cultivation on the Northwest Coast of North America*, edited by Douglas Deur and Nancy J. Turner, 67–100. Seattle: University of Washington Press.

Ames, Kenneth M., and Herbert G. D. Maschner. 1999. *Peoples of the Northwest Coast: Their Archaeology and Prehistory*. New York: Thames and Hudson.

Amoss, Pamela. 1978. *Coast Salish Spirit Dancing: The Survival of an Ancestral Religion*. Seattle: University of Washington Press.

Andela, Niels, and Guido R. van der Werf. 2014. "Recent Trends in African Fires Driven by Cropland Expansion and El Niño to La Niña Transition." *Nature Climate Change* 4 (9): 791–195.

Anderson, M. Kat. 2009. "The Ozette Prairies of Olympic National Park: Their Former Indigenous Uses and Management: Final Report to Olympic National Park." Davis, CA: National Plant Data Center, USDA Natural Resources Conservation Service.

Angassa, Ayana, and Gufu Oba. 2008. "Herder Perceptions on Impacts of Range Enclosures, Crop Farming, Fire Ban and Bush Encroachment on the Rangelands of Borana, Southern Ethiopia." *Human Ecology* 36(2): 201–215.

Aragón, Roxana, Juan Manuel Morales, and E. Ezcurra. 2003. "Species Composition and Invasion in NW Argentinian Secondary Forests: Effects of Land Use History, Environment and Landscape." *Journal of Vegetation Science* 14 (2): 195–204.

Archibald, Sally, A. Carla Staver, and Simon A. Levin. 2012. "Evolution of Human-Driven Fire Regimes in Africa." *Proceedings of the National Academy of Sciences* 109 (3): 847–852.

Bach, Andrew, and Dave Conca. 2004. "Final Report: Natural History of the Ahlstrom's and Roose's Prairies, Olympic National Park, Washington." Bellingham: Western Washington University.

Bai, Xuemei, Sander van der Leeuw, Karen O'Brien, Frans Berkhout, Frank Biermann, Eduardo S. Brondizio, Christophe Cudennec, John Dearing, Anantha Duraiappah, Marion Glaser, Andrew Revkin, Will Steffen, and James Syvitski. 2016. "Plausible and Desirable Futures in the Anthropocene: A New Research Agenda." *Global Environmental Change* 39 (July): 351–362.

Beerling, David J., and Colin P. Osborne. 2006. "The Origin of the Savanna Biome." *Global Change Biology* 12 (11): 2023–2031.

Benson, J. S., and P. A. Redpath. 1997. "The Nature of Pre-European Native Vegetation in South-Eastern Australia: A Critique of Ryan, D.G., Ryan, J.R. and Starr, B.J. (1995) The Australian Landscape—Observations of Explorers and Early Settlers." *Cunninghamia* 5 (2): 285–328.

Berkes, Fikret. 2012. *Sacred Ecology*. New York: Routledge.

Berna, Francesco, Paul Goldberg, Liora Kolska Horwitz, James Brink, Sharon Holt, Marion Bamford, and Michael Chazan. 2012. "Microstratigraphic Evidence of in Situ Fire in the Acheulean Strata of Wonderwerk Cave, Northern Cape Province, South Africa." *Proceedings of the National Academy of Sciences* 109 (20): E1215–1220.

Bilbao, Bibiana A., Alejandra V. Leal, and Carlos L. Méndez. 2010. "Indigenous Use of Fire and Forest Loss in Canaima National Park, Venezuela: Assessment of and Tools for Alternative Strategies of Fire Management in Pemón Indigenous Lands." *Human Ecology* 38 (5): 663–673.

Bird, Douglas W., Rebecca Bliege Bird, and Brian F. Codding. 2016. "Pyrodiversity and the Anthropocene: The Role of Fire in the Broad Spectrum Revolution." *Evolutionary Anthropology: Issues, News, and Reviews* 25 (3): 105–116.

Bird, Douglas W., Rebecca Bliege Bird, and Christopher H. Parker. 2005. "Aboriginal Burning Regimes and Hunting Strategies in Australia's Western Desert." *Human Ecology* 33 (4): 443–464.

Bird, Michael I., and J. A. Cali. 1998. "A Million-Year Record of Fire in Sub-Saharan Africa." *Nature* 394 (6695): 767–769.

Bliege Bird, Rebecca, Douglas W. Bird, C. H. Parker, and J. H. Jones. 2008. "The 'Fire Stick Farming' Hypothesis: Australian Aboriginal Foraging Strategies, Biodiversity, and Anthropogenic Fire Mosaics." *Proceedings of the National Academy of Sciences* 105 (39): 14796–14801.

Bliege Bird, Rebecca, Brian F. Codding, Peter G. Kauhanen, and Douglas W. Bird. 2012. "Aboriginal Hunting Buffers Climate-Driven Fire-Size Variability in Australia's Spinifex Grasslands." *Proceedings of the National Academy of Sciences* 109 (26): 10287–10292.

Bond, William J., and Jon E. Keeley. 2005. "Fire as a Global 'Herbivore': The Ecology and Evolution of Flammable Ecosystems." *Trends in Ecology & Evolution* 20 (7): 387–394.

Bond, William J., and Catherine L. Parr. 2010. "Beyond the Forest Edge: Ecology, Diversity and Conservation of the Grassy Biomes." *Biological Conservation* 143 (10): 2395–2404.

Bond, William J., F. I. Woodward, and G. F. Midgley. 2005. "The Global Distribution of

Ecosystems in a World without Fire." *New Phytologist* 165 (2): 525–538.

Bowman, David M. J. S. 1998. "The Impact of Aboriginal Landscape Burning on the Australian Biota." *New Phytologist* 140 (3): 385–410.

———. 2014. "What Is the Relevance of Pyrogeography to the Anthropocene?" *The Anthropocene Review* 1 (2): 73–76.

Bowman, David M. J. S., Jennifer K. Balch, Paulo Artaxo, William J. Bond, Jean M. Carlson, Mark A. Cochrane, Carla M. D'Antonio, Ruth S. DeFries, John C. Doyle, Sandy P. Harrison, Fay H. Johnston, Jon E. Keeley, Meg A. Krawchuk, Christian A. Kull, J. Brad Marston, Max A. Moritz, I. Colin Prentice, Christopher I. Roos, Andrew C. Scott, Thomas W. Swetnam, Guido R. van der Werf, and Stephen J. Pyne. 2009. "Fire in the Earth System." *Science* 324 (5926): 481–484.

Bowman, David M. J. S., M. Garde, and A. Saulwick. 2001. "Kunj-Ken Makka Man-Wurrk Fire Is for Kangaroos: Interpreting Aboriginal Accounts of Landscape Burning in Central Arnhem Land." In *Land, Histories of Old Ages: Essays in Honour of Rhys Jones*, edited by A. Anderson, I. Lilley, and S. O'Connor, 61–78. Canberra: Pandanus Books.

Bowman, David M. J. S., Jessica A. O'Brien, and Johann G. Goldammer. 2013. "Pyrogeography and the Global Quest for Sustainable Fire Management." *Annual Review of Environment and Resources* 38 (1): 57–80.

Bowman, David M. J. S., Angie Walsh, and L. D. Prior. 2004. "Landscape Analysis of Aboriginal Fire Management in Central Arnhem Land, North Australia." *Journal of Biogeography* 31 (2): 207–223.

Boyd, Robert T. 1990. *The Coming of the Spirit of Pestilence: Introduced Infectious Diseases and Population Decline among Northwest Coast Indians, 1774-1874*. Seattle: University of Washington Press.

———, ed. 1999. *Indians, Fire and the Land in the Pacific Northwest*. Corvallis: Oregon State University Press.

Bradstock, Ross A., M. M. Boer, G. J. Cary, O. F. Price, R. J. Williams, D. Barrett, G. Cook, A.M. Gill, L.B.W. Hutley, H. Keith, S.W. Maier, M. Meyer, S.H. Roxburgh, and J. Russell-Smith. 2012. "Modelling the Potential for Prescribed Burning to Mitigate Carbon Emissions from Wildfires in Fire-Prone Forests of Australia." *International Journal of Wildland Fire* 21 (6): 629–639.

Bradstock, Ross A., Richard J. Williams, and A. Malcolm Gill. 2012. "Future Fire Regimes of Australian Ecosystems: New Perspectives on Enduring Questions of Management." In *Flammable Australia: Fire Regimes, Biodiversity and Ecosystems in a Changing World*, edited by Ross A. Bradstock, A. Malcolm Gill, and Richard J. Williams, 307–324. Collingwood: CSIRO.

Braithwaite, Richard W. 1991. "Aboriginal Fire Regimes of Monsoonal Australia in the 19th Century." *Search* 22 (7): 247–249.

Brondizio, Eduardo S, Karen O'Brien, Xuemei Bai, Frank Biermann, Will Steffen, Frans Berkhout, Christophe Cudennec, Maria Carmen Lemos, Alexander Wolfe, Jose Palma-Oliveira, and Chen-Tung Arthur Chen. 2016. "Re-Conceptualizing the Anthropocene: A Call for Collaboration." *Global Environmental Change* 39: 318-327.

Brown, K. J., and R. J. Hebda. 2003. "Coastal Rainforest Connections Disclosed through a Late Quaternary Vegetation, Climate, and Fire History Investigation from the Mountain Hemlock Zone on Southern Vancouver Island, British Colombia, Canada." *Review of Palaeobotany and Palynology* 123 (3–4): 247–269.

Burrows, Neil D., Andrew A. Burbidge, Phillip J. Fuller, and Graeme Behn. 2006. "Evidence of Altered Fire Regimes in the Western Desert Region of Australia." *Conservation Science Western Australia* 5 (3): 272–284.

Burtchard, Greg C. 2009. "Archaeology and Prehistory of Buck Lake, Mt. Rainier National Park." Auburn, WA: Preservation Department, Muckleshoot Indian Tribe.

Busse, Walter Carl Otto. 1908. "Die periodischen Grasbrände im tropischen Afrika ihr Einfluss auf die Vegetation und ihre Bedeutung für die Landeskultur." *Schutzgebieten* 2: 113–139.

Butler, Virginia L., and Sarah K. Campbell. 2004. "Resource Intensification and Resource Depression in the Pacific Northwest of North America: A Zooarchaeological Review." *Journal of World Prehistory* 18 (4): 327–405.

Butz, Ramona J. 2009. "Traditional Fire Management: Historical Fire Regimes and Land Use Change in Pastoral East Africa." *International Journal of Wildland Fire* 18 (4): 442–450.

———. 2013. "Changing Land Management: A Case Study of Charcoal Production among a Group of Pastoral Women in Northern Tanzania." *Energy for Sustainable Development* 17 (2): 138–145.

Campbell, Sarah, and Virginia Butler. 2010. "Archaeological Evidence for Resilience of Pacific Northwest Salmon Populations and the Socioecological System over the Last ~7,500 Years." *Ecology and Society* 15 (1): 17.

Carranza, Tharsila, Andrew Balmford, Valerie Kapos, and Andrea Manica. 2014. "Protected Area Effectiveness in Reducing Conversion in a Rapidly Vanishing Ecosystem: The Brazilian Cerrado." *Conservation Letters* 7 (3): 216–223.

Carvalho, Fábio M. V., Paulo De Marco, and Laerte G Ferreira. 2009. "The Cerrado into Pieces: Habitat Fragmentation as a Function of Landscape Use in the Savannas of Central Brazil." *Biological Conservation* 142 (7): 1392–1403.

Cashdan, Elizabeth A. 1985. "Coping with Risk: Reciprocity Among the Basarwa of Northern Botswana." *Man* 20 (3): 454–474.

Castro, Elmar Andrade de, and J. Boone Kauffman. 1998. "Ecosystem Structure in the Brazilian Cerrado: A Vegetation Gradient of Aboveground Biomass, Root Mass and Consumption by Fire." *Journal of Tropical Ecology* 14 (3): 263–283.

Cavalcanti, Roberto B., and Maria Alice S. Alves. 1997. "Effects of Fire on Savanna Birds in Central Brazil." *Ornitologia Neotropical* 8: 85–87.

Cerling, Thure E., Jonathan G. Wynn, Samuel A. Andanje, Michael I. Bird, David Kimutai Korir, Naomi E. Levin, William Mace, Anthony N. Macharia, Jay Quade, and Christopher H. Remien. 2011. "Woody Cover and Hominin Environments in the Past 6 Million Years." *Nature* 476 (7358): 51–56.

Clarkson, Chis, Zenobia Jacobs, Ben Marwick, Richard Fullagar, Lynley Wallis, Mike Smith, Richard G. Roberts, Elspeth Hayes, Kelsey Lowe, Xavier Carah, S. Anna Florin, Jessica McNeil, Delyth Cox, Lee J. Arnold, Quan Hua, Jillian Huntley, Helen E. A. Brand, Tiina Manne, Andrew Fairbairn, James Shulmeister, Lindsey Lyle, Makiah Salinas, Mara Page, Kate Connell, Gayoung Park, Kasih Norman, Tessa Murphy, and Colin Pardoe. 2017. "Human occupation of northern Australia by 65,000 years ago." *Nature* 547: 306–310.

Clean Energy Regulator. 2016. "Emissions Reduction Fund Project Register." Accessed on February 11, 2016. http://www.cleanenergyregulator.gov.au/ERF/project-and-contracts-registers/project-register.

Cohen, Tim J., John D. Jansen, Luke A. Gliganic, Joshua R. Larsen, Gerald C. Nanson, Jan-Hendrik May, Brian G. Jones, and David M. Price. 2015. "Hydrological Transformation Coincided with Megafaunal Extinction in Central Australia." *Geology* 43 (3): 195–98.

Colfer, Carol J. Pierce, Janis B. Alcorn, and Diane Russell. 2015. "Swiddens and Fallows: Reflections on the Global and Local Values of 'Slash and Burn.'" In *Shifting Cultivation and Environmental Change: Indigenous People, Agriculture and Forest Conservation*, edited by Malcolm Cairns, 62–86. Abingdon: Routledge.

Collins, June McCormick. 1974. *Valley of the Spirits: The Upper Skagit Indians of Western*

Washington. Seattle: University of Washington Press.

Coupland, Gary. 1998. "Maritime Adaptation and Evolution of the Developed Northwest Coast Pattern on the Central Northwest Coast." *Arctic Anthropology* 35 (1): 36–56.

Coutinho, Leopoldo Magno. 1990. "Fire in the Ecology of the Brazilian Cerrado." In *Fire in the Tropical Biota: Ecosystem Processes and Global Challenges*, edited by J.G. Goldammer, 82–105. New York: Columbia University Press.

———. 2002. "O Bioma do Cerrado." In *Eugene Warming e o Cerrado Brasileiro: Um Século Depois*, edited by Aldo Luiz Klein, 77–91. São Paulo: Editora UNESP.

Crowley, G. M., and S. T. Garnett. 2000. "Changing Fire Management in the Pastoral Lands of Cape York Peninsula of Northeast Australia, 1623 to 1996." *Australian Geographical Studies* 38 (1): 10–26.

Denevan, William M. 1992. "The Pristine Myth: The Landscape of the Americas in 1492." *Annals of the Association of American Geographers* 82 (3): 369–385.

———. 2001. *Cultivated Landscapes of Native Amazonia and the Andes*. New York: Oxford University Press.

Department of the Environment and Energy, Australia. 2016a. "Indigenous Protected Areas." Accessed on May 9, 2016. http://www.environment.gov.au/land/indigenous-protected-areas.

———. 2016b. "Working on Country." Accessed on May 9, 2016. http://www.environment.gov.au/indigenous/workingoncountry/.

Derr, Kelly M. 2014. "Anthropogenic Fire and Landscape Management on Valdes Island, Southwestern BC." *Canadian Journal of Archaeology* 38 (1): 250–279.

Deur, Douglas, and Nancy J. Turner, eds. 2005. *Keeping It Living: Traditions of Plant Use and Cultivation on the Northwest Coast of North America*. Seattle: University of Washington Press.

Dickson, Louise. 2016. "Planned Burn on Tumbo Island Has Saturna Residents Fuming." *Times Colonist*, August 6. Accessed on February 8, 2016. http://www.timescolonist.com/news/local/planned-burn-on-tumbo-island-has-saturna-residents-fuming-1.2317721.

Director of National Parks, Australia. 2016. "Kakadu National Park Management Plan 2016-2026." Canberra, ACT. http://www.environment.gov.au/resource/kakadu-national-park-management-plan-2016-2026.

Dore, Jeremy, Christine Michael, Jeremy Russell-Smith, Maureen Tehan, and Lisa Caripis. 2014. "Carbon Projects and Indigenous Land in Northern Australia." *Rangeland Journal* 36 (4): 389–402.

Dupont, Lydie M., Florian Rommerskirchen, Gesine Mollenhauer, and Enno Schefuß. 2013. "Miocene to Pliocene Changes in South African Hydrology and Vegetation in Relation to the Expansion of C4 Plants." *Earth and Planetary Science Letters* 375 (August): 408–417.

Durigan, Giselda, and James A. Ratter. 2016. "The Need for a Consistent Fire Policy for Cerrado Conservation." *Journal of Applied Ecology* 53 (1): 11–15.

Eiten, George. 1972. "The Cerrado Vegetation of Brazil." *Botanical Review* 38 (2): 201–341.

———. 1994. "Vegetação." In *Cerrado: Caracterização, Ocupação E Perspectivas*, edited by Maria Novae Pinto. Brasília: Editora Universidade de Brasília.

Eriksen, Christine. 2007. "Why Do They Burn the 'Bush'? Fire, Rural Livelihoods, and Conservation in Zambia." *Geographical Journal* 173 (3): 242–256.

Eriksson, Ove, Sara A. O. Cousins, Hans Henrik Bruun, and S. Díaz. 2002. "Land-Use History and Fragmentation of Traditionally Managed Grasslands in Scandinavia." *Journal of Vegetation Science* 13 (5): 743–748.

Erni, Christian. 2009. "Shifting the Blame? Southeast Asia's Indigenous Peoples and Shifting Cultivation in the Age of Climate Change." *Indigenous Affairs* 1: 38–49.

Fairhead, James, and Melissa Leach. 1996. *Misreading the African Landscape: Society and Ecology in a Forest-Savanna Mosaic*. Cambridge: Cambridge University Press.

Falleiro, Rodrigo de Moraes. 2011. "Resgate do manejo tradicional do Cerrado com fogo para proteção das Terras Indígenas do oeste do Mato Grosso: Um estudo de caso." *Biodiversidade Brasileira* (2): 86–96.

Falleiro, Rodrigo de Moraes, Marcelo Trindade Santana, and Cendi Ribas Berni. 2016. "As contribuições do Manejo Integrado do Fogo para o controle dos incêndios florestais nas Terras Indígenas do Brasil." Biodiversidade Brasileira 6 (2):88–105.

Fensham, R. J. 1997. "Aboriginal Fire Regimes in Queensland, Australia: Analysis of the Explorers' Record." *Journal of Biogeography* 24 (1): 11–22.

Flannery, Tim F. 1994. *The Future Eaters: An Ecological History of the Australasian Lands and People*. Chatswood, N.S.W.: Reed Books.

Forsyth, Tim. 2003. *Critical Political Ecology: The Politics of Environmental Science*. London: Routledge.

Fowler, Cynthia T., and James R. Welch. 2015. "Introduction: Special Issue on Fire Ecology and Ethnobiology." *Journal of Ethnobiology* 35 (1): 1–3.

França, H., G. J. Oliveira, and A. Pereira. 2004. "Mapeamento de queimadas antrópicas no cerrado durante a primeira metade do século XIX com base em relatos de naturalistas europeus." In *Anais do XV Congresso da Sociedade de Botânica de São Paulo*. São Paulo: Sociedade de Botânica de São Paulo

Françoso, Renata D., Reuber Brandão, Cristiano C. Nogueira, Yuri B. Salmona, Ricardo Bomfim Machado, and Guarino R. Colli. 2015. "Habitat Loss and the Effectiveness of Protected Areas in the Cerrado Biodiversity Hotspot." *Natureza & Conservação* 13 (1): 35–40.

Franklin, Jerry F., and C. T. Dyrness. 1988. *Natural Vegetation of Oregon and Washington*. Corvallis: Oregon State University Press.

Frizzo, Tiago L. Massochini, Camila Bonizário, and Heraldo L. Vasconcelos. 2011. "Revisão dos efeitos do fogo sobre a fauna de formações savânicas do Brasil." *Oecologia Australis* 15 (2): 365–379.

Frost, Peter G. H. 1999. "Fire in Southern African Woodlands: Origins, Impacts, Effects and Control." In *FAO Meeting on Public Policies Affecting Forest Fires: Food and Agriculture Organization of the United Nations, Rome 1998*, 181–205. Rome: Forestry Department, Food and Agriculture Organization of the United Nations (FAO).

Furley, Peter A. 1999. "The Nature and Diversity of Neotropical Savanna Vegetation with Particular Reference to the Brazilian Cerrados." *Global Ecology and Biogeography* 8 (3–4): 223–241.

Fynn, Richard W. S., David J. Augustine, Michael J. S. Peel, and Michel de Garine-Wichatitsky. 2016. "Strategic Management of Livestock to Improve Biodiversity Conservation in African Savannahs: A Conceptual Basis for Wildlife-livestock Coexistence." *Journal of Applied Ecology* 53 (2): 388–397.

Gammage, Bill. 2011. *The Biggest Estate on Earth: How Aborigines Made Australia*. Crows Nest, N.S.W.: Allen and Unwin.

Garde, M., B. L. Nadjamerrek, M. Kolkkiwarra, J. Kalarriya, J. Djandjomerr, B. Birriyabirriya, R. Bilindja, M. Kubarkku, P. Biless, J. Russell-Smith, P. J. Whitehead, and P. Cooke. "The Language of Fire: Seasonally, Resources and Landscape Burning on the Arnhem Land Plateau." In *Culture, Ecology and Economy of Fire Management in North Australian Savannas: Rekindling the Wurrk Tradition*, edited by Jeremy Russell-Smith, Peter Whitehead, and Peter Cooke, 85–164. Collingwood: CSIRO.

Gavin, Daniel G., David M. Fisher, Erin M. Herring, Ariana White, and Linda B. Brubaker.

2013. "Paleoenvironmental Change on the Olympic Peninsula, Washington: Forests and Climate from the Last Glaciation to the Present." Final report to Olympic National Park, Contract J8W07100028.

Gibbs, George. 1978. *Indian Tribes of Washington Territory*. Fairfield, WA: Galleon.

Giglio, Louis, James T. Randerson, and Guido R. van der Werf. 2013. "Analysis of Daily, Monthly, and Annual Burned Area Using the Fourth-Generation Global Fire Emissions Database (GFED4)." *Journal of Geophysical Research: Biogeosciences* 118 (1): 317–328.

Gil-Romera, Graciela, Henry F. Lamb, David Turton, Miguel Sevilla-Callejo, and Mohammed Umer. 2010. "Long-Term Resilience, Bush Encroachment Patterns and Local Knowledge in a Northeast African Savanna." *Global Environmental Change* 20 (4): 612–626.

Goland, Carol. 1991. "The Ecological Context of Hunter-Gatherer Storage: Environmental Predictability and Environmental Risk." *Michigan Discussions in Anthropology* 10: 107–120.

Gómez-Baggethun, Erik, Esteve Corbera, and Victoria Reyes-García. 2013. "Traditional Ecological Knowledge and Global Environmental Change: Research Findings and Policy Implications." *Ecology and Society: A Journal of Integrative Science for Resilience and Sustainability* 18 (4).

Gould, Richard A. 1971. "Uses and Effects of Fire among the Western Desert Aborigines of Australia." *Mankind* 8 (1): 14–24.

Gouveia, S. E. M., L. C. R. Pessenda, R. Aravena, R. Boulet, R. Scheel-Ybert, J. A. Bendassoli, A. S. Ribeiro, and H. A. Freitas. 2002. "Carbon Isotopes in Charcoal and Soils in Studies of Paleovegetation and Climate Changes during the Late Pleistocene and the Holocene in the Southeast and Centerwest Regions of Brazil." *Global and Planetary Change* 33 (1–2): 95–106.

Gowlett, John A. J., and Richard W. Wrangham. 2013. "Earliest Fire in Africa: Towards the Convergence of Archaeological Evidence and the Cooking Hypothesis." *Azania: Archaeological Research in Africa* 48 (1): 5–30.

Green, W. 1991. "Lutte contre les feux de Brousse." Report for project DERIK (Développement Rural Intégré de Kissidougou). Conakry: République de Guinée.

Grégoire, J. M., H. D. Eva, A. S. Belward, I. Palumbo, D. Simonetti, and A. Brink. 2013. "Effect of Land-Cover Change on Africa's Burnt Area." *International Journal of Wildland Fire* 22 (2): 107–120.

Haeberlin, Hermann Karl, and Erna Gunther. 1930. *The Indians of Puget Sound*. Seattle: University of Washington Press.

Hallam, Sylvia J. 1975. *Fire and Hearth: A Study of Aboriginal Usage and European Usurpation in South-Western Australia*. Canberra: Australian Institute of Aboriginal Studies.

Hamman, Sarah T., Peter W. Dunwiddie, Jason L. Nuckols, and Mason McKinley. 2011. "Fire as a Restoration Tool in Pacific Northwest Prairies and Oak Woodlands: Challenges, Successes, and Future Directions." *Northwest Science* 85 (2): 317–328.

Harris, Grant, Simon Thirgood, J. Grant, C. Hopcraft, Joris P. G. M. Cromsigt, and Joel Berger. 2009. "Global Decline in Aggregated Migrations of Large Terrestrial Mammals." *Endangered Species Research* 7 (1): 55–76.

Harris, John W. K., and Glynn L. Isaac. 1997. "Sites in the Upper KBS, Okote, and Chari Members: Reports." In *Koobi Fora Research Project*, 5: 115–236. Edited by Glynn Llywelyn Isaac and Barbara Isaac. Oxford: Clarendon.

Haynes, Christopher David. 1985. "The Pattern and Ecology of Munwag: Traditional Aboriginal Fire Regimes in North-Central Arnhemland." *Ecology of the Wet Dry Tropics: Proceedings of the Ecological Society of Australia* 13: 203–214.

Hebda, Richard J., and Rolf W. Mathewes. 1984. "Holocene History of Cedar and Native Indian Cultures of the North American Pacific Coast." *Science* 225 (4663): 711–713.

Hecht, Susanna B. 2009. "Kayapó Savanna Management: Fire, Soils, and Forest Islands in a Threatened Biome." In *Amazonian Dark Earths: Wim Sombroek's Vision*, edited by William I. Woods, Wenceslau G. Teixeira, Johannes Lehmann, Christoph Steiner, Antoinette WinklerPrins, and Lilian Rebellato, 143–162. Berlin: Springer.

Hemstrom, Miles A., and Jerry F. Franklin. 1982. "Fire and Other Disturbances of the Forests in Mount Rainier National Park." *Quaternary Research* 18 (1): 32–51.

Higgins, Steven I., William J. Bond, and Winston S. W. Trollope. 2000. "Fire, Resprouting and Variability: A Recipe for Grass-tree Coexistence in Savanna." *Journal of Ecology* 88 (2): 213–229.

Hoffman, Kira M., Daniel G. Gavin, and Brian M. Starzomski. 2016. "Seven Hundred Years of Human-Driven and Climate-Influenced Fire Activity in a British Columbia Coastal Temperate Rainforest." *Royal Society Open Science* 3 (10): 160608.

Homewood, Katherine. 2008. *Ecology of African Pastoralist Societies*. Oxford: James Currey.

Homewood, Katherine, and Daniel Brockington. 1999. "Biodiversity, Conservation and Development in Mkomazi Game Reserve, Tanzania." *Global Ecology and Biogeography* 8 (3–4): 301–313.

Hunt, J. 2012. "North to South?" In *People on Country: Vital Landscapes, Indigenous Futures*, edited by Jon C. Altman and Seán Kerins, 94–114. Annandale, N.S.W.: Federation.

Ickowitz, Amy. 2006. "Shifting Cultivation and Deforestation in Tropical Africa: Critical Reflections." *Development and Change* 37 (3): 599–626.

Ingalsbee, Timothy. 2010. *Getting Burned: A Taxpayer's Guide to Wildfire Suppression*. Eugene, OR: Firefighters United for Safety, Ethics, and Ecology.

Jeffery, Kathryn Jane, Lisa Korte, Florence Palla, Gretchen M. Walters, Lee White, and Katharine Abernethy. 2014. "Fire Management in a Changing Landscape: A Case Study from Lopé National Park, Gabon." *PARKS: The International Journal of Protected Areas and Conservation* 20 (1): 39–52.

Johnson, C. N., J. Alroy, N. J. Beeton, Michael I. Bird, B. W. Brook, A. Cooper, R. Gillespie, S. Herrando-Pérez, Z. Jacobs, G. H. Miller, G. J. Prideaux, R. G. Roberts, M. Rodríguez-Rey, F. Saltré, C. S. M. Turney, and C. J. A. Bradshaw. 2016. "What Caused Extinction of the Pleistocene Megafauna of Sahul?" *Proceedings of the Royal Society B: Biological Sciences* 283 (1824): 2015–2399.

Johnson, Chris. 2006. *Australia's Mammal Extinctions: A 50,000 Year History*. Cambridge: Cambridge University Press.

Jones, Rhys. 1969. "Fire-Stick Farming." *Australian Natural History* 16 (7): 224–228.

———. 1975. "The Neolithic, Palaeolithic and the Hunting Gardeners: Man and Land in the Antipodes." In *Quaternary Studies: Selected Papers from IX INQUA Congress, Christchurch, New Zealand 2-10 December, 1973*, edited by R. P. Suggate and M. M. Cresswell, 21–34. Wellington: Royal Society of New Zealand.

———. 1980. "Hunters in the Australian Coastal Savanna." In *Human Ecology in Savanna Environments*, 107–146. London: Academic Press.

Kaiser, Sandra. 2016. "Washington's Brutal Wildfires: A Call to Action." Invited address presented at Western Washington University, Pullman, January 26.

Kakadu National Park Board of Management. 2016. Kakadu National Park Management Plan 2016–2026. Canberra, ACT: Director of National Parks.

Kamau, Peter Ngugi, and Kimberly E. Medley. 2014. "Anthropogenic Fires and Local Livelihoods at Chyulu Hills, Kenya." *Landscape and Urban Planning* 124 (April): 76–84.

Kayll, A. J. 1974. "Use of Fire in Land Management." In *Fire and Ecosystems*, edited by T. T.

Kozlowski and C. E. Ahlgren, 483–511. New York: Academic.

Kennedy, A. D., and A. L. F. Potgieter. 2003. "Fire Season Affects Size and Architecture of Colophospermum Mopane in Southern African Savannas." *Plant Ecology* 167 (2): 179–192.

Kimber, Richard. 1983. "Black Lightning: Aborigines and Fire in Central Australia and the Western Desert." *Archaeology in Oceania* 18 (1): 38–45.

Kingsbury, Nancy. 2001. "Impacts of Land Use and Cultural Change in a Fragile Environment: Indigenous Acculturation and Deforestation in Kavanayén, Gran Sabana-Venezuela." *Interciencia* 26 (8): 327–336.

Klink, Carlos A., and Ricardo B. Machado. 2005. "Conservation of the Brazilian Cerrado." *Conservation Biology* 19 (3): 707–713.

Klink, Carlos A., and Adriana G. Moreira. 2002. "Past and Current Human Occupation, and Land Use." In *The Cerrados of Brazil: Ecology and Natural History of a Neotropical Savanna*, edited by Paulo S. Oliveira and Robert J. Marquis, 69–88. New York: Columbia University Press.

Krawchuk, Meg A., Max A. Moritz, Marc-André Parisien, Jeff Van Dorn, and Katharine Hayhoe. 2009. "Global Pyrogeography: The Current and Future Distribution of Wildfire." *PLOS ONE* 4 (4): e5102.

Kroeber, Alfred L. 1939. *Cultural and Natural Areas of Native North America*. Berkeley: University of California Press.

Kronik, Jakob, and Dorte Verner. 2010. *Indigenous Peoples and Climate Change in Latin America and the Caribbean*. Washington DC: World Bank Publications.

Kull, Christian A. 2000. "Deforestation, Erosion, and Fire: Degradation Myths in the Environmental History of Madagascar." *Environment and History* 6 (4): 423–450.

———. 2002. "Madagascar Aflame: Landscape Burning as Peasant Protest, Resistance, or a Resource Management Tool?" *Political Geography* 21 (7): 927–953.

Lambin, Eric F., B. L. Turner, Helmut J. Geist, Samuel B. Agbola, Arild Angelsen, John W. Bruce, Oliver T. Coomes, Rodolfo Dirzo, Günter Fischer, Carl Folke, P. S. George, Katherine Homewood, Jacques Imbernon, Rik Leemans, Xiubin Li, Emilio F. Moran, Michael Mortimore, P. S. Ramakrishnan, John F. Richards, Helle Skånes, Will Steffen, Glenn D. Stone, Uno Svedin, Tom A. Veldkamp, Coleen Vogel, and Jianchu Xu. 2001. "The Causes of Land-Use and Land-Cover Change: Moving beyond the Myths." *Global Environmental Change* 11 (4): 261–269.

Laris, Paul. 2002. "Burning the Seasonal Mosaic: Preventative Burning Strategies in the Wooded Savanna of Southern Mali." *Human Ecology* 30 (2): 155–186.

Laris, Paul, Sebastien Caillault, Sepideh Dadashi, and Audrey Jo. 2015. "The Human Ecology and Geography of Burning in an Unstable Savanna Environment." *Journal of Ethnobiology* 35 (1): 111–139.

Laris, Paul, and David Andrew Wardell. 2006. "Good, Bad or 'Necessary Evil'? Reinterpreting the Colonial Burning Experiments in the Savanna Landscapes of West Africa." *Geographical Journal* 172 (4): 271–290.

Latz, Peter. 1995. *Bushfires and Bushtucker: Aboriginal Plant Use in Central Australia*. Alice Springs: IAD.

Lawes, Michael J., Brett P. Murphy, Alaric Fisher, John C. Z. Woinarski, Andrew C. Edwards, and Jeremy Russell-Smith. 2015. "Small Mammals Decline with Increasing Fire Extent in Northern Australia: Evidence from Long-Term Monitoring in Kakadu National Park." *International Journal of Wildland Fire* 24 (5): 712–722.

LeCompte-Mastenbrook, Joyce. 2014. "Historical Ecologies of Swətixʷtəd in the Duwamish-Green-White River of Washington State." Unpublished fellowship report. Seattle: Burke Museum of Natural History. Accessed on February 26, 2017. https://ethnobiology.org/conference/abstracts/38.

———. 2015. "Restoring Coast Salish Foods and Landscapes: A More-Than-Human Politics of Place, History and Becoming." Ph.D. dissertation, University of Washington.

Ledru, Marie-Pierre. 2002. "Late Quaternary History and Evolution of the Cerrados as Revealed by Palynological Records." In *The Cerrados of Brazil: Ecology and Natural History of a Neotropical Savanna*, edited by Paulo S. Oliveira and Robert J. Marquis, 33–50. New York: Columbia University Press.

Leichhardt, Ludwig. 1847. *Journal of an Overland Expedition in Australia from Moreton Bay to Port Essington, a Distance of Upwards of 3000 Miles, during the Years, 1844-1845*. London: T & W Boone.

Leopold, Estella, and Robert T. Boyd. 1999. "An Ecological History of Old Prairie Areas in Southwestern Washington." In *Indians, Fire, and the Land in the Pacific Northwest*, edited by Robert T. Boyd, 139–163. Corvallis: Oregon State University Press.

Lepofsky, Dana, Douglas Hallett, Ken Lertzman, Rolf Mathewes, Albert McHalsie, and Kevin Washbrook. 2005. "Documenting Precontact Plant Management on the Northwest Coast: An Example of Prescribed Burning in the Central and Upper Fraser Valley, British Columbia." In *Keeping It Living: Traditions of Plant Use and Cultivation on the Northwest Coast*, edited by Douglas Deur and Nancy J. Turner, 218–239. Seattle: University of Washington Press.

Li, Peng, Zhiming Feng, Luguang Jiang, Chenhua Liao, and Jinghua Zhang. 2014. "A Review of Swidden Agriculture in Southeast Asia." *Remote Sensing* 6 (2): 1654–1683.

Linder, H. Peter. 2014. "The Evolution of African Plant Diversity." *Frontiers in Ecology and Evolution* 2: 38.

Loarie, Scott R., David B. Lobell, Gregory P. Asner, Qiaozhen Mu, and Christopher B. Field. 2011. "Direct Impacts on Local Climate of Sugar-Cane Expansion in Brazil." *Nature Climate Change* 1 (2): 105–109.

Lövbrand, Eva, Silke Beck, Jason Chilvers, Tim Forsyth, Johan Hedrén, Mike Hulme, Rolf Lidskog, and Eleftheria Vasileiadou. 2015. "Who Speaks for the Future of Earth? How Critical Social Science Can Extend the Conversation on the Anthropocene." *Global Environmental Change* 32: 211–218.

Lynch, Amanda H., Jason Beringer, Peter Kershaw, Andrew Marshall, Scott Mooney, Nigel Tapper, Chris Turney, and Sander van der Kaars. 2007. "Using the Paleorecord to Evaluate Climate and Fire Interactions in Australia." *Annual Review of Earth and Planetary Sciences* 35: 215–239.

Mann, Charles C. 2005. *1491: New Revelations of the Americas before Columbus*. New York: Knopf.

Mbow, C., T. T. Nielsen, and K. Rasmussen. 2000. "Savanna Fires in East-Central Senegal: Distribution Patterns, Resource Management and Perceptions." *Human Ecology* 28 (4): 561–583.

McGowan, Hamish, Samuel Marx, Patrick Moss, and Andrew Hammond. 2012. "Evidence of ENSO Mega-Drought Triggered Collapse of Prehistory Aboriginal Society in Northwest Australia." *Geophysical Research Letters* 39 (22): L22702.

Melo, Monica Martins de, and Carlos Hiroo Saito. 2013. "The Practice of Burning Savannas for Hunting by the Xavante Indians Based on the Stars and Constellations." *Society and Natural Resources* 26 (4): 478–487.

Mertz, Ole. 2009. "Trends in Shifting Cultivation and the REDD Mechanism." *Current Opinion in Environmental Sustainability* 1 (2): 156–160.

Miller, Gifford H., Marilyn L. Fogel, John W. Magee, Michael K. Gagan, Simon J. Clarke, and Beverly J. Johnson. 2005. "Ecosystem Collapse in Pleistocene Australia and a Human

Role in Megafaunal Extinction." *Science* 309 (5732): 287–290.

Miller, Gifford H., Jennifer Mangan, David Pollard, Starley Thompson, Benjamin Felzer, and John Magee. 2005. "Sensitivity of the Australian Monsoon to Insolation and Vegetation: Implications for Human Impact on Continental Moisture Balance." *Geology* 33 (1): 65–68.

Miranda, Heloisa S., Mercedes M. C. Bustamante, and Antonio C. Miranda. 2002. "The Fire Factor." In *The Cerrados of Brazil: Ecology and Natural History of a Neotropical Savannah*, edited by Paulo S. Oliveira and Robert J. Marquis, 51–68. New York: Columbia University Press.

Miranda, Heloisa S., Margarete Naomi Sato, Walter Nascimento Neto, and Felipe Salvo Aires. 2009. "Fires in the Cerrado, the Brazilian Savanna." In *Tropical Fire Ecology: Climate Change, Land Use and Ecosystem Dynamics*, edited by Mark A. Cochrane, 427–450. Berlin: Springer.

Miranda, Izildinha Souza, and Maria Lúcia Absy. 2000. "Physiognomy of the Savannas of Roraima, Brazil." *Acta Amazonica* 30 (3): 423–440.

Mistry, Jayalaxshmi, Andrea Berardi, Valeria Andrade, Txicaprô Krahô, Phocrok Krahô, and Othon Leonardos. 2005. "Indigenous Fire Management in the Cerrado of Brazil: The Case of the Krahô of Tocantíns." *Human Ecology* 33 (3): 365–386.

Mistry, Jayalaxshmi, Bibiana A. Bilbao, and Andrea Berardi. 2016. "Community Owned Solutions for Fire Management in Tropical Ecosystems: Case Studies from Indigenous Communities of South America." *Philosophical Transactions of the Royal Society B: Biological Sciences* 371 (1696): 20150174.

Mitchell, T. L. 1848. *Journal of an Expedition into the Interior of Tropical Australia in Search of a Route from Sydney to the Gulf of Carpentaria*. London: Longman, Brown, Green and Longmans.

MMA/IBAMA. 2011. "Monitoramento do desmatamento nos biomas brasileiros por satélite—Acordo de cooperação técnica MMA/IBAMA: Monitoramento do bioma cerrado 2009-2010." Brasilia: Ministério do Meio Ambiente, Accessed on December 27, 2017. http://www.mma.gov.br/estruturas/sbf_chm_rbbio/_arquivos/relatoriofinal_cerrado_2010_final_72_1.pdf.

Mooney, S. D., S. P. Harrison, P. J. Bartlein, A. L. Daniau, J. Stevenson, K. C. Brownlie, S. Buckman et al. 2011. "Late Quaternary Fire Regimes of Australasia." *Quaternary Science Reviews* 30 (1–2): 28–46.

Mooney, S. D., S. P. Harrison, P. J. Bartlein, and J. Stevenson. 2012. "The Prehistory of Fire in Australia." In *Flammable Australia: Fire Regimes, Biodiversity and Ecosystems in a Changing World*, edited by Ross A. Bradstock, A. Malcolm Gill, and Richard J. Williams. Collingwood: CSIRO.

Moss, Madonna. 2011. *Northwest Coast: Archaeology as Deep History*. Washington, DC: SAA Press.

Mulvaney, John, and Johan Kamminga. 1999. *Prehistory of Australia*. Sydney: Allen and Unwin.

Neke, Kirsten S., and Morné A. Du Plessis. 2004. "The Threat of Transformation: Quantifying the Vulnerability of Grasslands in South Africa." *Conservation Biology* 18 (2): 466–477.

Nimuendaju, Curt. 1939. *The Apinayé*. Washington, DC: Catholic University of America Press.

Norton, Helen H. 1979. "The Association Between Anthropogenic Prairies and Important Food Plants in Western Washington." *Northwest Anthropological Research Notes* 13 (2): 175–200.

———. 1985. "Women and Resources of the Northwest Coast: Documentation from the 18th and Early 19th Centuries." Ph.D. dissertation, University of Washington.

O'Connell, J. F., and J. Allen. 2015. "The Process, Biotic Impact, and Global Implications of the Human Colonization of Sahul about 47,000

Years Ago." *Journal of Archaeological Science* 56: 73–84.

O'Connor, Christopher D., Gregg M. Garfin, Donald A. Falk, and Thomas W. Swetnam. 2011. "Human Pyrogeography: A New Synergy of Fire, Climate and People Is Reshaping Ecosystems across the Globe." *Geography Compass* 5 (6): 329–350.

Odum, Eugene P. 1964. "The New Ecology." *BioScience* 14 (7): 14–16.

———. 1977. "The Emergence of Ecology as a New Integrative Discipline." *Science* 195 (4284): 1289–1293.

Park, Mi Sun, and Daniela Kleinschmit. 2016. "Framing Forest Conservation in the Global Media: An Interest-Based Approach." *Forest Policy Analysis: Advancing the Analytical Approach* 68: 7–15.

Parr, Catherine L., and B. H. Brockett. 1999. "Patch-Mosaic Burning: A New Paradigm for Savanna Fire Management in Protected Areas?" *Koedoe* 42 (2): 117–130.

Parr, Catherine L., John C. Z. Woinarski, and Danie J. Pienaar. 2009. "Cornerstones of Biodiversity Conservation? Comparing the Management Effectiveness of Kruger and Kakadu National Parks, Two Key Savanna Reserves." *Biodiversity and Conservation* 18 (13): 3643.

Parrotta, John A., and Mauro Agnoletti. 2012. "Traditional Forest-Related Knowledge and Climate Change." In *Traditional Forest-Related Knowledge*, edited by John A. Parrotta and Ronald L. Trosper, 12:491–533. Dordrecht: Springer Netherlands.

Pascoe, Bruce. 2014. *Dark Emu: Black Seeds Agriculture or Accident?* Broome, W.A.: Magabala Books.

Peasley, W. J. 1983. *The Last of the Nomads.* Fremantle, W. A.: Fremantle.

Pellatt, Marlow G., Marian M. McCoy, and Rolf W. Mathewes. 2015. "Paleoecology and Fire History of Garry Oak Ecosystems in Canada: Implications for Conservation and Environmental Management." *Biodiversity and Conservation* 24 (7): 1621–1639.

Pereira Júnior, Alfredo C., Sofia L. J. Oliveira, José M. C. Pereira, and Maria Antónia Amaral Turkman. 2014. "Modelling Fire Frequency in a Cerrado Savanna Protected Area." *PLOS ONE* 9 (7): e102380.

Perrot, Radhika. 2015. "The Trojan Horses of Global Environmental and Social Politics." In *Earth, Wind and Fire: Unpacking the Political, Economic and Security Implications of Discourse on the Green Economy. Green Economy Research Report, Research and Policy Development to Advance a Green Economy in South Africa*, edited by L. Mytelka, V. Msimang, R. Perrot, and Mapungubwe Institute for Strategic Reflection, 21–36. Johannesburg: Real African Publishers.

Peter, D. H., and D. Shebitz. 2006. "Historic Anthropogenically Maintained Bear Grass Savannas of the Southeastern Olympic Peninsula." *Restoration Ecology* 14 (4): 605–615.

Petty, Aaron M., Vanessa deKoninck, and Ben Orlove. 2015. "Cleaning, Protecting, or Abating? Making Indigenous Fire Management 'Work' in Northern Australia." *Journal of Ethnobiology* 35 (1): 140–162.

Phillips, John F. V. 2012. "'Fire: Its Influence on Biotic Communities and Physical Factors in South and East Africa.'" *Fire Ecology* 8 (2): 2–16.

Pivello, Vânia R. 2006. "Fire Management for Biological Conservation in the Brazilian Cerrado." In *Savannas and Dry Forests: Linking People with Nature*, edited by Jayalaxshmi Mistry and Andrea Berardi, 129–154. Aldershot, England: Ashgate.

———. 2011. "The Use of Fire in the Cerrado and Amazonian Rainforests of Brazil: Past and Present." *Fire Ecology* 7 (1): 24–39.

Posey, Darrell A. 1985. "Indigenous Management of Tropical Forest Ecosystems: The Case of the Kayapó Indians of the Brazilian Amazon." *Agroforestry Systems* 3 (2): 139–158.

———. 1986. "Manejo da floresta secundária, capoeiras, campos e cerrados (Kayapó)." In *Suma Etnológica Brasileira 1: Etnobiologia*, edited by Berta G. Ribeiro, 173–185. Petrópolis: FINEP.

Prada, Manrique, Jader S. Marinho-Filho, and Peter W. Price. 1995. "Insects in Flower Heads of Aspilia Foliacea (Asteraceae) after a Fire in a Central Brazilian Savanna: Evidence for the Plant Vigor Hypothesis." *Biotropica* 27: 513–518.

Preece, Noel. 2002. "Aboriginal Fires in Monsoonal Australia from Historical Accounts." *Journal of Biogeography* 29 (3): 321–336.

———. 2007. "Traditional and Ecological Fires and Effects of Bushfire Laws in North Australian Savannas." *International Journal of Wildland Fire* 16 (4): 378–389.

Preston, D. 2005. "Tribes Foster Plants for Traditional Uses." *NWIFC News* (Spring): 8–9.

———. 2016. "Prairie Burns Return to Quinault." *Northwest Treaty Tribes*. June 24. Accessed on February 11, 2017. http://nwtreatytribes.org/prairie-burning-returns-quinault/.

Price, Gilbert J., Gregory E. Webb, Jian-xin Zhao, Yue-xing Feng, Andrew S. Murray, Bernard N. Cooke, Scott A. Hocknull, and Ian H. Sobbe. 2011. "Dating Megafaunal Extinction on the Pleistocene Darling Downs, Eastern Australia: The Promise and Pitfalls of Dating as a Test of Extinction Hypotheses." *Quaternary Science Reviews* 30 (7–8): 899–914.

Pyne, Stephen J. 2007. "Problems, Paradoxes, Paradigms: Triangulating Fire Research." *International Journal of Wildland Fire* 16 (3): 271–276.

Ramos-Neto, Mário Barroso, and Vânia Regina Pivello. 2000. "Lightning Fires in a Brazilian Savanna National Park: Rethinking Management Strategies." *Environmental Management* 26 (6): 675–684.

Ratnam, Jayashree, William J. Bond, Rod J. Fensham, William A. Hoffmann, Sally Archibald, Caroline E. R. Lehmann, Michael T. Anderson, Steven I. Higgins, and Mahesh Sankaran. 2011. "When Is a 'Forest' a Savanna, and Why Does It Matter?" *Global Ecology and Biogeography* 20 (5): 653–660.

Ratter, James A., José Felipe Ribeiro, and Samuel Bridgewater. 1997. "The Brazilian Cerrado Vegetation and Threats to Its Biodiversity." *Annals of Botany* 80 (3): 223–230.

Ratter, James A., P. W. Richards, G. Argent, and D. R. Gifford. 1973. "Observations on the Vegetation of Northeastern Mato Grosso: I. The Woody Vegetation Types of the Xavantina-Cachimbo Expedition Area." *Philosophical Transactions of the Royal Society B: Biological Sciences* 266 (880): 449–492.

Reyes-García, Victoria. 2015. "The Values of Traditional Ecological Knowledge." In *Handbook of Ecological Economics*, edited by Joan Martínez-Alier and Roldan Muradian, 283–306. Cheltenham, UK: Edward Elgar.

Rizzini, Carlos Toledo, and Ezechias Paulo Heringer. 1961. "Underground Organs of Plants from Some Southern Brazilian Savannas with Special Reference to the Xylopodium." *Phyton* 17 (1): 105–124.

Roberts, Richard G., Rhys Jones, and M. A. Smith. 1990. "Thermoluminescence Dating of a 50,000-Year-Old Human Occupation Site in Northern Australia." *Nature* 345 (6271): 153–156.

Roos, Christopher I., David M. J. S. Bowman, Jennifer K. Balch, Paulo Artaxo, William J. Bond, Mark Cochrane, Carla M. D'Antonio, Ruth DeFries, Michelle Mack, Fay H. Johnston, Meg A. Krawchuk, Christian A. Kull, Max A. Moritz, Stephen Pyne, Andrew C. Scott, and Thomas W. Swetnam. 2014. "Pyrogeography, Historical Ecology, and the Human Dimensions of Fire Regimes." *Journal of Biogeography* 41 (4): 833–836.

Roos, Christopher I., Andrew C. Scott, Claire M. Belcher, William G. Chaloner, Jonathan Aylen, Rebecca Bliege Bird, Michael R. Coughlan, Bart R. Johnson, Fay H. Johnston, Julia McMorrow, Toddi Steelman, and the Fire and

Mankind Discussion Group. 2016. "Living on a Flammable Planet: Interdisciplinary, Cross-Scalar and Varied Cultural Lessons, Prospects and Challenges." *Philosophical Transactions of the Royal Society B: Biological Sciences* 371 (1696).

Ruggiero, Patricia Guidão Cruz, Marco Antônio Batalha, Vânia Regina Pivello, and Sérgio Tadeu Meirelles. 2002. "Soil-Vegetation Relationships in Cerrado (Brazilian Savanna) and Semideciduous Forest, Southeastern Brazil." *Plant Ecology* 160 (1): 1–16.

Rule, Susan, Barry W. Brook, Simon G. Haberle, Chris S. M. Turney, A. Peter Kershaw, and Christopher N. Johnson. 2012. "The Aftermath of Megafaunal Extinction: Ecosystem Transformation in Pleistocene Australia." *Science* 335 (6075): 1483–1486.

Russell-Smith, Jeremy, Garry D. Cook, Peter M. Cooke, Andrew C. Edwards, Mitchell Lendrum, C. P. (Mick) Meyer, and Peter J. Whitehead. 2013. "Managing Fire Regimes in North Australian Savannas: Applying Aboriginal Approaches to Contemporary Global Problems." *Frontiers in Ecology and the Environment* 11 (s1): e55–63.

Russell-Smith, Jeremy, Diane Lucas, Minnie Gapindi, Billy Gunbunuka, Nipper Kapirigi, George Namingum, Kate Lucas, Pina Giuliani, and George Chaloupka. 1997. "Aboriginal Resource Utilization and Fire Management Practice in Western Arnhem Land, Monsoonal Northern Australia: Notes for Prehistory, Lessons for the Future." *Human Ecology* 25 (2): 159–195.

Russell-Smith, Jeremy, and Peter J. Whitehead. 2015. "Reimagining Fire Management in Fire-Prone North Australian Savannas." In *Carbon Accounting and Savanna Fire Management*, edited by Brett P. Murphy, Andrew C. Edwards, and Jeremy Russell-Smith, 1–22. Clayton: CSIRO.

Russell-Smith, Jeremy, Cameron Yates, Andrew Edwards, Grant E. Allan, Garry D. Cook, Peter Cooke, Ron Craig, Belinda Heath, and Richard Smith. 2003. "Contemporary Fire Regimes of Northern Australia, 1997–2001: Change since Aboriginal Occupancy, Challenges for Sustainable Management." *International Journal of Wildland Fire* 12 (4): 283–297.

Russell-Smith, Jeremy, Cameron P. Yates, Peter J. Whitehead, Richard Smith, Ron Craig, Grant E. Allan, Richard Thackway, Ian Frakes, Shane Cridland, Mick C. P. Meyer, and A. Malcolm Gill. 2007. "Bushfires Down Under: Patterns and Implications of Contemporary Australian Landscape Burning." *International Journal of Wildland Fire* 16 (4): 361–377.

Saamak, Christopher F. 2001. "A Shift from Natural to Human-Driven Fire Regime: Implications for Trace-Gas Emissions." *The Holocene* 11 (3): 373–375.

Sakaguchi, Shota, David M. J. S. Bowman, Lynda D. Prior, Michael D. Crisp, Celeste C. Linde, Yoshihiko Tsumura, and Yuji Isagi. 2013. "Climate, Not Aboriginal Landscape Burning, Controlled the Historical Demography and Distribution of Fire-Sensitive Conifer Populations across Australia." *Proceedings of the Royal Society B: Biological Sciences* 280 (1773).

Saltré, Frédérik, Marta Rodríguez-Rey, Barry W. Brook, Christopher N. Johnson, Chris S. M. Turney, John Alroy, Alan Cooper, Nicholas Beeton, Michael I. Bird, Damien A. Fordham, Richard Gillespie, Salvador Herrando-Pérez, Zenobia Jacobs, Gifford H. Miller, David Nogués-Bravo, Gavin J. Prideaux, Richard G. Roberts, and Corey J. A. Bradshaw. 2016. "Climate Change Not to Blame for Late Quaternary Megafauna Extinctions in Australia." *Nature Communications* 7: 10511.

Sankaran, M., and J. Ratnam. 2013. "African and Asian Savannas." In *Encyclopedia of Biodiversity*, 2nd ed., edited by Simon A. Levin, 1:58–74. Amsterdam: Academic Press.

Scholes, R. J., and S. R. Archer. 1997. "Tree-Grass Interactions in Savannas." *Annual Review of Ecology and Systematics* 28 (1): 517–544.

Scoones, I. 1999. "New Ecology and the Social Sciences: What Prospects for a Fruitful Engagement?" *Annual Review of Anthropology* 28: 479–507.

Scott, Andrew C., William G. Chaloner, Claire M. Belcher, and Christopher I. Roos. 2016. "The Interaction of Fire and Mankind: Introduction." *Philosophical Transactions of the Royal Society B: Biological Sciences* 371 (1696).

Senut, Brigitte, Martin Pickford, and Loïc Ségalen. 2009. "Neogene Desertification of Africa." *Comptes Rendus Geoscience* 341 (8–9): 591–602.

Shaffer, L. Jen. 2010. "Indigenous Fire Use to Manage Savanna Landscapes in Southern Mozambique." *Fire Ecology* 6 (2): 43–59.

Shea, Ronald W., Barbara W. Shea, J. Boone Kauffman, Darold E. Ward, Craig I. Haskins, and Mary C. Scholes. 1996. "Fuel Biomass and Combustion Factors Associated with Fires in Savanna Ecosystems of South Africa and Zambia." *Journal of Geophysical Research: Atmospheres* 101 (D19): 23551–23568.

Sheuyange, Asser, Gufu Oba, and Robert B. Weladji. 2005. "Effects of Anthropogenic Fire History on Savanna Vegetation in Northeastern Namibia." *Journal of Environmental Management* 75 (3): 189–198.

Silva, J. F., M. R. Farinas, J. M. Felfili, and C. A. Klink. 2006. "Spatial Heterogeneity, Land Use and Conservation in the Cerrado Region of Brazil." *Journal of Biogeography* 33 (3): 536–548.

Silvério, Divino Vicente, Oriales Rocha Pereira, Henrique Augusto Mews, Leonardo Maracahipes-Santos, Josias Oliveira dos Santos, and Eddie Lenza. 2015. "Surface Fire Drives Short-Term Changes in the Vegetative Phenology of Woody Species in a Brazilian Savanna." *Biota Neotropica* 15 (3): 1–9.

Sim, T. R. 1907. *The Forests and Forest Flora of the Colony of the Cape of Good Hope*. Aberdeen, Scotland: Taylor and Henderson.

Simon, Marcelo F., Rosaura Grether, Luciano P. de Queiroz, Cynthia Skema, R. Toby Pennington, and Colin E. Hughes. 2009. "Recent Assembly of the Cerrado, a Neotropical Plant Diversity Hotspot, by in Situ Evolution of Adaptations to Fire." *Proceedings of the National Academy of Sciences* 106 (48): 20359–20364.

Singh, G., A. P. Kershaw, and R. Clark. 1981. "Quaternary Vegetation and Fire History in Australia." In *Fire and the Australian Biota*, edited by A. M. Gill, R. H. Groves, and I. R. Noble, 23–54. Canberra: Australian Academy of Science.

Sletto, Bjørn. 2008. "The Knowledge That Counts: Institutional Identities, Policy Science, and the Conflict Over Fire Management in the Gran Sabana, Venezuela." *World Development* 36 (10): 1938–1955.

Sletto, Bjørn, and Iokiñe Rodriguez. 2013. "Burning, Fire Prevention and Landscape Productions among the Pemon, Gran Sabana, Venezuela: Toward an Intercultural Approach to Wildland Fire Management in Neotropical Savannas." *Journal of Environmental Management* 115: 155–166.

Smith, Allan H. 2006. *Takhoma: Ethnography of Mount Rainier National Park*. Pullman: Washington State University Press.

Smith, Bruce D. 2005. "Low-Level Food Production and the Northwest Coast." In *Keeping It Living: Traditions of Plant Use and Cultivation on the Northwest Coast*, edited by Douglas Deur and Nancy J. Turner, 37–66. Seattle: University of Washington Press.

Smith, Linda Tuhiwai. 1999. *Decolonizing Methodologies: Research and Indigenous Peoples*. London: Zed Books.

Smith, Marian W. 1940. *The Puyallup-Nisqually*. New York: Columbia University Press.

Smith, Mike A. 2013. *The Archaeology of Australia's Deserts*. Cambridge: Cambridge University Press.

Smith, Natalie. 2001. "Are Indigenous People Conservationists? Preliminary Results from the Machiguenga of the Peruvian Amazon." *Rationality and Society* 13 (4): 429–461.

Sorrensen, Cynthia. 2009. "Potential Hazards of Land Policy: Conservation, Rural Development and Fire Use in the Brazilian Amazon." *Land Use Policy* 26 (3): 782–791.

Spoon, Jeremy, Richard Arnold, Brian J. Lefler, and Christopher Milton. 2015. "Nuwuvi (Southern Paiute), Shifting Fire Regimes, and the Carpenter One Fire in the Spring Mountains National Recreation Area, Nevada." *Journal of Ethnobiology* 35 (1): 85–110.

Staver, A. Carla, Sally Archibald, and Simon A. Levin. 2011a. "Tree Cover in Sub-Saharan Africa: Rainfall and Fire Constrain Forest and Savanna as Alternative Stable States." *Ecology* 92 (5): 1063–1072.

———. 2011b. "The Global Extent and Determinants of Savanna and Forest as Alternative Biome States." *Science* 334 (6053): 230–232.

Steiner, Christoph, Wenceslau Geraldes Teixeira, and Wolfgang Zech. 2004. "Slash and Char: An Alternative to Slash and Burn Practiced in the Amazon Basin." In *Amazonian Dark Earths: Explorations in Space and Time*, edited by Dr. Bruno Glaser and Professor William I. Woods, 183–193. Berlin, Heidelberg: Springer.

Stephenson, Eleanor S., and Peter H. Stephenson. 2016. "The Political Ecology of Cause and Blame: Environmental Health Inequities in the Context." In *A Companion to the Anthropology of Environmental Health*, edited by Merrill Singer, 302–324. Chichester, UK: John Wiley and Sons.

Stewart, Omer C. 2002. "The Effects of Burning of Grasslands and Forests by Aborigines the World Over." In *Forgotten Fires: Native Americans and the Transient Wilderness*, edited by Henry T. Lewis and M. Kat Anderson, 65–312. Norman: University of Oklahoma Press.

Storm, Linda E. 2002. "Patterns and Processes of Indigenous Burning: How to Read Landscape Signatures of Past Human Practices." In *Ethnobiology and Biocultural Diversity: Proceedings of the Seventh International Congress of Ethnobiology*, 496–508. Athens: University of Georgia Press.

Storm, Linda E., and Daniela Shebitz. 2006. "Evaluating the Purpose, Extent, and Ecological Restoration Applications of Indigenous Burning Practices in Southwestern Washington." *Ecological Restoration* 24 (4): 256–268.

Suttles, Wayne P. 1987. "Coping with Abundance: Subsistence on the Northwest Coast." In *Coast Salish Essays*, edited by Wayne P. Suttles, 45–66. Vancouver: Talonbooks.

———. 2005. "Coast Salish Resource Management: Incipient Agriculture?" In *Keeping It Living: Traditions of Plant Use and Cultivation on the Northwest Coast*, edited by Douglas Deur and Nancy J. Turner, 181–193. Seattle: University of Washington Press.

Tacconi, Luca, and Yayat Ruchiat. 2006. "Livelihoods, Fire and Policy in Eastern Indonesia." *Singapore Journal of Tropical Geography* 27 (1): 67–81.

Tacconi, Luca, and Andrew P. Vayda. 2006. "Slash and Burn and Fires in Indonesia: A Comment." *Ecological Economics* 56 (1): 1–4.

Thomas, S. M., and M. W. Palmer. 2007. "The Montane Grasslands of the Western Ghats, India: Community Ecology and Conservation." *Community Ecology* 8 (1): 67–73.

Thompson, Laurence C., and Dale M. Kinkade. 1990. "Languages." In *Northwest Coast*, edited by Wayne Suttles, 30–51. *Handbook of North American Indians*, Vol. 7, William C. Sturtevant, general editor. Washington, DC: Smithsonian Institution.

Thomson, Donald F. 1949. *Economic Structure and the Ceremonial Exchange Cycle in Arnhem Land*. Melbourne: Macmillan.

Tiffen, Mary. 2006. "Urbanization: Impacts on the Evolution of 'Mixed Farming' Systems in Sub-Saharan Africa." *Experimental AgricultureResearchGate* 42 (3): 259–287.

Trant, Andrew J., Wiebe Nijland, Kira M. Hoffman, Darcy L. Mathews, Duncan McLaren, Trisalyn A. Nelson, and Brian M. Starzomski.

2016. "Intertidal Resource Use over Millennia Enhances Forest Productivity." *Nature Communications* 7: 12491.

Turner, Matthew D. 2011. "The New Pastoral Development Paradigm: Engaging the Realities of Property Institutions and Livestock Mobility in Dryland Africa." *Society and Natural Resources* 24 (5): 469–484.

Turner, Nancy J. 1999. "'Time to Burn': Traditional Use of Fire to Enhance Resource Production by Aboriginal Peoples in British Columbia." In *Indians, Fire and the Land in the Pacific Northwest*, edited by Robert Boyd, 185–218. Corvallis: Oregon State University Press.

Turner, Nancy J., and Sandra Peacock. 2005. "Solving the Perennial Paradox: Ethnobotanical Evidence for Plant Resource Management on the Northwest Coast and Elsewhere." In *Keeping It Living: Traditions of Plant Use and Cultivation on the Northwest Coast*, edited by Douglas Deur and Nancy J. Turner, 101–150. Seattle: University of Washington Press.

Tweiten, Michael. 2007. "The Interaction of Changing Patterns of Land Use, Sub-Alpine Forest Composition at Buck Lake, Mount Rainier National Park, USA." Unpublished report prepared for the Columbia Cascades Support Office, National Park Service, Seattle, Washington. Honolulu: International Archaeological Research Institute.

US Court of Claims. 1947. The Duwamish, Lummi, Whidbey Island, Skagit, Upper Skagit, Swinomish, Kikiallus, Snohomish, Snoqualmie, Stillaguamish, Suquamish, Samish, Puyallup, Squaxin, Skokomish, Upper Chehalis, Muckleshoot, Nooksack, Chinook and San Juan Island Tribes of Indians, Claimants, v. the United States of America, Defendant. Consolidated Petition No. F-275. United States Court of Claims, Washington, DC

USDA. 2013. "Fiscal Year 2014 Budget Overview." Unpublished report. Washington, DC: Forest Service, US Department of Agriculture. Accessed on July 8, 2016. https://www.fs.fed.us/aboutus/budget/2014/FY2014ForestServiceBudgetOverview041613.pdf.

Van der Schijff, H. P. 1958. "Inleidende verslag oor veldbrandnavorsing in die Nasionale Krugerwildtuin." *Koedoe* 1 (1): 60–94.

van der Werf, G. R., J. T. Randerson, L. Giglio, G. J. Collatz, M. Mu, P. S. Kasibhatla, D. C. Morton, R. S. DeFries, Y. Jin, and T. T. van Leeuwen. 2010. "Global Fire Emissions and the Contribution of Deforestation, Savanna, Forest, Agricultural, and Peat Fires (1997–2009)." *Atmospheric Chemistry and Physics* 10 (23): 11707–11735.

van Vliet, Nathalie, O. Mertz, T. Birch-Thomsen, and B. Schmook. 2013. "Is There a Continuing Rationale for Swidden Cultivation in the 21st Century?." *Human Ecology* 41 (1): 1–5.

van Vliet, Nathalie, Ole Mertz, Andreas Heinimann, Tobias Langanke, Unai Pascual, Birgit Schmook, Cristina Adams, Dietrich Schmidt-Vogt, Peter Messerli, Stephen Leisz, Jean-Christophe Castella, Lars Jørgensen, Torben Birch-Thomsen, Cornelia Hett, Thilde Bech-Bruun, Amy Ickowitz, Kim Chi Vu, Kono Yasuyuki, Jefferson Fox, Christine Padoch, Wolfram Dressler, and Alan D. Ziegler. 2012. "Trends, Drivers and Impacts of Changes in Swidden Cultivation in Tropical Forest-Agriculture Frontiers: A Global Assessment." *Global Environmental Change* 22 (2): 418–429.

van Wilgen, Brian W. 2009. "The Evolution of Fire Management Practices in Savanna Protected Areas in South Africa." *South African Journal of Science* 105 (9–10): 343–349.

van Wilgen, Brian W., C. S. Everson, and W. S. W. Trollope. 1990. "Fire Management in Southern Africa: Some Examples of Current Objectives, Practices, and Problems." In *Fire in the Tropical Biota: Ecosystem Processes and Global Challenges*, edited by Johann Georg Goldammer, 179–215. Berlin, Heidelberg: Springer.

van Wilgen, Brian W., Navashni Govender, and Sandra MacFadyen. 2008. "An Assessment of the Implementation and Outcomes of Recent

Changes to Fire Management in the Kruger National Park." *Koedoe* 50 (1): 22–31.

Vigilante, Tom. 2001. "Analysis of Explorers' Records of Aboriginal Landscape Burning in the Kimberley Region of Western Australia." *Australian Geographical Studies* 39 (2): 135–155.

Vitt, Laurie J., and Janalee P. Caldwell. 1993. "Ecological Observations on Cerrado Lizards in Rondônia, Brazil." *Journal of Herpetology* 27 (1): 46–52.

Walsh, Megan K., Jennifer R. Marlon, Simon J. Goring, Kendrick J. Brown, and Daniel G. Gavin. 2015. "A Regional Perspective on Holocene Fire–Climate–Human Interactions in the Pacific Northwest of North America." *Annals of the Association of American Geographers* 105 (6): 1135–1157.

Walters, Gretchen. 2012. "Customary Fire Regimes and Vegetation Structure in Gabon's Bateke Plateaux." *Human Ecology* 40 (6): 943–955.

———. 2015. "Changing Fire Governance in Gabon's Plateaux Bateke Savanna Landscape." *Conservation and Society* 13 (3): 275.

Warming, Eugene. 1892. *Lagoa Santa: Et Bidrag Til Den Biologiske Plantegeografi*. Copenhagen: Bianco Luno.

Waters, Michael R., Charlotte D. Pevny, David L. Carlson, William A. Dickens, Ashley M. Smallwood, Scott A. Minchak, E. Bartelink, Jason M. Wiersema, James E. Wiederhold, and Heidi M. Luchsinger. 2011. *A Clovis Workshop in Central Texas: Archaeological Investigations of Excavation Area 8 at the Gault Site*. College Station: Texas A&M University Press.

Weiser, Andrea, and Dana Lepofsky. 2009. "Ancient Land Use and Management of Ebey's Prairie, Whidbey Island, Washington." *Journal of Ethnobiology* 29 (2): 184–212.

Welch, James R. 2013. *Sprouting Valley: Historical Ethnobotany of the Northern Pomo from Potter Valley, California*. Denton, TX: Society of Ethnobiology.

———. 2014. "Xavante Ritual Hunting: Anthropogenic Fire, Reciprocity, and Collective Landscape Management in the Brazilian Cerrado." *Human Ecology* 42 (1): 47–59.

———. 2015. "Learning to Hunt by Tending the Fire: Xavante Youth, Ethnoecology, and Ceremony in Central Brazil." *Journal of Ethnobiology* 35 (1): 183–208.

Welch, James R., Eduardo S. Brondízio, Scott S. Hetrick, and Carlos E. A. Coimbra. 2013. "Indigenous Burning as Conservation Practice: Neotropical Savanna Recovery amid Agribusiness Deforestation in Central Brazil." Edited by Brock Fenton. *PLOS ONE* 8 (12): e81226.

Werner, Patricia A., Brian H. Walker, and P. A. Stott. 1991. "Introduction." In *Savanna Ecology and Management : Australian Perspectives and Intercontinental Comparisons*, edited by Patricia A. Werner, xi–xii. Oxford: Blackwell Scientific.

White, Tim D., Berhane Asfaw, Yonas Beyene, Yohannes Haile-Selassie, C. Owen Lovejoy, Gen Suwa, and Giday WoldeGabriel. 2009. "Ardipithecus Ramidus and the Paleobiology of Early Hominids." *Science* 326 (5949): 64–86.

Whitehead, Peter J., D. M. J. S. Bowman, Noel Preece, Fiona Fraser, and Peter Cooke. 2003. "Customary Use of Fire by Indigenous Peoples in Northern Australia: Its Contemporary Role in Savanna Management." *International Journal of Wildland Fire* 12 (4): 415–425.

Whitehead, Peter J., Jeremy Russell-Smith, and Cameron P. Yates. 2014. "Carbon Markets and Improved Management of Fire in North Australian Savannas: Identifying Sites for Productive Targeting of Emissions Reductions." *Rangelands Journal* 36: 371–388.

Wigley, B. J., William J. Bond, and M. T. Hoffman. 2009. "Bush Encroachment under Three Contrasting Land-Use Practices in a Mesic South African Savanna." *African Journal of Ecology* 47: 62–70.

Williams, Alan N. 2013. "A New Population Curve for Prehistoric Australia." *Proceedings* of the Royal Society B: Biological Sciences 280 (1761).

Williams, Alan N., Scott D. Mooney, Scott A. Sisson, and Jennifer Marlon. 2015. "Exploring the Relationship between Aboriginal Population Indices and Fire in Australia over the Last 20,000 Years." *Palaeogeography, Palaeoclimatology, Palaeoecology* 432: 49–57.

Williams, Alan N., Sean Ulm, Chris S. M. Turney, David Rohde, and Gentry White. 2015. "Holocene Demographic Changes and the Emergence of Complex Societies in Prehistoric Australia." *PLOS ONE* 10 (6): e0128661.

Wilson, Michael C., Stephen M. Kenady, and Randall F. Schalk. 2009. "Late Pleistocene Bison Antiquus from Orcas Island, Washington, and the Biogeographic Importance of an Early Postglacial Land Mammal Dispersal Corridor from the Mainland to Vancouver Island." *Quaternary Research* 71 (1): 49–61.

Woinarski, John C. Z., Sarah Legge, James A. Fitzsimons, Barry J. Traill, Andrew A. Burbidge, Alaric Fisher, and Ron S. C. Firth. 2011. "The Disappearing Mammal Fauna of Northern Australia: Context, Cause, and Response." *Conservation Letters* 4 (3): 192–201.

Wolverton, Steve. 2013. "Ethnobiology 5: Interdisciplinarity in an Era of Rapid Environmental Change." *Ethnobiology Letters* 4: 21–25.

Wong, Carmen, Holger Sandmann, and Brigitte Dorner. 2003. *Historical Variability of Natural Disturbances in British Columbia: A Literature Review*. Kamloops, B.C.: Forest Research Extension Partnership.

Wroe, Stephen, Judith H. Field, Michael Archer, Donald K. Grayson, Gilbert J. Price, Julien Louys, J. Tyler Faith, Gregory E. Webb, Iain Davidson, and Scott D. Mooney. 2013. "Climate Change Frames Debate over the Extinction of Megafauna in Sahul (Pleistocene Australia-New Guinea)." *Proceedings of the National Academy of Sciences* 110 (22): 8777–8781.

Yibarbuk, D. 1998. "Notes on Traditional Use of Fire on the Upper Cadell River." In *Burning Questions: Ongoing Environmental Issues for Indigenous Peoples in Northern Australia*, edited by Marcia Langton, 1–16. Darwin: Northern Territory University Press.

Yibarbuk, D., P. J. Whitehead, J. Russell-Smith, D. Jackson, C. Godjuwa, A. Fisher, P. Cooke, D. Choquenot, and D. M. J. S. Bowman. 2001. "Fire Ecology and Aboriginal Land Management in Central Arnhem Land, Northern Australia: A Tradition of Ecosystem Management." *Journal of Biogeography* 28 (3): 325–343.

Zander, Kerstin K., Desleigh R. Dunnett, Christine Brown, Otto Campion, and Stephen T. Garnett. 2013. "Rewards for Providing Environmental Services—Where Indigenous Australians' and Western Perspectives Collide." *Ecological Economics* 87: 145–154.

Zhang, Quanfa, Christopher O. Justice, and Paul V. Desanker. 2002. "Impacts of Simulated Shifting Cultivation on Deforestation and the Carbon Stocks of the Forests of Central Africa." *Agriculture, Ecosystems and Environment* 90 (2): 203–209.

Ziegler, Alan D., Jacob Phelps, Jia QI Yuen, Edward L. Webb, Deborah Lawrence, Jeff M. Fox, and Thilde B. Bruun. 2012. "Carbon Outcomes of Major Land-Cover Transitions in SE Asia: Great Uncertainties and REDD+ Policy Implications." *Global Change Biology* 18 (10): 3087–3099.

Ziembicki, Mark R., John C. Z. Woinarski, Jonathan K. Webb, Eric Vanderduys, Katherine Tuft, James Smith, and Euan G. Ritchie. 2014. "Stemming the Tide: Progress towards Resolving the Causes of Decline and Implementing Management Responses for the Disappearing Mammal Fauna of Northern Australia." *Therya* 6 (1): 169–225.

Zimmerer, Karl S. 1994. "Human Geography and the 'New Ecology': The Prospect and Promise of Integration." *Annals of the Association of American Geographers* 84 (1): 108–125.

CHAPTER 3

Fire in the African Savanna

Identifying Challenges to Traditional Burning Practices in Tanzania and Malawi

RAMONA J. BUTZ

Fire is an integral component of many African ecosystems, including savanna grasslands and woodlands where it is used by local populations to promote diversity, health, and function. Yet both tribal peoples and land managers encounter numerous challenges in maintaining traditional fire regimes on the landscape. Fire suppression policies, food insecurity, unpredictable rainfall patterns and climate change, increasing population pressure, poaching, alternative agricultural practices, and an increasing number of catastrophic accidental fires are just a handful of the obstacles practitioners routinely face.

Tropical savannas comprise almost a third of the world's land surface, including more than half of the land surface of Africa (Kennedy and Potgieter 2003; Werner et al. 1991). Savanna landscapes are noted for their inherent dynamism (Sharp and Bowman 2004). Dramatic shifts in species composition and relative abundances of trees and grasses or forbs that characterize savannas are a response to complex ecological factors (Scholes and Archer 1997; Sharp and Bowman 2004; Skarpe 1991). Savannas have existed in Africa for at least 30 million years (van Wilgen 2009) and they are, in most cases, highly modified by humans.

The ratio of woody to herbaceous plants in arid and semiarid savannas is determined by a combination of climate, water and nutrient availability, herbivory, and fire (van Langevelde et al. 2002; Cole 1986; Frost 1996; Higgins et al. 2000; Skarpe 1992; van Wilgen et al. 2009). Today, African savannas burn more frequently and extensively than any other region on earth (Dwyer et al. 2000; Laris 2002a). Fire regimes can direct pathways of vegetation change (Bradstock and Kenny 2003). Such changes result from the interaction between fire intensity, frequency, season, type, and life history characteristics of species with attributes that provide a basis for predicting species and vegetation dynamics in fire-prone environments (Bond and van Wilgen 1996: Noble and Slatyer 1980). Given this complexity, management and conservation of savanna systems are challenging undertakings with little methodological agreement among academics, land managers, and traditional land inhabitants.

Fire and herbivores modify vegetation structures and levels of rangeland productivity that are otherwise determined by plant-available moisture and nutrients (Boone et al. 2002; Higgins et al. 2000). Rangeland systems such as *Themeda triandra* Forssk. fire-climax grasslands and wooded grasslands in the northern savanna region of Tanzania, where fire has been a regular feature for centuries, have correspondingly fire-adapted species compositions (Sinclair and Arcese 1995). Temporary protection against fire in the absence of compensatory grazing allows accumulation of standing plant biomass with low nutritive value and an increase of fire-sensitive and often unpalatable herbaceous species (Homewood and Rodgers 1991). In the absence of periodic fires, savanna vegetation shifts from herbaceous dominance toward an increase in shrub cover (Trollope 1982; Knoop and Walker 1985; Oba et al. 2000; Sheuyange et al. 2005). In addition, a change in timing or frequency of fire can alter the response of herbaceous or woody species, thereby producing a change in species composition (Dyer et al. 1997; Jacobs and Schloeder 2002).

Yet despite anthropogenic fires accounting for more than 70 percent of annual fires in African savannas (van Wilgen et al. 1990), the roles that anthropogenic fires play in vegetation dynamics are poorly documented (Sheuyange et al. 2005). Vegetation changes in savanna landscapes can affect both ecosystem productivity and conservation value. The interactions between human burning practices, fire regimes, and savanna landscapes have received limited research attention (Laris 2002b). Understanding the influences of anthropogenic fire and site conditions on savanna vegetation composition and structure can help clarify how changes in fire frequency and intensity may significantly impact these diverse communities.

The persistence of functioning African savannas is critical to support human populations, natural resource use, economic activities, and conservation of dynamically functioning ecosystems (Sporton and Thomas 2002; Bassett and Crummey 2003; Amanor and Moyo 2008). Yet climate change, privatization of resources, reallocation of land and localized conflicts, increasing water stress, food insecurity, and increasing demand for biofuels are a few of the many factors threatening the sustainability of savanna ecosystems and the people who rely on them (Amigun et al. 2008; Leichenko and O'Brien 2002; Tyson et al. 2002; Boko et al. 2007; Conway 2012; Muller et al. 2008; Casale et al. 2010). As these factors change, potentially so do fire regimes and fire management. Are traditional fire management systems in East Africa changing? If so, why should we care?

This chapter explores and contrasts the contemporary challenges of fire management and driving factors of change in two East African savannas. The first is a rural village in northern Tanzania, home to a seminomadic group of Maasai pastoralists. The second is Nyika National Park in northern Malawi, managed by a federal agency and supported by a foreign trust. Although seemingly divergent in location, ecology, and management, both face similar challenges in maintaining appropriate fire regimes for their respective landscapes. Substantial modifications to historical anthropogenic fire regimes are having cascading consequences for ecosystem health in savanna systems by altering species composition and fuel loads, and reducing overall species diversity.

A Rural Tanzanian Village

Tanzania, one of the larger countries in Africa (947,300 square kilometers), is home to roughly 45 million people, a density of 47.5 people per square kilometer (Tanzania 2012). Nearly a third of the country's total area is protected to varying degrees as national parks, game reserves, marine parks, forest reserves, and conservation areas.

Engikareti is a Maasai village in northern Tanzania along the main road between Arusha and Nairobi, Kenya. The village is located within the bounds of the Longido Game Controlled Area in Longido District, Arusha Region, northeastern Tanzania (03°00.9'S, 036°42.4'E) (Figure 3.1). Village land is bounded by the town of Oldonyo Sambu to the south, the Maasai villages of Kiserian to the north, Emuseregi to the west, and Engarenanyuki to the east. Engikareti covers roughly 200 square kilometers and is home to approximately 3,000 seminomadic Maasai pastoralists. Within the village, residents live in *bomas*, small communities often composed of extended families that range in size from approximately 20 to 200 people and their livestock. In 2005 and 2006 there were approximately 80 bomas in the village.

Specialized herders have occupied the savannas of northern Tanzania and southern Kenya for more than 3,000 years. The Maasai, a large and loosely defined ethnic group of seminomadic pastoralists and agropastoralists, have lived in this region since approximately 1850 (for a detailed ethnography see Beidelman 1960; Fosbrooke 1948; Homewood and Rodgers 1991; Mitzlaff 1988; Spear and Waller 1993; Spencer 1988). Similar to other pastoralists in East Africa (Kijazi et al. 1999; Oba 2001; Oba and Kotile 2001), oral and pictorial traditions of the Maasai describe in rich detail both past and present management practices for local and useful plants as well as a wealth of knowledge on the ecology and environmental history of the local region. Yet the balance between human occupation and wildlife concerns is continually being reassessed, resulting in increasing levels of tension and conflict between wildlife managers and resident Maasai (Coast 2002; McCabe 1998). One of the main reasons for increasing tensions has to do with the traditional burning practices employed by Maasai pastoralists.

Pastoralists are people whose livelihoods depend mainly on raising domestic animals including cattle, goats, sheep, donkeys, and camels, which are used for milk, meat, transport, and trade. Pastoralists typically occupy large tracts of communally shared land and use kinship ties for mutual herding and defense (Fratkin 2001). The Maasai, as with many other nomadic and seminomadic pastoral groups worldwide, are politically disempowered and economically marginalized by national politics oriented toward development, agricultural, and conservation interests. Privatization of land for ranching, agriculture, urban and rural development, and conservation areas has curtailed pastoral mobility and encouraged sedentarization (Kwashirai 2013).

The region encompassing Engikareti is arid to semiarid with mean annual rainfall ranging from 200 to 600 millimeters per year. Rainfall is bimodal, and interannual variability is high. The short rains normally fall in November to December, and the long rains begin as early as March and terminate by the end of May or June. The short rains are highly unreliable, overall rainfall is poorly distributed, and droughts are frequent (Ngailo et al. 2001). A severe drought from 1999 to early 2006 resulted in a 50 to 80 percent reduction in livestock numbers and the starvation deaths of two children in early 2006 before the late onset of the long rains. Food security in the village is a major concern.

The landscape is characterized by undulating treeless, short-grass savanna interspersed with *Acacia-Commiphora* scrub, and savanna woodlands dominated by umbrella acacia (*Acacia tortilis* [Forssk.] Hayne) (Butz 2009). It is highly arid, and plant cover is low even in the most densely vegetated areas. Low plant biomass, a high percentage of bare ground, and virtually no litter layer create an environment where—particularly with the high fire-return interval common in savanna sys-

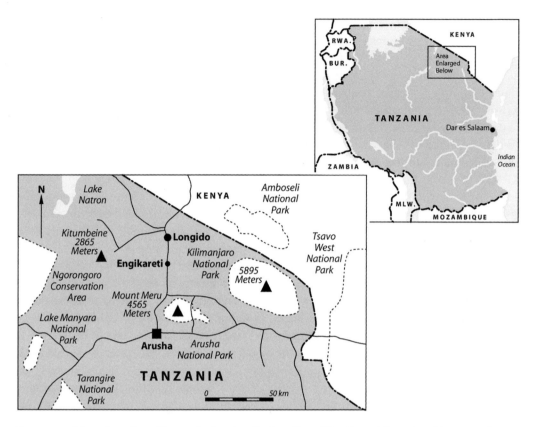

FIGURE 3.1. Map of northern Tanzania showing the location of Engikareti (based on Homewood et al. 2006).

tems—rapid, low-temperature fires are most common. The village is located within the western rain shadow of Mount Kilimanjaro and along the windward side of Mount Meru. The region has complex landforms resulting from folding, faulting, volcanic eruptions, and erosion (Oba and Kaitira 2006). Elevations generally vary from 1,200 to 1,700 meters. There is no permanent water located on village lands. The Longido Game Controlled Area effectively forms a southern extension of the Amboseli plains (Kenya) and is used extensively by large mammals when Amboseli is dry. It is an area of considerable importance for a large number of migratory wildlife, and more than 400 species of birds have been recorded in the area (Fishpool and Evans 2001).

A public primary school, a Catholic secondary prep school, housing barracks for teachers, a village office, and three churches are the only infrastructure other than earthen family homes located within the village confines of Engikareti. The Maasai pastoralists living in this region engage mostly in livestock husbandry. Cattle, goats, sheep, and donkeys are grazed on communal pastures. Small garden plots have been created near a handful of bomas in the southwestern corner of the village due to influences from the neighboring WaArusha tribe, but no large-scale agriculture exists, due in part to the lack of available arable land. In the Longido Game Controlled Area in Tanzania, all human activities are permitted except the exploitation of wildlife (Kaiza-Boshe et al. 1998).

Fire Use among the Maasai

Maasai pastoralists have an elaborate system of traditional fire management. They use a progression of small fires throughout the dry season to create a checkerboard or patch-mosaic (Parr and Brockett 1999) pattern of burns to prevent large, catastrophic late-season fires. These early dry-season fires are small and contained. In addition to natural or constructed firebreaks, such as a dry riverbed or a constructed trench, as well as back-burns to constrain the area to be burned, pasture areas grazed down by livestock also serve to prevent fires from escaping by containing minimal fuel loads. Green tree branches are cut and used to beat out small fires and fires that jump a firebreak into an undesired area. Sand is also used to put out fires when careful containment is necessary. Heavily forested areas or areas near occupied bomas are typically not burned to prevent loss of tree cover (used as building material and firewood by the Maasai) and to protect existing structures from accidentally escaped fires (for more detail on traditional burning practices, see Butz 2009).

Burning is conducted on a one- to eight-year rotation depending on the health of the existing vegetation, the perceived prospect of a good rain year based on the current drought status of village lands, and the percentage of adequate grazing land remaining were the rains to fail.

Although fires vary in size, most are small. They are most frequently started in the evenings when the wind is minimal or non-existent, the temperatures are cool, and the humidity is slightly higher. If the area to be burned is experiencing strong winds at night, men ignite the fire(s) very early in the morning and assign a team of people, often other warriors and young boys, to carefully watch the progression of the fire and keep livestock out of harm's way.

During interviews about historical and current practices, reasons for burning, the history of land use, and their perceptions of fire, the Maasai identified eight major reasons for using fire on a landscape scale in savannas (Butz 2009, 2013). The single most important reason mentioned in individual and group interviews for the use of fire was to promote new, diverse, high-quality forage for cattle and small livestock. Other frequently mentioned reasons include the prevention of shrub encroachment, opening up landscapes to allow freer movement of livestock, killing disease-carrying ticks, eliminating cover for dangerous predators of livestock (such as lions, leopards, and cheetahs), preventing late-season catastrophic fires, removing overgrown or matted grasses to more readily identify holes where livestock could be injured or killed, and to create suitable habitat for plants of edible, medicinal, or construction value.

Currently there is little active vegetation management using fire, largely due to a combination of federal fire suppression policies, unpredictable rainfall patterns, increasing population pressures, and a subsequent increase in the number of catastrophic accidental fires. Although the villagers who were interviewed disagreed on the exact timing of this departure from traditional practices, they agreed that the shift has been most dramatic in the last ten to twenty years.

Nyika National Park, Malawi

The Republic of Malawi is a landlocked country in southeast Africa that was formerly known as Nyasaland. Malawi, in stark contrast to Tanzania, is one of the smallest African countries (118,484 square kilometers). The 2013 estimated population density of Malawi was more than two and a half times that of Tanzania, at 128.8 people per square kilometer, with a total population of roughly the contour marking the zone of transition from woodland to montane grassland and forest (Donovan et al. 2002; Dowsett-Lemaire 1985; Lemon 1968; Meadows 1984; Meadows and Linder 1993).

The first formal protection of the Nyika Plateau was the establishment in 1948 of a forest reserve to protect the southernmost population of juniper trees (*Juniperus procera* Hoscht. ex Endl.) in Africa. In 1951 the grasslands of the high plateau were protected under the Natural Resources Rules, and hunting was prohibited. In the 1950s the Colonial Development Corporation established a plantation of 542 hectares of pines, blue gums, and wattle at Chelinda. The plantations were transferred to the Department of Forestry in 1958. In 1965 (after independence) the Malawi National Park, covering 940 square kilometers of the Nyika Plateau, was established. The name was changed to Nyika National Park in 1969. In 1978 the park was considerably enlarged to its present size of 3,134 square kilometers with the addition of the foothills lying south and north of the plateau. This expansion was due to the plateau's critical role as the water catchment area that much of northern Malawi depends on for domestic consumption, irrigation, and hydroelectric power. *Nyika* means "where the water comes from," and the plateau is one of the most important water catchments in Malawi and contains the sources for four large rivers that drain into Lake Malawi (McCracken 2006). The expansion of the park boundaries also took account of the winter migration of large mammals—particularly roan antelope, eland, and zebra—from the plateau to the adjacent woodlands, and their return to the grasslands in the spring (Chitaukali et al. 2001; Hall-Martin et al. 2007).

Approximately 80 square kilometers of Nyika National Park are in Zambia but are accessible by road only from the Malawi side. The park is located 480 kilometers by road

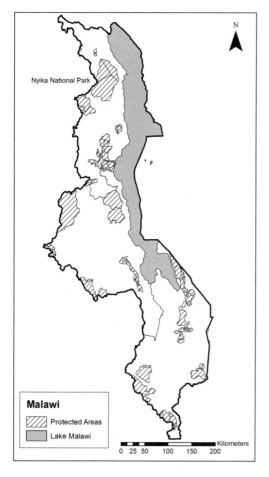

FIGURE 3.2. Map of Malawi showing the location of Nyika National Park and other protected areas (map by Ramona J. Butz).

14 million. Malawi has a large rural population, and the economy is heavily based on agriculture.

Located to the south of Engikareti and Tanzania, within the same chain of isolated mountain groups running parallel to the Indian Ocean from Ethiopia to the Drakensberg Mountains, is the Nyika Plateau (Figure 3.2). This mountain island, bounded on all sides by steep scarps, lies between 10°15'–10°50'S and 33°35'–34°05'E, predominantly in northern Malawi. It is the largest montane complex in south-central Africa with approximately 1,800 square kilometers above 1,800 meters, north of Lilongwe and 35 kilometers west of Livingstonia. The plateau is a biotite-granite monolith rising over 2,500 meters above sea level. The highest point is Nganda Hill (2,607 meters). Summers are cloudy and cool with frequent misty conditions, and cold, dry winters have frost during several months. The mean monthly temperature in November, the warmest month, is only 16.5°C, and maximum daily temperatures seldom rise above 25°C. Prevailing easterly winds blow over Lake Malawi, causing orographic rainfall on the eastern escarpment of about 1,700 millimeters per year. The central grasslands have an average rainfall of 1,000 to 1,200 millimeters per year, while the northern (Malawi) and western (Zambia) slopes average less than 900 millimeters per year.

The lower slopes of the plateau consist of steep foothills covered with *Brachystegia* woodland and divided into deep gullies (McCracken 2006). These are replaced at higher elevations by short, open grassland dominated by *Loudetia simplex* (Nees) C. E. Hubb, *T. triandra*, and *Exotheca abyssinica* (A. Rich) Anderss. Relict evergreen forest covers 2 to 4 percent of the upper plateau, occurring mainly on valley heads, valley slopes, and in hollows. Soils are rather poor and tend to be acidic. More than 200 species of terrestrial orchids grow there, making it the richest orchid area in south-central Africa (McCracken 2006). Zebras, elephants, warthogs, jackals, hyenas, lions, leopards, and a variety of antelopes are among the mammals present. More than 435 species of birds have been recorded, including the rare Denham's bustard, wattled crane, and endemic red-winged francolin. Nyika National Park was added to the UNESCO World Heritage Tentative List on May 17, 2000, in the Mixed (Cultural and Natural) category.

Nyika National Park is under the authority of the Department of National Parks and Wildlife, with management assistance granted to the Nyika-Vwaza Trust. Prior to colonial settlement and the creation of the park, the Nyika Plateau and surrounding areas were inhabited by the Phoka, a small group of subsistence farmers and hunters (McCracken 2006). They remain among the large numbers of people settled just outside the park boundaries, and a small handful are hired each year by the Nyika-Vwaza Trust to help with park management. Little has been written about Phoka traditional management of the plateau.

The grasslands in Nyika National Park have been experiencing very high fire-return intervals for several decades, and possibly since the creation of the park in the 1960s. A large proportion of the plateau currently burns every year. During the 2010 fire season, it was estimated that more than 90 percent of the plateau grasslands experienced fire through a combination of early-season area burning conducted by the Nyika-Vwaza Trust (with authority granted from the Department of National Parks and Wildlife) and late-season fires set by poachers, beekeepers, and other park users. Numerous early reports list both natural and human fire ignitions as important disturbance events on the plateau, with an estimated one- to three-year fire-return interval in the grasslands (Lemon 1964, 1968).

The Nyika-Vwaza Trust has implemented an early controlled-burn program to help prevent large late-season fires. Throughout March and early April, firebreaks around sensitive areas and structures are cleared. Next, roads and tracks are cleared of long grass to act as additional firebreaks. Prescribed burning usually runs from the end of May through July, depending on weather conditions. The aim of this program is to burn roughly one-third of the grassland on the plateau every year, establishing a three-year rotation. When combined with fires set by other park users, however, the area of the park burned annually often exceeds 50 percent and, in some recent years, nears 100 percent.

These additional fires may be accidental or purposeful, and come from a variety of sources. Escaped fires from agricultural burning and beekeeping are frequently cited. Hunters, deemed "poachers" by park staff, also start fires to flush out game and reduce vegetation cover. In many years, the areas not burned early in the season through the prescribed burn program are subsequently burned in the dry season by a combination of these other ignition sources. The so-called poachers have also burned wooden bridges throughout the park to restrict the movement of game scouts and antipoaching patrols. Poaching is largely attributed to subsistence activities of hungry nearby residents, not to the black-market sale of bush meat.

Miombo and evergreen woodlands are scattered in drainages throughout the plateau. Many of these are shrinking in size or disappearing altogether due to the high frequency of late dry-season fires. While many of the species in these woodlands are fire-tolerant or fire-resistant, frequent higher-intensity, late-season burning can kill trees and decrease the overall resistance of the woodland patches to fire. A more natural fire-return interval for miombo woodlands is approximately every five to fifteen years, although more frequent low-intensity, early-season burns are often tolerated quite well. Once trees reach a height of approximately 3 meters and higher, they are also generally less susceptible to fire (Furley et al. 2008), but fire tolerance can occur when trees are as little as 12 centimeters diameter at breast height.

A small number of forest patches within the park are currently being protected from fire by the implementation of firebreaks. One of these patches, Zovochipolo Forest, is expanding, and the firebreak was modified in 2010 by the Nyika-Vwaza Trust to capture some of this regeneration. Another of these patches, the Juniper Forest (*J. procera*), has been protected from fire since approximately 1940. Determining appropriate fire-return intervals for this vegetation type will require further research, but total fire exclusion is not recommended. The possibility of fire entering the forest in a particularly dry year late in the fire season could be very damaging to the juniper if earlier fuels mitigation is not undertaken.

Yellow Himalayan raspberry (*Rubus ellipticus*); several types of pine, predominantly including Mexican pine (*Pinus patula*); and blue gum (*Eucalyptus* sp.) are exotic invasive species that have been introduced in the park. Yellow Himalayan raspberry spreads rapidly by root suckers and regenerates from underground shoots after fire or cutting. Its seeds are dispersed by fruit-eating birds and mammals. This species was observed throughout the park, but not in large, impenetrable thickets typical of the species.

A plantation of Mexican pine was established at Chelinda in the 1950s and is in the process of being removed through logging. This highly invasive species has been shown to cause both environmental and economic impacts in southern Africa, including reduction in stream flow and available water, loss of grazing potential, increased cost of fire protection, siltation of dams, change in soil nutrient status, threats to native plant species, change in habitat suitability for native animal species, and replacement of native tree species (Nyoka 2003). Mexican pines that have regenerated from the plantation are invading grasslands and riverine, evergreen, and miombo woodlands throughout the park and will continue to spread if not actively managed. The species can produce viable seed within five years and abundant cones by eight years. The cones are serotinous, and seed can remain viable for at least seven years. Seedlings and saplings have been observed in the park in Zovochipolo Forest, in wetlands near Dam 2, around Lake Kaulime, and in grassland near the juniper forest. The presence of

FIGURE 3.3. Bracken fern on the Nyika Plateau. Although it is native to Malawi, some park managers believe it is spreading in an invasive manner (photo by Ramona J. Butz).

this non-native pine is altering fire intensities and rates of spread, which can exceed the suppression capabilities of the local park firefighting staff.

Bracken fern, a native species, is also considered by many park managers to be spreading in an invasive manner (Figure 3.3). It responds very well to fire and thrives under frequent fire-return conditions. It may be aided in its spread due to the simplified species composition of annually burned grasslands providing little competition. A reduction in both the number of late dry-season fires and the overall fire-return interval for the plateau will aid in the containment of bracken fern.

The current levels of fire experienced annually on the Nyika Plateau are shrinking native miombo and evergreen woodlands, further encouraging the spread of non-native invasive species, and appear to be simplifying grassland composition (Figure 3.4). A shift from grasslands dominated by perennials to an increasing dominance by annuals was noted by many of the park's long-term staff during a recent visit.

Discussion

Regular disturbances such as fire and grazing of varying intensity and frequency prevent savanna systems from developing into either pure grasslands or woodlands, thereby serving to create patchy habitats and potentially leading to greater species richness (Oba, Vetaas, and Stenseth 2001; Scholes and Walker 1993; Skarpe 1992). Infrequency of occurance of species in the landscape (low alpha diversity) is common in savanna systems and has been positively correlated with drought resistance in temperate grasslands (Frank and McNaughton 1991).

FIGURE 3.4. Contraction of Miombo woodland in Nyika National Park (photo by Ramona J. Butz).

It is assumed that the structure and productivity of plant communities in arid and semiarid ecosystems are mainly driven by external factors such as moisture and soil composition (Oba, Vetaas, and Stenseth 2001; Noy-Meir 1973; DeAngelis and Waterhouse 1987; Ellis and Swift 1988; Fernandez-Gimenez and Allen-Diaz 1999; Fynn and O'Connor 2000; Oba, Stenseth, and Lusigi 2000), fire (Wiegand et al. 2006; Sankaran et al. 2005; Laris 2002b), and herbivory (Illius and O'Connor 1999; Oba, Mengistu, and Stenseth 2000; Oba 1998). While primary production is positively correlated with rainfall throughout the region (Mwalyosi 1992), moisture is certainly not the only determinant of plant community structure and composition.

Long-term persistence of savanna sites requires that a dynamic balance be maintained between tree mortality caused by too short of a fire-return interval and the encroachment of trees and shrubs resulting from too long of a fire-return interval, such that neither trees nor grasses become locally extinct (Peterson and Reich 2001). In Engikareti, woody plant establishment is more severely limited by droughts or competition from grasses due to site-specific soil composition and nutrient availability than in less arid savanna systems such as those found in Nyika National Park. Woody plant establishment is also strongly controlled by fire, an important source of disturbance in this system. Fire frequency creates changes in structural measures in the community such as growth form shape, size, and age.

In the absence of frequent fires, accumulation of standing plant biomass of low nutritive value and a shift in dominance in favor of increased shrub cover can occur (Figure 3.5). As fire becomes more rare, the

Figure 3.5. Adjacent landscapes in Engikareti that were burned one and eight years prior: (*top*) a grassland burned to increase forage for grazing; (*bottom*) a woodland burned to clear the understory for easier passage of humans and animals (photos by Ramona J. Butz).

transition in the plant community toward woody plant dominance will select for different foragers than grass-dependent ungulates. An increase in woody vegetation will also change the intensity of future fires, resulting in a potentially profound impact on postburn plant communities.

Maasai elders cite high grazing pressure, what they perceive as an increasing frequency of droughts, and a government-legislated ban on burning as negatively affecting range condition in terms of botanical composition, bush encroachment, forage production, and soil erosion (Butz 2009). Degradation, frequently called "green desertification," involves a nearly complete replacement of perennial grasses by annual grasses and the widespread encroachment of acacia thornbush into grassland communities (Bollig and Schulte 1999). Great care must be taken, however, when reviewing degradation narratives, as they have often served colonial and postcolonial critiques of traditional African land use practices (Kwashirai 2013). Although no observations of the replacement of perennial grasses by annual grasses were observed in Engikareti, encroachment of *A. tortilis* seedlings and saplings was readily visible at all but the most recently burned sites. Managers and long-term staff in Nyika did, however, cite changing species composition as a concern. Both decreased fire frequency and increased grazing intensity have been shown to increase shrub encroachment (Dougill and Trodd 1999; Perkins and Thomas 1993a, 1993b).

Reductions in fire frequency caused by landscape fragmentation, land use change, fire prevention activities, and active fire suppression can lead to significant structural changes—including increased tree density, basal area, and canopy cover (Stout 1944; Cooper 1960; Agee 1996; Gillson 2004; Kennedy and Potgieter 2003; Trollope 1982)—and succession toward more fire-sensitive and shade-tolerant overstory species (Agee 1996; Kennedy et al. 2003; Langaas 1992; Mistry et al. 2005; Trollope 1982). These changes may have important implications for pastoralism, wildlife habitat, biotic diversity, and the risk of future catastrophic disturbance (Cooper 1960; Homewood and Rodgers 1991; Wiegand et al. 2006; Laris 2002a; Homewood et al. 2001; Dublin et al. 1990; Bilbao et al. 1996).

Fire has probably always been a major factor in Tanzanian grassland ecology, but the total area burned each year is in decline. Fire control programs initiated by national park personnel have contributed to this decline (McNaughton 1983). Patchy grazing by migratory herds at the beginning of the dry season (McNaughton 1976) also serves to produce extensive firebreaks that can prevent widespread grass fires later in the dry season. High grazing pressure, water pond development, and the banning of fire use are believed to negatively affect range conditions in terms of botanical composition, bush encroachment, forage production, and soil erosion (van Vegten 1984; Angassa and Baars 2000; Roques et al. 2001; Ghermandi et al. 2004).

The limitation of fire use is also causing intense hardship for the Maasai, one of the world's last remaining nomadic pastoralists (Longiporo, personal communication). Decreased fire use by the Maasai, driven in part by fear of imprisonment, limits their ability to manage landscapes for livestock production, their primary source of income, and food security. Restrictions on fire are also accelerating the rate at which traditional tribal knowledge of burning practices are being lost. This loss is particularly concerning because traditional knowledge has been shown to contribute to conservation (Gadgil et al. 1993; Turner and Berkes 2006; Gómez-Baggethun et al. 2010), provide multiple benefits to the holders of the traditional knowledge (e.g., McDade et al. 2007), and reflect an important aspect of human cultural diversity (Maffi 2005).

In the face of similar challenges, Nyika National Park appears to be experiencing a rather dramatic increase in the total area burned over recent years. This increase can be attributed, in large part, to changes in land tenure and management, increasing population pressures, and competing resource needs. Attempts by park management to limit the annual area burned through their prescribed fire program are rendered less effective by both escaped and purposefully set fires by adjacent land owners for activities related to food security such as agriculture, beekeeping, and hunting. Consequently, the traditional occupants of this land are now viewed as detrimental to conservation efforts in the park.

In the cases of the Maasai in Tanzania and that of Nyika National Park, changes in fire management practices over time can be both directly and indirectly attributed to land dispossession and elitist conservation rhetoric. Colonial intrusion cut Maasai land in Kenya and Tanzania in half in the mid- to late 1800s (Fratkin 2001). Further reduction of Maasai grazing land occurred with the creation of large game reserves and national parks in Tanzania from the late 1940s through mid-1960s and beyond. *Ujamaa*, a socialist policy enacted after colonial independence from the British in 1964, led to the burning of Maasai homesteads, the confiscation of livestock, and resettlement of many Maasai into contained villages that controlled grazing and water resources (Hodgson 2001).

Similarly, the Nyika Plateau has often been incorrectly described in the literature as one of the last remaining untouched natural habitats in Africa (McCracken 2006). Human habitation on this landscape dates back at least 3,000 years, and the ancestors of the current-day Phoka people are thought to have settled there as early as the fourteenth century (see McCracken 2006). Colonial intervention beginning in the 1940s sought to establish the plateau as a source of industrial timber production, and its establishment as a national park was announced in 1965. While some local Phoka were retained as temporary park staff, most were forced out of the area by 1975, when the park's boundaries were extended from 367 to 1,209 square miles. All hunting, honey collecting, grazing, and firewood collection were outlawed at that time. Although there is very little literature available on the Phoka's traditional fire practices, all traditional burning practices are assumed to have ceased with their exclusion from their homelands on the Nyika Plateau.

The fire regimes of both Engikareti and Nyika are now dominated by unplanned wildfires. Traditional fire management by local people has been severely limited or is prohibited altogether. This trend, if it continues, could have detrimental effects on many dominant plant species and exacerbate the problem of invasive alien plants. It also directly impacts the lives and livelihoods of local peoples and contributes to the loss of traditional ecological knowledge. The sustainability of African savanna systems is critical for human health and welfare, natural resource use, and ecosystem functioning (Amanor and Moyo 2008; Bassett and Crummey 2003). The potentially serious consequences of this loss of savanna warrant further study and will require a more pragmatic approach to fire management.

References Cited

Agee, James K. 1993. *Fire Ecology of Pacific Northwest Forests*. Washington, DC: Island.

Amanor, Kojo Sebastian, and Sam Moyo. 2008. *Land and Sustainable Development in Africa*. London: Zed Books.

Amigun, Bamikole, Rovani Sigamoney, and Harro von Blottnitz. 2008. "Commercialisation of Biofuel Industry in Africa: A Review."

Renewable and Sustainable Energy Reviews 12 (3): 690–711.

Angassa, Ayana, and Robert M. T. Baars. 2000. "Ecological Condition of Encroached and Non-Encroached Rangelands in Borana, Ethiopia." *African Journal of Ecology* 38 (4): 321–328.

Bassett, Thomas J., and Donald Crummey. 2003. *African Savannas: Global Narratives and Local Knowledge of Environmental Change.* Portsmouth, RI: James Currey.

Beidelman, Thomas O. 1960. "The Baraguyu." *Tanganyika Notes and Records* 54: 245–278.

Bilbao, Bibiana, Richard Braithwaite, Christiane Dall'Aglio, Adriana Moreira, Paulo E. Oliviera, José Felipe Ribeiro, and Philip Stott. 1996. "Biodiversity, Fire, and Herbivory in Tropical Savannas." In *Biodiversity and Savanna Ecosystem Processes: A Global Perspective,* edited by Otto T. Solbrig, Ernesto Medina, and Juan F. Silva, 197–203. Berlin, Heidelberg: Springer-Verlag.

Boko, Michel, Isabelle Niang, Anthony Nyong, Coleen Vogel, Andrew Githeko, Mahmoud Medany, Balgis Osman-Elasha, Ramadjita Tabo, and Pius Yanda. 2007. "Africa." In *Climate Change 2007: Impacts, Adaptation and Vulnerability. Contribution of Working Group II to the Fourth Assessment Report of the Intergovernmental Panel on Climate Change (IPCC),* edited by Martin. L. Parry, Osvaldo F. Canziani, Jean. P. Palutikof, Paul J. van der Linden, and Clair E. Hanson, 433–467. Cambridge: Cambridge University Press.

Bollig, Michael, and Anja Schulte. 1999. "Environmental Change and Pastoral Perceptions: Degradation and Indigenous Knowledge in Two African Pastoral Communities." *Human Ecology* 27 (3): 493–514.

Bond, William J., and Brian W. van Wilgen. 1996. *Fire and Plants.* London: Chapman and Hall.

Boone, Randall B., Michael B. Coughenour, Kathleen A. Galvin, and James E. Ellis. 2002. "Addressing Management Questions for Ngorongoro Conservation Area, Tanzania, Using the SAVANNA Modelling System." *African Journal of Ecology* 40 (2): 138–150.

Bradstock, Ross A., and Belinda J. Kenny. 2003. "An Application of Plant Functional Types to Fire Management in a Conservation Reserve in Southeastern Australia." *Journal of Vegetation Science* 14 (3): 345–354.

Butz, Ramona J. 2009. "Traditional Fire Management: Historical Fire Regimes and Land Use Change in Pastoral East Africa." *International Journal of Wildland Fire* 18 (4): 442–450.

———. 2013. "Changing Land Management: A Case Study of Charcoal Production Among a Group of Pastoral Women in Northern Tanzania." *Energy for Sustainable Development* 17 (2): 138–145.

Casale, Marisa, Scott Drimie, Timothy Quinlan, and Gina Ziervogel. 2010. "Understanding Vulnerability in Southern Africa: Comparative Findings Using a Multiple-Stressor Approach in South Africa and Malawi." *Regional Environmental Change* 10 (2): 157–168.

Chitaukali, Wilbert N., Hynek Burda, and Dieter Kock. 2001. "On Small Mammals of the Nyika Plateau, Malawi." In *African Small Mammals,* edited by C. Denys, L. Granjon, and A. Poulet, 415–426. Paris: IRD.

Coast, Ernestina. 2002. "Maasai Socio-Economic Conditions: A Cross-Border Comparison." *Human Ecology* 30 (1): 79–105.

Cole, Monica M. 1986. *The Savannas: Biogeography and Geobotany.* London: Academic.

Conway, Gordon. 2012. *One Billion Hungry: Can We Feed the World?* Ithaca, NY: Cornell University Press.

Cooper, Charles F. 1960. "Changes in Vegetation, Structure, and Growth of Southwestern Pine Forests Since White Settlement." *Ecological Monographs* 30: 129–164.

DeAngelis, Donald L., and J. C. Waterhouse. 1987. "Equilibrium and Non-equilibrium Concepts in Ecological Models." *Ecological Monographs* 57 (1): 1–21.

Donovan, Sarah E., Paul Eggleton, and Andy Martin. 2002. "Species Composition of Termites of the Nyika Plateau Forests, Northern Malawi, Over an Altitudinal Gradient." *African Journal of Ecology* 40 (4): 379–385.

Dougill, Andrew, and Nigel Trodd. 1999. "Monitoring and Modelling Open Savannas Using Multisource Information: Analyses of Kalahari Studies." *Global Ecology and Biogeography* 8 (3–4): 211–221.

Dowsett-Lemaire, Francoise. 1985. "The Forest Vegetation of the Nyika Plateau (Malawi-Zambia): Ecological and Phenological Studies." *Bulletin du Jardin Botanique National de Belgique / Bulletin van de Nationale Plantentuin van Belgie* 55(3/4): 301–392.

Dublin, Holly T., Alan R. E. Sinclair, and Jacqueline M. McGlade. 1990. "Elephants and Fire as Causes of Multiple Stable States in the Serengeti-Mara Tanzania Woodlands." *Journal of Animal Ecology* 59 (3): 1147–1164.

Dwyer, Edward, Simon Pinnock, Jean-Marie Gregoire, and Jose M. C. Pereira. 2000. "Global Spatial and Temporal Distribution of Vegetation Fire as Determined from Satellite Observations." *International Journal of Remote Sensing* 21 (6–7): 1289–1302.

Dyer, Rodd, Michael Cobiac, Linda Café, and T. Stockwell. 1997. *Developing Sustainable Pasture Management Practices for the semiarid Tropics of the Northern Territory*. Final Report, MRC Project NT 1022. Katherine, Northern Territory, Australia: Northern Territory Department of Primary Industries and Fisheries.

Ellis, James E., and David M. Swift. 1988. "Stability of African Pastoral Ecosystems: Alternate Paradigms and Implications for Development." *Journal of Range Management* 41 (6): 450–459.

Fernandez-Gimenez, Maria E., and Barbara Allen-Diaz. 1999. "Testing a Non-Equilibrium Model of Range Dynamics in Mongolia." *Journal of Applied Ecology* 36 (6): 871–885.

Fishpool, Lincoln D. C., and Michael I. Evans. 2001. *Important Bird Areas in Africa and Associated Islands*. Cambridge: Pisces Publications and BirdLife International.

Fosbrooke, Henry A. 1948. "An Administrative Survey of the Masai Social System." *Tanganyika Notes and Records* 26: 1–50.

Frank, Doug A., and Sam J. McNaughton. 1991. "Stability Increases with Diversity in Plant Communities: Empirical Evidence from the 1988 Yellowstone Drought." *Oikos* 62 (3): 360–362.

Fratkin, Elliot. 2001. "East African Pastoralism in Transition: Maasai, Boran, and Rendille Cases." *African Studies Review* 44 (3): 1–25.

Frost, Peter. 1996. "The Ecology of Miombo Woodlands." In *The Miombo in Transition: Woodlands and Welfare in Africa*, edited by Bruce M. Campbell, 11–57. Bogor, Indonesia: Center for International Forestry Research.

Furley, Peter A., Robert M. Rees, Casey M. Ryan, and Gustavo Saiz. 2008. "Savanna Burning and the Assessment of Long-term Fire Experiments with Particular Reference to Zimbabwe."*Progress in Physical Geography* 32 (6): 611–634.

Fynn, Richard W. S., and Timothy G. O'Connor. 2000. "Effect of Stocking Rate and Rainfall on Rangeland Dynamics and Cattle Performance in a semiarid Savanna, South Africa." *Journal of Applied Ecology* 37 (3): 491–507.

Gadgil, Madhav, Fikret Berkes, and Carl Folke. 1993. "Indigenous Knowledge for Biodiversity Conservation." *Ambio* 22 (2/3): 151–156. http://www.jstor.org/stable/4314060.

Ghermandi, Luciana, Nadia Guthmann, and Donaldo Bran. 2004. "Early Post-Fire Succession in Northwestern Patagonia grasslands." *Journal of Vegetation Science* 15 (1): 67–76.

Gillson, Lindsey. 2004. "Evidence of Hierarchical Patch Dynamics in an East African Savanna?" *Landscape Ecology* 19 (8): 883–894.

Gómez-Baggethun, Erik, Sara Mingorria, Victoria Reyes-García, Laura Calvet, and Carlos Montes. 2010. "Traditional Ecological Knowledge Trends in the Transition to a Market

Economy: Empirical Study in the Doñana Natural Areas." *Conservation Biology* 24 (3): 721–729.

Hall-Martin, Anthony, Humphrey E. Nzima, Werner Myburgh, and Willem van Riet Jr. 2007. *Establishment and Development of Malawi-Zambia Transfrontier Conservation Areas: Nyika Transfrontier Conservation Area Joint Management Plan*. Stellenbosch, South Africa: Peace Parks Foundation.

Higgins, Steven I., William J. Bond, and Winston S. W. Trollope. 2000. "Fire, Resprouting and Variability: A Recipe for Grass-Tree Coexistence in Savanna." *Journal of Ecology* 88 (2): 213–229.

Hodgson, Dorothy Louise. 2001. *Once Intrepid Warriors: Gender, Ethnicity, and the Cultural Politics of Maasai Development*. Bloomington: Indiana University Press.

Homewood, Katherine, Eric F. Lambin, Ernestina Coast, A. Kariuki, Idris Kikula, Julius Kivelia, M. Said, S. Serneels, and Mick Thompson. 2001. "Long-Term Changes in Serengeti-Mara Wildebeest and Land Cover: Pastoralism, Population, or Policies?" *Proceedings of the National Academy of Sciences of the United States of America* 98 (22): 12544–12549.

Homewood, Katherine., Pippa Trench, Sara Randall, Godelieve Lynen, and Beth Bishop. 2006. "Livestock health and socio-economic impacts of a veterinary intervention in Maasailand: Infection-and-treatment vaccine against East Coast fever." *Agricultural Systems* 89 (2–3): 248–271.

Homewood, Katherine M., and W. Alan Rodgers. 1991. *Maasailand Ecology: Pastoralist Development and Wildlife Conservation in Ngorongoro, Tanzania*. Cambridge: Cambridge University Press.

Illius, Andrew W., and Timothy G. O'Connor. 1999. "On the Relevance of Non-equilibrium Concepts to Arid and semiarid Grazing Systems." *Ecological Applications* 9 (3): 798–813.

Jacobs, Michael J., and Catherine A. Schloeder. 2002. "Fire Frequency and Species Associations in Perennial Grasslands of Southwest Ethiopia." *African Journal of Ecology* 40 (1): 1–9.

Kaiza-Boshe, Theonestina, Byarugaba Kamara, and John Mugabe. 1998. "Biodiversity Management in Tanzania." In *Managing Biodiversity: National Systems of Conservation and Innovation in Africa,* edited by John Mugabe and Norman Clark, 155–184. Nairobi: African Centre for Technology Studies (ACTS).

Kennedy, Andrew David, Harry C. Biggs, and Nick Zambatis. 2003. "Relationship Between Grass Species Richness and Ecosystem Stability in Kruger National Park, South Africa." *African Journal of Ecology* 41 (2): 131–140.

Kennedy, Andrew David, and Andre L. F. Potgieter. 2003. "Fire Season Affects Size and Architecture of Colophospermum Mopane in Southern African Savannas." *Plant Ecology* 167 (2): 179–192.

Kijazi, Allan, Samson Mkumbo, and D. Michael Thompson. 1999. "Human and Livestock Population Trends." In *Multiple Land-Use: The Experience of the Ngorongoro Conservation Area, Tanzania,* edited by D. Michael Thompson, 169–180. Gland, Switzerland: International Union for the Conservation of Nature.

Knoop, Warren T., and Brian H. Walker. 1985. "Interactions of Woody and Herbaceous Vegetation in a Southern Africa Savanna." *Journal of Ecology* 73(1): 235–253.

Kwashirai, Vimbai C. 2013. "Environmental Change, Control and Management in Africa." *Global Environment* 6 (12): 166–196.

Langaas, Sindre. 1992. "Temporal and Spatial Distribution of Savanna Fires in Senegal and the Gambia, West Africa, 1989–90, Derived from Multi-Temporal AVHRR Night Images." *International Journal of Wildland Fire* 2 (1): 21–36.

Laris, Paul. 2002a. "Burning the Savanna Mosaic: Fire Patterns, Indigenous Burning Regimes, and Ecology in the Savanna of Mali." PhD dissertation, Clark University, 2002.

———. 2002b. "Burning the Seasonal Mosaic: Preventative Burning Strategies in the Wooded Savannah of Southern Mali." *Human Ecology* 30 (2): 155–186.

Leichenko, Robin M., and Karen L. O'Brien. 2002. "The Dynamics of Rural Vulnerability to Global Change: The Case of Southern Africa." *Mitigation and Adaptation Strategies for Global Change* 7 (1): 1–18.

Lemon, Paul C. 1964. "The Nyika Wild Life Frontier." *Nyasaland Journal* 17 (2): 29–41.

———. 1968. "Effects of Fire on an African Plateau Grassland." *Ecology* 49 (2): 316–322.

Maffi, Luisa. 2005. "Linguistic, Cultural, and Biological Diversity." *Annual Review of Anthropology* 34: 599–617.

McCabe, J. Terrence. 1998. "Risk and Uncertainty Among the Maasai of the Ngorongoro Conservation Area in Tanzania: A Case Study in Economic Change." *Nomadic Peoples* 1 (1): 54–65.

McCracken, John. 2006. "Imagining the Nyika Plateau: Laurens van der Post, the Phoka and the Making of a National Park." *Journal of Southern African Studies* 32 (4): 807–821.

McDade, Thomas W., Victória Reyes-García, P. Blackinton, Susan Tanner, T. Huanca, and William R. Leonard. 2007. "Ethnobotanical Knowledge Is Associated with Indices of Child Health in the Bolivian Amazon." *Proceedings of the National Academy of Sciences of the United States of America* 104 (15): 6134–6139.

McNaughton, Samuel J. 1976. "Serengeti Migratory Wildebeest: Facilitation of Energy Flow by Grazing." *Science* 191(4222): 92–94.

———. 1983. "Serengeti Grassland Ecology: The Role of Composite Environmental Factors and Contingency in Community Organization." *Ecological Monographs* 53 (3): 291–320.

Meadows, Michael E. 1984. "Late Quaternary Vegetation History of the Nyika Plateau, Malawi." *Journal of Biogeography* 11 (3): 209–222.

Meadows, Michael E., and H. Peter Linder. 1993. "Special Paper: A Palaeoecological Perspective on the Origin of Afromontane Grasslands." *Journal of Biogeography* 20 (4): 345–355.

Mistry, Jayalaxshmi, Andrea Berardi, Valeria Andrade, Txicapro Kraho, Phocrok Kraho, and Othon Leonardos. 2005. "Indigenous Fire Management in the Cerrado of Brazil: The Case of the Kraho of Tocantins." *Human Ecology* 33 (3): 365–386.

Mitzlaff, Ulricke von. 1988. *Maasai Women: Life in a Patriarchial Society: Field Research Among the Parakuyo, Tanzania.* Munich: Trickster Verlag.

Muller, Alexander, Josef Schmidhuber, Jippe Hoogeveen, and Pasquale Steduto. 2008. "Some Insights in the Effect of Growing Bio-Energy Demand on Global Food Security and Natural Resources." *Water Policy* 10: 83–94.

Mwalyosi, Raphael B. B. 1992. "Influence of Livestock Grazing on Range Condition in South-West Masailand, Northern Tanzania." *Journal of Applied Ecology* 29 (3): 581–588.

Ngailo, Jerry A., Fidelis B. S. Kaihura, Frederick Baijukya, and Barnabas J. Kiwambo. 2001. "Land Use Changes and Their Impact on Agricultural Biodiversity in Arumeru, Tanzania." Paper presented at the Land Management and Ecosystem Conservation Conference.

Noble, Ian R., and Ralph O. Slatyer. 1980. "The Use of Vital Attributes to Predict Successional Changes in Plant Communities Subject to Recurrent Disturbances." In *Succession*, edited by Eddy van der Maarel, 5–21. Nijmegen: Springer Netherlands.

Noy-Meir, Imanuel. 1973. "Desert Ecosystems: Environment and Producers." *Annual Review of Ecology and Systematics* 4: 25–51.

Nyoka, Betserai Isaac. 2003. "Biosecurity in Forestry: A Case Study on the Status of Invasive Forest Tree Species in Southern Africa." Forest Biosecurity Working Paper, Food and Agriculture Organization of the United Nations,

Rome. http://www.fao.org/docrep/005/AC846E/AC846E00.HTM.

Oba, Gufu. 1998. "Effects of Excluding Goat Herbivory on *Acacia tortilis* Woodland Around Pastoralist Settlements in Northwest Kenya." *Acta Oecologica—International Journal of Ecology* 19 (4): 395–404.

———. 2001. "Indigenous Ecological Knowledge of Landscape Change in East Africa." *IALE Bulletin* 19: 1–3.

Oba, Gufu, and L. M. Kaitira. 2006. "Herder Knowledge of Landscape Assessments in Arid Rangelands in Northern Tanzania." *Journal of Arid Environments* 66 (1): 168–186.

Oba, Gufu, and Dido G. Kotile. 2001. "Assessments of Landscape Level Degradation in Southern Ethiopia: Pastoralists Versus Ecologists." *Land Degradation and Development* 12 (5): 461–475.

Oba, Gufu, Zelalem Mengistu, and Nils C. Stenseth. 2000. "Compensatory Growth of the African Dwarf Shrub *Indigofera spinosa* Following Simulated Herbivory." *Ecological Applications* 10 (4): 1133–1146.

Oba, Gufu, Eric Post, Per Ole Syvertsen, and Nils C. Stenseth. 2000. "Bush-Cover and Range Condition Assessment in Relation to Landscape and Grazing in Ethiopia." *Landscape Ecology* 15 (6): 535–546.

Oba, Gufu, Nils C. Stenseth, and Walter J. Lusigi. 2000. "New Perspectives on Sustainable Grazing Management in Arid Zones of Sub-Saharan Africa." *BioScience* 50 (1): 35–51.

Oba, Gufu, Ole R. Vetaas, and Nils C. Stenseth. 2001. "Relationships Between Biomass and Plant Species Richness in Arid-Zone Grazing Lands." *Journal of Applied Ecology* 38 (4): 836–845.

Parr, Catherine L., and Bruce H. Brockett. 1999. "Patch-Mosaic Burning: A New Paradigm for Savanna Fire Management in Protected Areas?" *Koedoe* 42 (2): 117–130.

Perkins, Jeremy S., and David S. G. Thomas. 1993a. "Environmental Response and Sensitivity to Permanent Cattle Ranching, semiarid Western Central Botswana." In *Landscape Sensitivity*, edited by David S. G. Thomas and R. J. Allison, 273–286. Chichester, UK: J. Wiley and Sons.

———. 1993b. "Spreading Deserts or Spatially Confined Environmental Impacts: Land Degradation and Cattle Ranching in the Kalahari Desert of Botswana." *Land Degradation and Rehabilitation* 4 (3): 179–194.

Peterson, David W., and Peter B. Reich. 2001. "Prescribed Fire in Oak Savanna: Fire Frequency Effects on Stand Structure and Dynamics." *Ecological Applications* 11 (3): 914–927.

Roques, Kim G., Timothy G. O'Connor, and Andrew R. Watkinson. 2001. "Dynamics of Shrub Encroachment in an African Savanna: Relative Influences of Fire, Herbivory, Rainfall and Density Dependence." *Journal of Applied Ecology* 38 (2): 268–280.

Sankaran, Mahesh, Niall P. Hanan, Robert J. Scholes, Jayashree Ratnam, David J. Augustine, Brian S. Cade, Jacques Gignoux, Steven I. Higgins, Xavier Le Roux, Fulco Ludwig, Jonas Ardo, Feetham Banyikwa, Andries Bronn, Gabriela Bucini, Kelly K. Caylor, Michael B. Coughenour, Alioune Diouf, Wellington Ekaya, Christie J. Feral, Edmund C. February, Peter G. Frost, Pierre Hiernaux, Halszka Hrabar, Kristine L. Metzger, Herbert H. T. Prins, Susan Ringrose, William Sea, Jorg Tews, Jeff Worden, and Nick Zambatis. 2005. "Determinants of Woody Cover in African Savannas." *Nature* 438 (7069): 846–849.

Scholes, Robert J., and Steven R. Archer. 1997. "Tree-Grass Interactions in Savannas." *Annual Review of Ecology and Systematics* 28: 517–544.

Scholes, Robert J., and Brian H. Walker. 1993. *An African Savanna: Synthesis of the Nylsvley Study*. Cambridge: Cambridge University Press.

Sharp, Ben R., and David M. J. S. Bowman. 2004. "Patterns of Long-Term Woody Vegetation Change in a Sandstone-Plateau Savanna

Woodland, Northern Territory, Australia." *Journal of Tropical Ecology* 20: 259–270.

Sheuyange, Asser, Gufu Oba, and Robert B. Weladji. 2005. "Effects of Anthropogenic Fire History on Savanna Vegetation in Northeastern Namibia." *Journal of Environmental Management* 75 (3): 189–198.

Sinclair, Alan R. E., and Peter Arcese, eds. 1995. *Serengeti II: Dynamics, Management, and Conservation of an Ecosystem*. Chicago: University of Chicago Press.

Skarpe, Christina. 1991. "Impact of Grazing in Savanna Ecosystems." *Ambio* 20 (8): 351–356.

———. 1992. "Dynamics of Savanna Ecosystems." *Journal of Vegetation Science* 3 (3): 293–300.

Spear, Thomas, and Richard Waller, eds. 1993. *Being Maasai: Ethnicity and Identity in East Africa*. London: James Currey.

Spencer, Paul. 1988. *The Maasai of Matapato: A Study of the Rituals of Rebellion*. London: Manchester University Press for the International African Institute.

Sporton, Deborah, and David S. G. Thomas. 2002. *Sustainable Livelihoods in Kalahari Environments: A Contribution to Global Debates*. Oxford: Oxford University Press.

Stout, Arlow B. 1944. "The Bur Oak Openings in Southern Wisonsin." *Transactions of the Wisconsin Academy of Sciences, Arts, and Letters* 36: 141–161.

Tanzania, National Bureau of Statistics. 2012. *Population and Housing Census: Population Distribution by Administrative Areas*. Dar es Salaam: Ministry of Finance.

Trollope, Winston S. W. 1982. "Ecological Effects of Fire in South African Savannas." In *Ecology of Tropical Savannas*, edited by Brian J. Huntley and Brian H. Walker, 292–306. Berlin: Springer-Verlag.

Turner, Nancy J., and Fikret Berkes. 2006. "Developing Resource Management and Conservation." *Human Ecology* 34 (4): 475–478.

Tyson, Peter, Eric Odada, Roland Schulze, and Coleen Vogel. 2002. "Regional-Global Change Linkages: Southern Africa." In *Global-Regional Linkages in the Earth System*, edited by Peter Tyson, Congbin Fu, Roland Fuchs, Louis Lebel, A. P. Mitra, Eric Odada, John Perry, Will Steffen and Hassan Virji, 3–73. Berlin, Heidelberg: Springer-Verlag.

van Langevelde, Frank, Claudius A. D. M. van de Vijver, Lalit Kumar, Johan van de Koppel, Nico de Ridder, Jelte van Andel, Andrew. K. Skidmore, John W. Hearne, Leo Stroosnijder, William J. Bond, Herbert H. T. Prins, and Max Rietkerk. 2002. "Effects of Fire and Herbivory on the Stability of Savanna Ecosystems." *Ecology* 84 (2): 337–350.

van Vegten, Jan A. 1984. "Thornbush Invasion in a Savanna Ecosystem in Eastern Botswana." *Vegetation* 56 (1): 3–7.

van Wilgen, Brian W. 2009. "The Evolution of Fire Management Practices in Savanna Protected Areas in South Africa." *South African Journal of Science* 105 (9–10): 343–349.

van Wilgen, Brian W., Chris S. Everson, and Winston S. W. Trollope. 1990. "Fire Management in Southern Africa: Some Examples of Current Objectives, Practices, and Problems." In *Fire in the Tropical Biota: Ecosystem Processes and Global Challenges*, edited by Johann G. Goldammer, 179–215. Berlin, Heidelberg: Springer-Verlag.

Werner, Patricia A., Brian H. Walker, and Philip A. Stott. 1991. "Introduction." In *Savanna Ecology and Management: Australian Perspectives and Intercontinental Comparisons*, edited by Patricia A. Werner. Oxford: Blackwell Scientific.

Wiegand, Kerstin, David Saltz, and David Ward. 2006. "A Patch-Dynamics Approach to Savanna Dynamics and Woody Plant Encroachment: Insights from an Arid Savanna." *Perspectives in Plant Ecology, Evolution, and Systematics* 7 (4): 229–242.

CHAPTER 4

Fire Management in Brazilian Savanna Wetlands

*New Insights from Traditional Swidden Cultivation
Systems in the Jalapão Region (Tocantins)*

LUDIVINE ELOY, SILVIA LAINE BORGES, ISABEL B. SCHMIDT,
AND ANA CAROLINA SENA BARRADAS

Fire has contributed significantly to the expansion of savannas (Bond and Keeley 2005), and today it is used more in these biomes than in any other in the world (Beerling and Osborne 2006). Since the 1950s, ecological research has indicated that fire is one of the determining factors for the existence and subsequent maintenance of savanna ecosystems (Jeltsch et al. 2000).

In recent decades, tropical savannas have been marked by fire regime transformations, with higher frequencies and greater severity of wildfires (Butz, chapter 3; Myers 2006). In the Brazilian *cerrado*, the largest tropical savanna in Latin America and a biodiversity hotspot, firefighting policies are justified by frequent wildfires that occur during the dry season. In 2009, it was estimated that approximately 24 percent of Brazil's CO_2 emissions due to land use between 2003 and 2005 came from the cerrado region, caused mainly by deforestation and burning (MCT 2009).

Fire management in protected areas in the cerrado has become a major challenge, with over 50 percent of the region affected by wildfires each year (Beatty 2014). Anthropogenic fire is generally identified as responsible for changes in the fire regime. Over the past fifty years, fires that occurred mainly at the beginning of the dry season (May–June) were replaced by late dry season fires (August–September), leading to large and intense wildfires, with several negative consequences for the environment (Coutinho 1990; Silva et al. 2011) and local communities.

Nevertheless, several researchers have shown that traditional fire management systems in tropical savannas play key roles in the maintenance of spatial heterogeneity, which prevents wildfires and supports important ecological processes, as in Africa (Brockett et al. 2001; Laris 2002), Australia (Bliege Bird et al. 2008; Russell-Smith et al. 1997), and Brazil (Welch et al. 2013). Such research, and the recurrence of wildfires in recent decades, has influenced fire management models, progressively turning fire suppression policies into fire management policies (Myers 2006). As Scott and his colleagues (2016, 5) assert, "It is important for policy-makers to accept that fire suppression is not the only mechanism and that sustainable fire management may be possible given a cooperative environment whereby all stakeholders have a say."

In the cerrado, this evolution of fire policies is beginning, but little research has gone into the rationales behind the agropastoral uses of fire and their transformations over time, especially in and around protected areas, where burning is usually prohibited (Borges et al. 2014). These productive activities depend mainly on the use of riparian vegetation, especially in upstream veredas (peat swamp palm forest), where soils are more fertile and humid, even during the long dry season (June to October).[1] Fire is a key element in the management of these activities: it clears and produces ash for agricultural fields in riparian forests and stimulates the regrowth of native pastures within the nearby humid grasslands (Eloy and Borges 2013). Yet fire is usually banned in riparian ecosystems because environmental managers are particularly sensitive about fire impacts.

However, a pioneering agreement signed in 2012 between the managers of a protected area of the Jalapão region (Tocantins State) and local people has led to an opportunity to negotiate fire uses as well as perform research on traditional fire management systems and their environmental impacts on wetland ecosystems.

The objectives of this chapter are to describe the past and present agricultural burning practices in the veredas of local farmers in Jalapão region, and introduce the preliminary elements of an environmental impact assessment of these practices. Swidden cultivation in the tropics is often blamed for deforestation and water resource degradation, without fully considering the role of traditional fire management practices for agrobiodiversity conservation, fallow succession, and landscape management (Padoch and Pinedo-Vasquez 2010). That is why, in the case of the wetlands ecosystems of the Jalapão, we propose to analyze the impacts of swidden cultivation on forest cover and water resources, and to understand its role in agrobiodiversity management. For these purposes, presented below is a literature review of the ecology and agricultural management of peat swamps in tropical landscapes and a description of local cultivation and fire management practices at Jalapão to provide a first assessment of their impact. Through this research, we hope to contribute to an improved dialogue between local communities and environmental managers on the topic of fire management in the cerrado, inside and outside legally protected areas.

Ecology and Agricultural Management of Peat Swamp Forests

Tropical peat lands contain some of the largest near-surface reserves of terrestrial organic carbon (Page et al. 2002), but their stability has been threatened by human activities since the early 1980s (Siegert et al. 2001).[2] Tropical peat swamp forest is a unique ecosystem, most extensive in Southeast Asia and under enormous threats from logging, fire, and land conversion. As a result, ecological research on peat swamps has been mostly carried out in these regions, especially in Indonesia and Malaysia (Posa et al. 2011; Usup et al. 2004; Yule 2010).

Fire in peat swamp forest vegetation causes effects similar to fire in other forest types: lower canopy cover, decreased species richness, and reduced tree and sapling density compared with areas of unburned forest (Yeager et al. 2003). Tropical peat swamp forests, however, are more vulnerable to destruction by fire than any other forest type because the soil substrate is extremely flammable when dry (Langner and Siegert 2009).

Moreover, controlling peat fires is particularly difficult in these areas because once fire has started, it can carry on for months in subterranean layers of the peat, potentially burning for the entire dry season. These subsurface fires can also cause the collapse of overlying material, increasing tree mortality. Frequent burning leads to rapid lowering of peat soils. Excessive drainage and uncontrolled burning also cause accelerated oxidation of peat soils and dramatic loss of organic material. Local communities know the importance of preventing these subsurface fires. One method is to dig a drain around the fire, down to the wet soil (Andriesse 1988). As has been noted in Indonesian peat swamp forests, the speed with which fire spreads in previously cleared areas is faster than in intact areas (Usup et al. 2004). This means that clearing vegetation and then waiting for the biomass to dry before burning may limit the progression of fire into deeper peat layers.

Compared to Southeast Asian peat swamp forests, little is known about fire uses and their impacts on savanna wetlands. These areas may function as refuge areas for wildlife (Redford and Fonseca 1986). They are often managed with fire and are culturally and economically important to local communities (Barbosa and Schmitz 1998; McGregor et al. 2010). The drainage of lowlands and humid soils, often with high fertility, is a common practice in tropical savanna regions marked by sharp differences between dry and rainy seasons (Lavigne-Delville 2003). In Latin America, seasonally flooded tropical savannas were used since prehistory for agriculture, through drainage and fire. Savanna formations in Brazil have been used for resource management and agricultural activities for at least 8,000 years (Feltran-Barbieri 2010). In several regions, practices such as the construction of raised fields created vast agricultural landscapes (Renard et al. 2012).

Fire Policies, Fire Ecology, and Traditional Fire Management Systems in the Peat Swamp Forest of the Brazilian Tropical Savanna (Cerrado)

The cerrado biome is composed of different vegetation types (physiognomies) ranging from grasslands to forests, which differ in terms of sensitivity and resistance to fire. These differences result in varying fire frequencies. Forest formations, such as gallery forests, tend to be more sensitive to fire than more open habitats (savannas and grasslands) (Walter and Ribeiro 2010).

Unlike Southeast Asian peat swamp forests, the veredas (palm swamps) of the cerrado are poorly studied. Their vegetation is visually dominated by *buriti* palm (*Mauritia flexuosa* L.f.) amid clusters of shrubs, and a subshrub layer dominated by grasses and forbs, with occurrence of Ciperacea and Xyridaceae species. The soil is hydromorphic, organic (peat), and flooded most of the year (Walter and Ribeiro, 2010; Sampaio et al. 2012). Veredas are geomorphologically young areas in the process of succession and formation (Ferreira 2008). In Brazil, this soil type covers about .03 percent of the country (IBGE and EMBRAPA 2001).

Given the seasonality that characterizes the savanna's climate, swampy palm forests and gallery forests function as islands of stability in terms of water availability and refuge areas for wildlife (Redford and Fonseca 1986; Cianciaruso and Batalha 2008). Paleoenvironmental studies in the veredas of Brazil (Ferraz-Vicentini and Salgado-Labouriau 1996; Barberi et al. 2000; Meneses et al. 2013) have reported the occurrence of fire in the prehistoric past, as evidenced by the presence of carbon in soil sediments. In Venezuela and Colombia, studies have shown the importance of fire for the maintenance of veredas (called *morichal*) (Vegas-Vilarrúbia et al. 2011; Montoya and Rull 2011)

Since veredas are considered sensitive to fire, research involving experimental fires in this type of vegetation has not been conducted, so the available information comes from studies on accidental fires. Late dry season fires devastate plants of the veredas (Bahia et al. 2009; Maillard et al. 2009). These fires are difficult to fight and can last for several days or even weeks (Maillard et al. 2009). The species composition of the swampy forests, in general, is such that they do not have protection mechanisms against fire. Due to the predominance of peat soil, subterranean fires are common, causing severe damage to soil structure and function (erosion and acidification). The removal of surface vegetation due to wildfires in veredas can begin an erosion process that can persist for decades (Wantzen et al. 2006). Biennial fire in the veredas can reduce populations of buriti, the dominant palm species, because there is not enough time for them to recover from the previous fire event. Because buriti populations take more than three years to recover after fire, Sampaio and his colleagues (2012) recommend a fire interval of at least ten years for the conservation of this species.

In the cerrado, management decisions are mostly oriented toward banning fire use and constructing firebreaks to prevent fire from entering forest formations. The farming systems of traditional populations[3] within or around the cerrado's protected areas, however, rely mainly on gallery forests and vereda environments, where soils are more fertile and humid even during the dry season (June–October). Palm swamps and the humid grasslands that surround them are often used for swidden cultivation; watering of livestock; extensive grazing on native grasses; and the harvesting of fibers, ornamental flowers, and palm leaves (for thatching) (Ribeiro 2008; Sampaio et al. 2012). Fire is a key element in the management of these activities: it clears new agricultural fields and stimulates the regrowth of native pastures during the dry season.

Most of the research on the impact of traditional fire management systems in the cerrado was conducted on Indigenous lands (Mistry et al. 2005; Welch et al. 2013; Falleiro 2011; Melo and Saito 2011). Except for one project in agricultural settlements (Mistry 1998), these studies explore the use of fire as it relates to resource protection, hunting, and harvesting, while little research has focused on the rationales behind the agropastoral uses of fire and their transformations over time. While swidden cultivation is well documented and gaining increasing recognition for its environmental and sociocultural importance in the Amazon and Atlantic Rain Forests (Adams et al. 2013; Emperaire and Eloy 2014; Padoch and Pinedo-Vasquez 2010; Steward, chapter 5), few such studies have been done in the cerrado, and those that do exist are very localized (Niemeyer 2011; Ribeiro 2005). Moreover, they lack environmental impact assessments. As a result, burning practices are usually banned in and around the cerrado's protected areas. In fact, in a context of rapid destruction of this biome,[4] environmental managers are particularly sensitive about fire use in and around the gallery forests and veredas. Preservation of the flagship palm species, buriti, is a priority, especially because of the role it plays in water regulation and biodiversity.

Methods
The Case Study

The Jalapão is a region in the eastern part of Tocantins State, bordering the states of Maranhão, Piauí, and Bahia. It has a population of 30,644, 37 percent of whom live in rural areas (MDA 2013).

The region hosts the largest extension of cerrado vegetation inside protected areas, representing a significant area of the remaining natural vegetation of this biome. One of the largest protected areas is the Serra Geral do Tocantins Ecological Station (SGTES) (Figure 4.1).

The SGTES covers 7,070 square kilometers, spanning two municipalities, Mateiros and Rio da Conceição (ICMBio 2014). It is a federal-level protected area managed by the

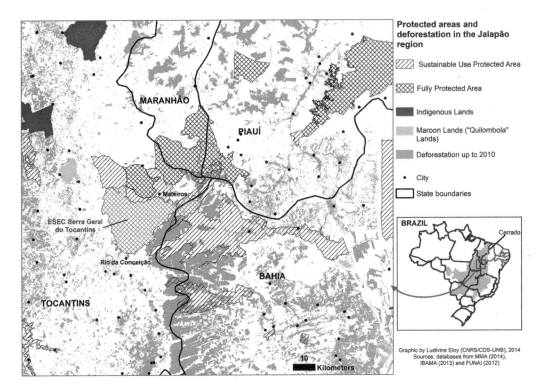

FIGURE 4.1. Map of the Jalapão region of Brazil showing the location of the Serra Geral do Tocantins Ecological Station, other protected areas, and areas of deforestation (map drawn by Ludivine Eloy).

Chico Mendes Institute of Biodiversity Conservation (ICMBio). In the Jalapão, farming systems rely mainly on swidden cultivation, extensive grazing on native grasses, and harvest of the ornamental flower *capim dourado* (*Syngonanthus nitens* [Bond.] Ruhland). While a body of collaborative research exists on fire ecology associated with the harvest and management for handicraft production (Schmidt et al. 2007; Schmidt et al. 2011; Viana 2013), the agricultural systems have so far been little explored.

The delimitation of SGTES took place in 2001, over an area occupied for centuries by Indigenous people and for nearly 120 years by families originating mainly from Bahia State, who identify themselves as quilombola communities (descendants of African slaves). They began to claim land rights in the mid-2000s when they were confronted by several resource use restrictions (Lindoso 2011).[5] The challenge of the ecological station managers is therefore to reconcile conservation objectives with the use of natural resources by local residents. In its quest for instruments of conflict mediation, the ICMBio signed a legal agreement (Termo de Compromisso) in 2012 with the local association, Ascolombas-Rios, permitting the traditional use of fire for agropastoral purposes, including in and around the veredas. According to this agreement, these activities are to be monitored and evaluated in collaboration with local residents. This is the context of our study.

Methods

The research focused on past and present swidden cultivation practices in the SGTES.

The combination of agroecological data obtained at the farmland scale and land cover data obtained from satellite imagery was used to investigate the farmers' production practices and their impacts on land cover. Data were obtained through semistructured interviews with farmers, field surveys, and remote sensing. The methodology sought to include members of the quilombola association who participated in the planning of fieldwork, data collection, and the validation of results.

Field visits were undertaken in the dry season between July and September 2013. We identified twelve families who had cultivated swidden fields and/or managed fallow forests within the SGTES. During the visits, farmers were asked to identify and interpret the different landscape components (native vegetation types, burning scars, rivers, cultivated fields, fallows, soil types, etc.) that were georeferenced. We used questionnaires to inquire about past agricultural activities in each swidden plot and studied fifteen swidden fields and eighteen fallow areas.

To assess the impact of swidden cultivation on land cover and on the water regime of the veredas, we used remote sensing tools to verify the type of vegetation within fallow areas and the dimension and location of cultivated plots with reference to river paths. We used ArcMap software (version 10.0) to map the different plots through GPS points projected on a satellite image acquired by ResourceSat-1 (IRS-P6) (LISS3do 08.07.2013, scene 328/084, resolution 23.5 meters).

To analyze the land cover within fallows areas, we used a vegetation classification map on Landsat TM images covering the SGTES, produced in 2010. Five vegetation types were identified in this map: open grassland (*campo limpo*), shrubby grassland (*campo sujo*), woodlands (*cerrado stricto sensu*), gallery forest, and converted areas (urbanized or converted to monoculture). Pixels contained in the fallow polygons were extracted in order to calculate the number of pixels corresponding to each class. To assess the impact of swidden cultivation on water regime in the veredas, we estimated the plot sizes and their distance to river sources. We also observed soil management practices and soil moisture levels in fallows.

Results

Population and Agricultural Practices

Agrarian History in the Jalapão

Prehistoric studies in Tocantins State show that the oldest settlements recorded date to circa 10,000 BP. In the Jalapão region, the first human populations maintained an economy based on hunting, fishing, and gathering. Archaeological registers of human occupation in Jalapão are located in smaller tributaries of the Sono River and indicate "a settlement pattern that prioritizes lower water flow areas" (Oliveira and Aguiar 2011, 20).

When European colonization began in the area now known as the state of Tocantins, Indigenous groups had production systems based on swidden agriculture (mainly corn, squash, and sweet potatoes) and hunting, fishing, and gathering of natural products.[6] There is strong evidence that these groups developed complex forms of agroforestry (Oliveira and Aguiar 2011). From the eighteenth century, contact between Indigenous peoples and European settlers was extremely contentious, mainly due to the expansion of mining (from the southern part of Goiás State), agriculture (expansion of rice farming from Maranhão State), and livestock (expansion of cattle from Bahia State) (Santos 2013). In this context, "the Jalapão became a refuge for many Indigenous peoples including Xerentes, Akroá, Krahôs, Timbiras, and even Xakriabá and Xavante" (Santos 2013, 250). This migration process may be attributed to the remoteness of the Jalapão region, as well as to the presence of abundant natural resources. Unlike other areas, the Jalapão is home to many important plant species of the central Brazilian savannas, as it is abundant in water (Oliveira and Aguiar 2011).

The nineteenth and early twentieth centuries were marked by the arrival of several families from the northeast region of the country who interacted and intermarried with the Indigenous population. Our interviews indicate that these newcomers were mostly Maroon slaves that escaped from large farms in Bahia or smallholder farmers who migrated to the region to escape severe drought and land conflicts in the Northeast. Many cowboys who drove livestock to Piauí and Bahia and used the abundant native pastures of the Jalapão during the dry season also established residence there (Eloy and Borges 2013). Today, part of the Jalapão population self-identifies as quilombola, but Indigenous influences in the local culture are numerous, expressed mainly in material culture, especially handicraft production (Schmidt et al. 2011), and agricultural and food practices. The complex techniques of wetland cultivation and fire management, however, may also result from the adaptation of specific skills brought to the Americas through the African diaspora (Carney 2002; Sluyter and Duvall 2016).

Our interviews indicate that the parents and grandparents of the families currently using the SGTES moved to the region in search of better land for cultivation and cattle raising (Eloy and Borges 2013). During the 1960s and 1970s, they concentrated near the Rio Novo River and its tributaries, forming the community Mata Verde, which received a public primary school. These families used natural resources in diverse ways. They used *gerais* (natural open grasslands) as common pasturelands and for hunting, as well as the forested valley for agriculture and the gathering of native fruits, fiber, and timber species.

Between 1950 and 1980, livestock were more numerous in the SGTES area than today because ranchers from Tocantins State used the native Jalapão pasture during the dry season to raise nearly 5,000 head of cattle. Many of the region's residents began raising their own livestock by receiving cattle in payment for tending the ranchers' animals (Lindoso and Parente 2013).

Until the 1980s, the local agricultural system was based on the cultivation of wetlands: flooded rice fields (*roça de brejo, roça de vazante*) opened in gallery forests during the rainy season, and drained vereda swidden cultivation (*roça de esgoto*) during the dry season. Rainfed swidden cultivation in dry soils (*roça de toco*) was used to complement this production. Our interviews indicate that flooded rice cultivation is older than drained peat swamp swidden cultivation, dating to at least the nineteenth century.

According to one of our informants, one of the early newcomers from Bahia knew how to cultivate peat swamp during the dry season by digging drainage ditches (*esgoto*) and using fire. He became locally famous for having developed and disseminated these techniques in the region.

Areas of roça de esgoto produced agricultural surpluses, especially of cassava, that were sold in the nearby cities. Production was greater and more diversified in these areas than in roças de toco.

> My father had a huge area of roça de esgoto. He used to grow pig and had a lot of sugarcane. He used to make rapadura [unrefined cane sugar]. He planted so much pumpkin and watermelon, we carried so much, that we had to call people from Riachão to come and fetch the products with their animals. He used to cultivate roça de toco fields, but they always had less than the roça de esgoto.[7] (Josefa Chagas dos Santos, interview)

Furthermore, the sizes of peat swamp fallow fields that we visited (from .5 to 14.6 hectares per household) indicate that some areas in the SGTES were more densely populated and cultivated over the last fifty years. For example, one informant stated that he and his father cultivated more than ten veredas during his lifetime in different locations within the northern portion of the SGTES.

Because peat swamp cultivation was particularly productive, its placement determined (and determines even today) the location of houses and motivated many families to regularly migrate every ten to twenty years within the same region and maintain at least two houses (an older house and a newer house). However, not all families had access to adequate places to establish peat swamp fields, which require specific soils at a certain inclination (see below).

Abandonment of flooded rice cultivation and the decline of peat swamp cultivation after the 1980s can be linked to three processes: increased consumption of industrialized food products, especially rice and beans; local rural exodus; and external imposition of new environmental norms and regulations.

In 1980, families from the Mata Verde community started building houses in the nearby towns of Mateiros and Rio da Conceição. Many young people moved to larger and more distant cities to attend school or find jobs. The growth of regional cities and job opportunities contributed to the gradual departure of youths and closure of the Mata Verde rural public school in 1997. Absence of a school led other families to move to nearby cities.

According to residents, creation of the SGTES protected area around 2001–2003 was another aggravating factor in the process of rural exodus because it created fear of expulsion and prohibition of most productive activities. Today, however, about fifteen families maintain productive activities in SGTES, often with dual residences (a house in Mateiros and a wattle and daub house within the SGTES).

Present Agricultural Practices

We identified three types of cultivation systems in the SGTES: rainfed swidden cultivation (roça de toco), drained peat swamp swidden cultivation (roça de esgoto), and planted livestock pastures (roça de pasto) with African grasses. Each of these types corresponds with a landscape component (Figure 4.2).

Rainfed swidden cultivation takes place on well-drained lands called *terra de cultura* (fertile clay soils) located in gallery forest formations (*capão*). The cultivation period of cassava and banana, especially, alternates with a long fallow period ranging from eight to twenty years. According to our informants, due to the depopulating rural exodus described above, as well as changes in cattle breeds and increased rainfall variability in recent decades, these swidden cultivation systems are used less than before. Nowadays, rainfed cassava cultivation areas are commonly converted to pastures planted with exotic African grass species, especially brachiaria (*Urochloa* spp. P. Beauv.), *Andropogon* spp. L., and *quicuio* (*Pennisetum clandestinum* Hochst. ex Chiov.).

Peat Swamp Cultivation

Drained peat swamp swidden cultivation (Figures 4.3 and 4.4) constitutes the basis of the local agricultural system. As mentioned above, its origins date from the nineteenth century, with the arrival of the first quilombola families, but might also go as far back as the Indigenous occupation of the Jalapão region. Such fields are cultivated in peat swamp forests (called *pantâmo* or *brejo* by farmers) which are dominated by buriti palm and correspond to veredas. The peat soil is very rich in organic matter. Farmers choose an area with a slight slope to allow for drainage. At the end of the rainy season (February–March), after digging drainage ditches around the plot (which is usually

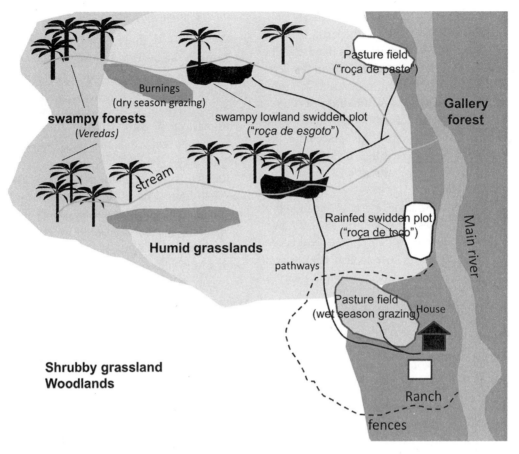

Figure 4.2 Representation of landscape diversity in the Rio Novo Valley, Serra Geral do Tocantins Ecological Station (drawing by Ludivine Eloy).

smaller than .5 hectare), the palm forest is felled and burned so cultivation can occur during the dry season (July–September).[8]

A field is burned only once, in June or July, which is the beginning of the production cycle (see Figure 4.4). To avoid starting wildfires, farmers open a firebreak a few days earlier by clearing fuels from field boundaries.

According to our informants, if well managed, this kind of field can be successfully cultivated continuously, without a fallow period, for ten to twenty years without additional burning or fertilization. Such an extended cultivation phase is followed by a forest fallow of at least ten years. Drainage ditches allow for control of soil moisture in the field. Thus ditches are enlarged during the rainy season to facilitate drainage, and their exits are usually blocked with mud during the late dry season to retain moisture.

According to farmers, controlling water levels within drainage ditches helps prevent soil drying and acidification, as well as reducing the danger of long-lasting subterranean fires with the potential to destroy peat soils.

A few days after burning, farmers plant fruit and vegetables (such as watermelon and squash) as well as beans. Cassava and banana are introduced later, in November or December, and remain in the field over several subsequent years. After the second or third year of cultivation on a plot, cassava becomes

FIGURE 4.3. Burning of a swidden plot in peat swamp forest (*vereda*) (photo by Ludivine Eloy).

FIGURE 4.4. The main ditch of a drained peat swamp swidden field (*roça de esgoto*) after burning (photo by Ludivine Eloy).

the dominant crop. Agrobiodiversity is high: among the forty-five cultivated varieties that we identified in the SGTES, twenty-eight were present exclusively in drained peat swamp swidden fields (Table 4.1).

It seems that peat swamp swidden fields function as agrobiodiversity repositories by ensuring the year-round survival of several cultivars propagated by vegetative reproduction (such as cassava, yams, and sugarcane). This is especially important to the rainfed swidden fields, which are cultivated only during the rainy season. Such circulation of cultivars relies on exchange networks among farmers. Farmers reopen a roça de esgoto after ten to twenty years of fallow by renovating the existing drainage ditches. A set of fields, fallows, and old drains therefore forms a familiar productive space that helps determine where houses are built for one to three decades (Figure 4.5).

Environmental Impacts of Swidden Cultivation in Peat Swamp Forests

Farmers strongly agree that "drains renovate the soils" ("o esgoto renova a terra"); that is, forest regeneration during fallow periods produces vegetation that is denser than before drained vereda cultivation. They further affirm that these secondary forests are richer in buriti palm than other swampy forests.

We observed species within these fallows that are typical of gallery forests, including *pindaíba-do-brejo* (*Xylopia emarginata* Mart.), buriti, *araruta* (*Maranta arundinacea* L.), and *embaúba* (*Cecropia* sp. Loefl.). This observation, however, requires confirmation by systematic botanical survey. In this study we used remote sensing tools to verify vegetation characteristics in drained peat swamp swidden fallows.

Figure 4.6 shows that fallows generated after drained peat swamp swidden cultivation coincide with the presence of dense vegetation, mainly classified as gallery forest or woodlands. Corroborating this observation, we found 82 percent of pixels in fallow polygons to be woody vegetation classes (gallery forests and woodlands) (Table 4.2). This evidence suggests that the observed swidden cultivation practices do not lead to deforestation of palm peat swamp, as they enable restoration of vegetation cover during the cultivation and fallow cycle.

The sizes of swidden fields were observed to be small (.4 hectares on average) and located downstream from river sources (Table 4.3). Whereas all drainage ditches observed in productive swidden fields were filled with water even in the late dry season (August), ditches in fallows were obstructed due to lack of maintenance, and consequently, soils were waterlogged. Thus this form of drainage is temporary.

Discussion

Given that vereda environments are recognized as essential to biodiversity conservation in the cerrado biome and are culturally and economically important to Indigenous, quilombola, and other traditional populations, surprisingly little research has been undertaken on traditional agricultural fire uses and impacts in these areas.

Others have described agricultural systems in Brazil similar to roça de esgoto. Traditional farmers in the north of Minas Gerais State cultivate swampy forests (brejos de pindaíba), known for their organic soils, during the dry season. Dayrell (1998) reports the use of drainage channels to cultivate rice, beans, corn, vegetables, sugar cane, banana, and cassava. Miller observed cultivation in buriti wetlands (Miller 2007) in Indigenous reserves in the lavrado savanna formation of Roraima State. Much like cassava cultivation in drained peat swamp swidden fields

TABLE 4.1 Cultivated plants associated with different field types

Species	Variety (local name)	Drained peat swamp swidden field (roça de esgoto)	Rainfed swidden field (roça de toco)	Pasture field (roça de pasto)
Watermelon (*Melancia*)	Comprida e rajada	x	x	
	Redonda	x	x	
Squash (*Abóbora*)	Comum	x		
	Cabutiá	x		
Beans (*Feijão*)	Carioca	x	x	
	Preto	x		
	Catador	x		
	Andu		x	
Banana	Nanica	x	x	
	Maçã	x		
	Roxa branca	x		
	Angola	x		
Cassava (*Mandioca*)	Amarelona	x	x	
	Quiri quiri	x	x	
	Branca roxa	x	x	
	Pimanê	x	x	
	Pé d'anta	x	x	
	Serrana	x	x	
Bur cucumber (*Maxixe*)	-		x	
Sweet potato (*Batata*)	Branca	x	x	
	Roxa	x		
Ginger (*Gengibre*)	-	x		
Turmeric (*Açafrão*)	-	x		
Ananas (*Abacaxi*)	-	x	x	
Ananas (*Ananás*)	-	x		
Fava Bean (*Fava*)	-	x		
Papaya (*Mamão*)	-	x		
Sugar cane (*Cana*)	Preta	x		
	Rajada	x		
	Caiana	x		
	Puba	x		
	Preta	x		
	Açucareira	x		
Capim santo	-	x		
Pepper (*Pimenta malagueta*)	-	x		
Taro (*Taioba*)	-	x		

continued on next page

TABLE 4.1.—*continued*

Species	Variety (local name)	Drained peat swamp swidden field (roça de esgoto)	Rainfed swidden field (roça de toco)	Pasture field (roça de pasto)
Yam (*Inhame*)	-		x	
Cocoa (*Cacau*)	-		x	
Orange (*Laranjeira*)	-		x	
Sorghum (*Sorgo*)	-		x	
Millet (*Milheto*)	-		x	
Brachiaria grass (*Capim brachiaria*)	-			x
Andropogon grass (*Capim andropogon*)	-			x
Quicuia grass (*Capim quicuia*)	-			x

Source: Field surveys (twelve farms, fifteen swidden fields, and five pasture fields).

in Jalapão, this farming practice ensures that farmers do "not lose the seed"; that is, the plants are kept alive during droughts, and cuttings are available for planting in the rainy season (Miller 2007, 126). In the northeast of Goiás State, roça de esgoto cultivation is practiced in remote agricultural settlements near the headwaters of rivers, frequently with the use of chemical fertilizers (Bosgiraud 2013). In the Jalapão region, fertilizers and chemical pesticides are not used in vereda swidden fields.

While this system is more labor intensive than rainfed swidden cultivation, it presents several advantages for farmers. The fields can ensure production throughout the year, especially during the dry season. They produce greater yields and host higher agrobiodiversity. They permit farmers to be less dependent on rainfed swidden fields, who therefore have less need to burn vegetation. Finally, these fields have limited vulnerability to rainfall variation and instability. These factors help explain this cultivation system's historical importance in Jalapão and its persistence into the present, even in the face of such recent threats as rural exodus.

Our results indicate that the observed use of agricultural fire in peat swamp does not lead to deforestation. On the contrary, it promotes the succession of fallow species after cultivation. The predominance of forest vegetation in fallows reinforces quilombola farmers' knowledge that "drainage ditches renovate the soils." Yet these farmers' observations go further still; as they reported, this traditional practice leaves a forest that is denser than before cultivation.

How might we explain this possible reforestation effect of traditional fire management in the veredas? After all, the use of agriculture fire directly contravenes normally strict conservation rules in these environments.

As mentioned above, paleoenvironmental studies in the veredas of Brazil indicate the occurrence of fire in prehistoric times. Ecology research shows that fires in veredas can affect the original vegetation, biodiversity, and water regimes depending on their frequencies and intensities (Bahia et al. 2009; Maillard et al. 2009). Fire intervals of at least ten years have been recommended for the conservation of buriti palm populations (Sampaio et al. 2012). We observed that tra-

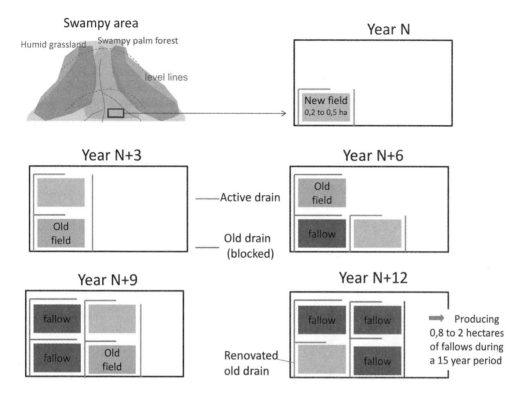

FIGURE 4.5. The common crop-fallow rotation cycle in a peat swamp (drawing by Ludivine Eloy).

TABLE 4.2 Vegetation cover in drained peat swamp swidden field fallows, Serra Geral do Tocantins Ecological Station

Vegetation class	Peat swamp swidden field fallows[a]	
Type	Number of pixels[b]	Percentage of pixels
Not identified	1	.2
Gallery forest	130	21.2
Woodland	373	60.8
Shrubby grassland	17	2.8
Open grassland	92	15
Total	613	100

Source: Supervised classification of vegetation cover based on a 2010 Landsat image. Mapping of fallows was done using field data (GPS points) collected in 2013 through farmer indications based on locations of old drains.
[a] Eleven plots, with an average fallow period of nineteen years and a range of six to fifty years.
[b] 1 pixel = 30 m /x 30 m.

ditional farmers in the Jalapão region use fire at intervals of ten to twenty years within veredas because the field is burned only once at the beginning of the cultivation cycle. In addition, fire is managed through diverse practices including burning in the early dry season, creating firebreaks, and controlling drainage ditch water levels to prevent un-

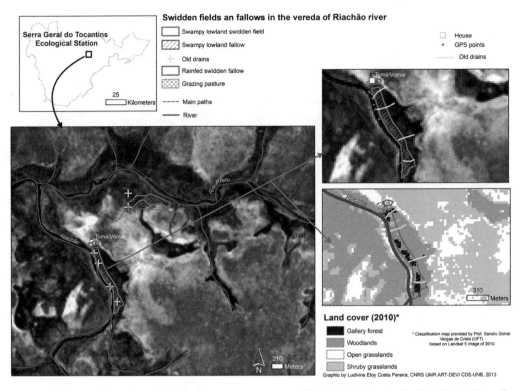

FIGURE 4.6. Map of two large fallows from peat swamp swidden cultivation in the Riachão Valley of the Serra Geral do Tocantins Ecological Station (graphic by Ludivine Eloy; classification map provided by Sandro Sidnei Vargas de Cristo).

TABLE 4.3 Locations of drained peat swamp swidden fields in relation to the mouths and sources of rivers

Field	Area (ha)	Distance (m) between field and mouth of river	Distance (m) between field and source of river
1	.267	16	5,154
2	.305	155	279
3	1.055	142	165
4	.068	146	1,109
5	.989	155	Not identified
6	.085	48	Not identified

Note: Distances were calculated from the centers of fields.
Source: Landsat satellite image from 2011 and field data.

derground fires and soil acidification. Our findings suggest that limiting the frequency (and perhaps the intensity) of fires may have been an important aspect of fallow management in cerrado wetlands, as is observed in seasonally flooded savannas in other regions (Renard et al. 2012).

Studies reporting devastating effects of fire in veredas usually refer to high-intensity and -frequency fire regimes and/or permanent cultivation scenarios (exotic pastures and eucalyptus) which cause soil compaction by removing native grasses and forbs, and trigger degradation processes. Such events are common in the cerrado, with the Rio dos Cochos in northern Minas Gerais State being one example (Ribeiro 2010).

Our empirical observations of Jalapão farmers lead us to believe that temporary drainage in combination with low-frequency (and maybe low-intensity) fire can transform vereda buriti formations (buritizais) into more diverse flooded gallery forests dominated by buriti by accelerating the natural succession process. Such succession occurs in the absence of fire and low-frequency fire contexts (Ferreira 2008). The observed traditional ways of managing fire and water may have a catalytic effect on this natural succession process in veredas because waterlogged soil restricts the growth of several species. When the soil becomes drier, many tree seeds already present in the soil seed bank and/or originating from gallery forest seed dispersal mechanisms may find a more suitable environment to germinate and grow. More research therefore needs to be carried out in the Jalapão to assess the density and variety of species in vereda swidden fallows as compared to those in noncultivated areas.

Verification of this catalytic effect on natural succession by swidden cultivation would suggest that part of the gallery forests in the Jalapão valleys have anthropogenic origins, especially considering the age of human occupancy and economic importance of wetland cultivation in the region. In the Llanos de Mojos savannas of the Bolivian Amazon, recognized for their high biodiversity, several pre-Columbian Indigenous practices led to the conversion of wetlands into productive agricultural areas, thus creating "domesticated landscapes" (Erickson 2006). Similar landscapes were found in the flooded savannas of Colombia, Ecuador, Venezuela, and Suriname, as well as in the Andean regions of Peru, Bolivia, Colombia, and Ecuador (Denevan 2001). Additionally, Carney (2002) calls attention to the importance of wetland rice cultivation in the Americas as a landscape legacy of the African diaspora, which suggests a possible alternative explanation of the historical origins of roça de esgoto in this region.

Conclusion

Our research, made possible by a pioneering agreement between environmental managers and a local quilombola association in a protected area of cerrado, describes traditional fire management practices that are usually prohibited in Brazil.[9] It was possible to verify farmers' knowledge of swidden fallow succession in swampy forests and affirm their practices' desirable outcomes for vegetation cover and landscape productivity. Our results contradict the idea that every kind of fire is devastating for such humid and fragile ecosystems. They also justify the creation of land management policies that effectively combine protection with productive uses of the cerrado biome by quilombola, Indigenous, and other traditional peoples.

It is possible that traditional peat swamp swidden cultivation systems involving draining and burning are more common than is recorded in the current cerrado literature. Fear of condemnation and fines may hinder identification of these practices by farmers and researchers since veredas are classified by national environmental legislation as Permanent Preservation Areas (APP) with strict use restrictions. Nevertheless, the use of fire continues to be the only viable alternative for many traditional farmers. In Brazil, most of the fire management rules are considered illegitimate by traditional communities because they fail to recognize or incorporate adequately local knowledge. In addition, these rules do not take the particular obstacles and constraints faced by these communities into account (Carmenta et al. 2013; Carvelheiro 2004). In conjunction with rural exodus, these measures tend to weaken collective management systems involving fire in the cerrado (Mistry and Bizerril 2011) and the Amazon (Toniolo 2004; Uriarte et al. 2012) by depreciating traditional agricultural systems. Undesirable occurrences of late dry season wildfires may also result from a sense of illegality and injustice, and even loss of traditional fire management knowledge, brought about in part by such public policies (Eloy et al. 2015; Mistry and Bizerril 2011).

In this study, dialogue between traditional and scientific knowledge holders within the framework of collaborative research undertaken in partnership with environmental managers and local populations has led to an improved understanding of the history of traditional fire management and the changes it has undergone over time.

Notes

1. Veredas are peat swamp forests dominated by buriti palm and characterized by peat soils. They occur around the headwaters of rivers and are surrounded by moist grasslands (campo limpo úmido) which locals use as native pastures and to gather flowers for handicraft production.
2. Peat lands are ecosystems characterized by the accumulation of partially decayed organic matter (peat) that forms from plant debris under waterlogged conditions (Andriesse 1988).
3. We define traditional populations (or traditional farmers) in a broad sense, independently of their political mobilizations and cultural affirmations: communities that are historical heirs of different forms of use, management, knowledge, and symbolic representations of the cerrado, characterized by their historical relationships with the territories they claim and by low-input production systems. This expression lumps together different identity groups, such as communities of Indigenous people or of quilombolas (descendants of Maroon groups or slaves), family farmers (in agricultural settlements or members of rural communities), artisans, fishermen, harvesters, and so on. Although these populations have been living in the cerrado since long before the advent of industrial agriculture, most of them do not have titles to the lands they live on.
4. Between 2002 and 2008, the average annual rate of deforestation in the cerrado was twice as high as in Amazonia. By 2008, 986,247 square kilometers, or 48.37 percent of the cerrado's original cover, was already cleared, compared to only 19 percent for Amazonia (IBGE 2010).
5. Article 68 of the Brazilian Constitution (1988) recognizes the land rights of descendants of Maroons (fugitive slaves) who established villages and occupied territories.
6. Mainly Karajá, Krahô, Nhyrkwãnje, Apinajé, Akroá, Xakriabá, Xavante, Xerente, and Avá-Canoeiro.
7. "Meu pai tinha um mundo de roça de esgoto. Ele criava muito porco, tinha muita cana. Fazia rapadura. Ele plantava tanta abóbora e melancia, a gente carregava tanto, que a gente mandava avisar o povo do Riachão, para eles vir buscar de animal. Ele fazia roça de toco, mas sempre era menos do que a roça de esgoto" (Josefa Chagas dos Santos, interview).
8. The felling operation is selective, with some individual trees of several species usually being protected, including pindaíba-do-brejo, buriti, pindaíba de capão (Xylopia sericea A. St.-Hil.) and pau d'óleo (Copaifera sp. Lindl.).
9. The Brazilian Constitution of 1988 recognizes the territorial rights of quilombola remnant communities, and Decree No. 6,040 from 2007 recognizes the importance of preserving their cultural practices and guaranteeing their access to the natural resources.

Acknowledgments

We thank the families of the association Ascolombas-Rios (Mateiros, Tocantins State) and ICMBio's environmental managers for welcoming us and taking part in the research. We also thank our academic research partners and the institutions that have supported this work, mainly Gesellschaft für Internationale Zusammenarbeit (GIZ) through the Cerrado Jalapão project, and the European Commission through the ENGOV project (FP7-2010, project no. SSH-CT-2010-266.710).

References Cited

Adams, C., L. Chamlian Munari, N. Vliet, R. Sereni Murrieta, B. Piperata, C. Futemma, N. Novaes Pedroso, Jr., C. Santos Taqueda, M. Abrahão Crevelaro, and V. Spressola-Prado. 2013. "Diversifying Incomes and Losing Landscape Complexity in Quilombola Shifting Cultivation Communities of the Atlantic Rainforest (Brazil)." *Human Ecology* 41 (1): 119–137.

Andriesse, J. P. 1988. *Nature and Management of Tropical Peat Soils*. Rome: Food and Agriculture Organization of the United Nations (FAO). http://www.fao.org/docrep/x5872e/x5872e00.htm.

Bahia, T. O., G. R. Luz, L. L. Braga, G. C. O. Menino, Y. R. F. Nunes, M. D. M. Veloso, W. V. Neves, and R. M. Santos. 2009. "Florística e fitossociologia de veredas em diferentese Estágios de conservação na APA do Rio Pandeiros, norte de Minas Gerais." *MG Biota* 2 (3): 14–21.

Barberi, M., M. L. Salgado-Labouriau, and K. Suguio. 2000. "Paleovegetation and Paleoclimate of 'Vereda de Aguas Emendadas', Central Brazil." *Journal of South American Earth Sciences* 13 (3): 241–254.

Barbosa, A. S., and P. I. Schmitz. 1998. "Ocupação indígena do cerrado: Esboço de uma história." In *Cerrado ambiente e flora*, edited by Sano and Almeida, 3–42. Brasilia: Empresa Brasileira de Pesquisa Agropecuária (Embrapa).

Beatty, Robin L. 2014. "Areas protegidas do cerrado brasileiro: Manejo integrado do fogo: Situação, estratégias, recomendações." Brasília: Cooperação Alemã para o Desenvolvimento.

Beerling, D. J., and C. P. Osborne. 2006. "The Origin of the Savanna Biome." *Global Change Biology* 12 (11): 2023–2031.

Bliege Bird, Rebecca, Douglas W. Bird, C. H. Parker, and J. H. Jones. 2008. "The 'Fire Stick Farming' Hypothesis: Australian Aboriginal Foraging Strategies, Biodiversity, and Anthropogenic Fire Mosaics." *Proceedings of the National Academy of Sciences* 105 (39): 14796–14801.

Bond, W. J., and J. E. Keeley. 2005. "Fire as a Global 'Herbivore': The Ecology and Evolution of Flammable Ecosystems." *Trends in Ecology and Evolution* 20 (7): 387–394.

Borges, S. L., L. Eloy, and T. Ludewigs. 2014. "O gado que circulava: Desafios da gestão participativa de unidades de conservação no Gerais de norte de Minas." *BioBrasil* 2014 (1): 130–156.

Bosgiraud, M. 2013. "Normes environnementales et transformation des pratiques de gestion des ressources dans le cerrado: L'exemple de L'Aire de Protection Environnementale (APA) nascentes do Rio Vermelho, Goiás." Master's thesis, Department of Agronomy and Development, Istom Ecole Supérieure D'Agro-Développement International, Paris.

Brockett, B. H., H. C. Biggs, and B. W. van Wilgen. 2001. "A Patch Mosaic Burning System for Conservation Areas in Southern African Savannas." *International Journal of Wildland Fire* 10 (2): 169–183.

Carmenta, R., S. Vermeylen, L. Parry, and J. Barlow. 2013. "Shifting Cultivation and Fire Policy: Insights from the Brazilian Amazon." *Human Ecology* 41 (4): 603–614.

Carney, J. 2002. *Black Rice: The African Origins of Rice Cultivation in the Americas*. Cambridge, MA: Harvard University Press.

Carvelheiro, K. 2004. "Community Fire Management in the Marana Region, Brazilian Amazonia." Master's thesis, University of Florida, Gainesville.

Cianciaruso, M., and M. Batalha. 2008. "A Year in a Cerrado Wet Grassland: A Non-Seasonal Island in a Seasonal Savanna Environment." *Brazilian Journal of Biology* 68: 495–501.

Coutinho, L. M. 1990. "O cerrado e a ecologia do Fogo." *Ciência Hoje* 12 (68): 23–30.

Dayrell, C. A. 1998. "Geraizeiros e biodiversidade no Norte de Minas: A contribuição da agroecologia e da etnoecologia nos estudos dos agroecossistemas tradicionais." Master's thesis, Sustainable Agroecology and Rural Development, Universidad Internacional De Andalucía, Huelva.

Denevan, W. M. 2001. *Cultivated Landscapes of Native Amazonia and the Andes*. Oxford: Oxford University Press.

Eloy, L., C. Aubertin, F. Toni, S. L. B. Lúcio, and M. Bosgiraud. 2015. "On the Margins of Soy Farms: Traditional Populations and Selective Environmental Policies in the Brazilian Cerrado." *Journal of Peasant Studies* 32 (2): 494–516.

Eloy, Ludivine, and S. L. Borges. 2013. "Caracterização agronômica e socioeconômica das roças de toco e de esgoto na Estação Ecológica Serra Geral do Tocantins." Brasilia: Cooperação Alemã para o Desenvolvimento (GIZ) and Instituto Chico Mendes de Conservação da Biodiversidade (ICMBio).

Emperaire, L., and Ludivine Eloy. 2014. "Amerindian Agriculture in an Urbanising Amazonia (Rio Negro, Brazil)." *Bulletin of Latin American Research* 34 (1): 70–84.

Erickson, C. L. 2006. "The Domesticated Landscapes of the Bolivian Amazon." In *Time and Complexity in Historical Ecology: Studies in the Neotropical Lowlands*, edited by William Balée and Clark L. Erickson, 235–278. New York: Columbia University Press.

Falleiro, R. d. M. 2011. "Resgate do manejo tradicional do cerrado com fogo para proteção das terras indígenas do oeste do Mato Grosso: Um estudo de caso." *Biodiversidade Brasileira* 2: 86–96.

Feltran-Barbieri, R. 2010. "Outro lado da fronteira agrícola: Breve história sobre a origem e declínio da agricultura autóctone no cerrado." *Ambiente e Sociedade* 13: 331–345.

Ferraz-Vicentini, K. R., and M. L. Salgado-Labouriau. 1996. "Palynological Analysis of a Palm Swamp in Central Brazil." *Journal of South American Earth Sciences* 9 (3): 207–219.

Ferreira, I. M. 2008. "Cerrado: Classificação geomorfológica de vereda." In *Anais do IX Simpósio Nacional do Cerrado e II Simpósio Internacional Savanas Tropicais*. Brasilia: Embrapa Cerrados.

IBGE (Instituto Brasileiro de Goegrafia e Estatistica). 2010. *Indicadores de Desenvolvimento Sustentável: Brasil 2010*. Brasilia: IBGE.

IBGE and EMBRAPA. 2001. *Mapa de solos do Brasil*. Rio de Janeiro: Instituto Brasileiro de Geografia e Estatística (IBGE) and Empresa Brasileira de Pesquisa Agropecuária (EMBRAPA).

ICMBio. 2014. *Plano de manejo para a Estação Ecologica Serra Geral do Tocantins*. Brasilia: ICMBio.

Jeltsch, F., G. Weber, and V. Grimm. 2000. "Ecological Buffering Mechanisms in Savannas: A Unifying Theory of Long-Term Tree-Grass Coexistence." *Plant Ecology* 150 (1–2): 161–171.

Langner, A., and F. Siegert. 2009. "Spatiotemporal Fire Occurrence in Borneo Over a Period of 10 years." *Global Change Biology* 15 (1): 48–62.

Laris, P. 2002. "Burning the Seasonal Mosaic: Preventive Burning Strategies in the Wooded Savanna of Southern Mali." *Human Ecology* 30 (2): 155–186.

Lavigne-Delville, P. 2003. "Le foncier et la gestion des ressources naturelles." In *Memento de l'Agronome*, 201–221. Paris: CIRAD-GRET-Ministère des Affaires Etrangères.

Lindoso, L. d. C. 2011. "Termo de ajustamento de conduta com população quilombola residente na Estação Ecológica Serra Geral do Tocantins." In *Anais do V seminário de áreas protegidas e inclusão social*. Manaus: Universidade Federal do Amazonas.

Lindoso, L. d. C., and T. G. Parente. 2013. "Fogo e liberdade nos Gerais do Jalapão: Uma análise à luz do conceito de recursos de uso comum." *Anais do VI seminário brasileiro sobre áreas protegidas e inclusão social*. Belo Horizonte: ICMBio.

MCT (Ministério da Ciência e Tecnologia). 2009. "Inventário Brasileiro das Emissões e Remoções Antrópicas de Gases de Efeito Estufa." Accessed March 1, 2017. http://ecen.com/eee75/eee75p inventario_emissoes_brasil.pdf

MDA (Ministerio do Desenvolvimento Agrario). 2013. "Território da Cidadania: Jalapão—TO." Accessed March 1, 2017. http://www.territoriosdacidadania.gov.br/dotlrn/clubs/territriosrurais/jalapoto/one-community?page_num=0.

Maillard, P., D. B. Pereira, and C. G. Souza. 2009. "Incêndios florestais em veredas: Conceitos e estudo de caso no Peruaçu." *Revista Brasileira de Cartografia* 61: 321–330.

McGregor, S., V. Lawson, P. Christophersen, R. Kennett, J. Boyden, P. Bayliss, A. Liedloff, B. McKaige, and A. Andersen. 2010. "Indigenous Wetland Burning: Conserving Natural and Cultural Resources in Australia's World Heritage-Listed Kakadu National Park." *Human Ecology* 38 (6): 721–729.

Melo, M. M. d., and C. H. Saito. 2011. "Regime de queima das caçadas com uso do fogo realizadas pelos Xavante no cerrado." *Revista Biodiversidade Brasileira* 2: 97–109.

Meneses, M. E. N. S., M. L. Costa, and H. Behling. 2013. "Late Holocene Vegetation and Fire Dynamics from a Savanna-Forest Ecotone in Roraima State, Northern Brazilian Amazon." *Journal of South American Earth Sciences* 42: 17–26.

Miller, R., ed. 2007. *Levantamento etnoambiental do complexo Macuxi-Wapixana*. Vol. 1: *Caracterização ambiental e antropológica*. Brasilia: Fundação Nacional do Índio e Projeto Integrado de Proteção às Populações e Terras Indígenas da Amazônia Legal (PPTAL).

Mistry, J. 1998. "Decision-Making for Fire Use Among Farmers in Savannas: An Exploratory Study in the Distrito Federal, Central Brazil." *Journal of Environmental Management* 54: 321–334.

Mistry, J., A. Berardi, V. Andrade, T. Krahô, P. Krahô, and O. Leonardos. 2005. "Indigenous Fire Management in the Cerrado of Brazil: The Case of the Krahô of Tocantíns." *Human Ecology* 33 (3): 365–386.

Mistry, J., and M. Bizerril. 2011. "Por que é importante entender as inter-relações entre pessoas, fogo e áreas protegidas?" *Revista Biodiversidade Brasileira* 2: 40–49.

Montoya, E., and V. Rull. 2011. "Gran Sabana Fires (SE Venezuela): A Paleoecological Perspective." *Quaternary Science Reviews* 30: 3430–3444.

Myers, R. L. 2006. *Living with Fire: Sustaining Ecosystems and Livelihoods Through Integrated Fire Management*. Arlington, VA: Nature Conservancy.

Niemeyer, F. D. 2011. "Cultura e agricultura: Resiliência e transformação do sistema agrícola Krahô." Master's thesis, Instituto de Filosofia e Ciências Humanas Universidade Estadual de Campinas, São Paulo.

Oliveira, J. E., and R. L. S. Aguiar. 2011. "Do megalitismo às gravuras rupestres: Contribuições para arqueologia da região do Jalapão, Tocantins, Brasil." *Revista Maracanan* 7 (7): 11–34.

Padoch, Christine, and M. Pinedo-Vasquez. 2010. "Saving Slash-and-Burn to Save Biodiversity." *Biotropica* 42 (5): 550–552.

Page, S. E., F. Siegert, J. O. Rieley, H.-D. V. Boehm, A. Jaya, and S. Limin. 2002. "The Amount of Carbon Released from Peat and Forest Fires in Indonesia during 1997." *Nature* 420 (6911): 61–65.

Posa, M. R. C., L. S. Wijedasa, and R. T. Corlett. 2011. "Biodiversity and Conservation of Tropical Peat Swamp Forests." *BioScience* 61 (1): 49–57.

Redford, K. H., and G. A. B. d. Fonseca. 1986. "The Role of Gallery Forests in the Zoogeography of the Cerrado's Non-Volant Mammalian Fauna." *Biotropica* 18 (2): 126–135.

Renard, D., J. Iriarte, J. J. Birk, S. Rostain, B. Glaser, and D. McKey. 2012. "Ecological Engineers Ahead of Their Time: The Functioning of Pre-Columbian Raised-Field Agriculture and Its Potential Contributions to Sustainability Today." *Ecological Engineering* 45: 30–44.

Ribeiro, E. M. 2010. *História dos Gerais*. Belo Horizonte: Editora da Universidade Federal de Minas Gerais.

Ribeiro, R. F. 2005. *Florestas anãs do Sertão—O cerrado na história de Minas Gerais*. Belo Horizonte: Autêntica Editora.

———. 2008. "Da Amazônia para o cerrado: As reservas extrativistas como estratégias socioambientais de conservação." *Sinapse Ambiental* 5 (1): 12–32.

Russell-Smith, Jeremy, D. Lucas, M. Gapindi, B. Gunbunuka, N. Kapirigi, G. Namingum, K. Lucas, P. Giuliani, and G. Chaloupka. 1997. "Aboriginal Resource Utilization and Fire Management Practice in Western Arnhem Land, Monsoonal Northern Australia: Notes for Prehistory, Lessons for the Future." *Human Ecology* 25 (2): 159–195.

Sampaio, M. B., T. Ticktin, C. S. Seixas, and F. A. M. dos Santos. 2012. "Effects of Socio-Economic Conditions on Multiple Uses of Swamp Forests in Central Brazil." *Human Ecology* 40 (6): 821–831.

Santos, R. M. D. 2013. "O Gê dos Gerais: Elementos de cartografia para a etno-história do Planalto Central: Contribuição à antropogeografia do cerrado." Master's thesis, Universidade de Brasília.

Schmidt, I., I. Figueiredo, and A. Scariot. 2007. "Ethnobotany and Effects of Harvesting on the Population Ecology of *Syngonanthus nitens* (Bong.) Ruhland (Eriocaulaceae), a NTFP from Jalapão region, central Brazil." *Economic Botany* 61 (1): 73–85.

Schmidt, I. B., M. B. Sampaio, I. B. Figueiredo, and T. Ticktin. 2011. "Fogo e artesanato de capim-dourado no Jalapão: Usos tradicionais e consequências ecológicas." *Biodiversidade Brasileira* 2: 67–85.

Scott, A. C., W. G. Chaloner, C. M. Belcher, and C. I. Roos. 2016. "The Interaction of Fire and Mankind: Introduction." *Philosophical Transactions of the Royal Society of London B* 371 (1696).

Siegert, F., G. Ruecker, A. Hinrichs, and A. A. Hoffmann. 2001. "Increased Damage from Fires in Logged Forests During Droughts Caused by El Nino." *Nature* 414 (6862): 437–440.

Silva, D. M., P. d. P. Loiola, N. B. Rosatti, I. A. Silva, M. V. Cianciaruso, and M. A. Batalha. 2011. "Os efeitos dos regimes de fogo sobre a vegetação de cerrado no Parque Nacional das Emas, GO: Considerações para a conservação da diversidade." *Biodiversidade Brasileira* 2: 26–39.

Sluyter, A., and C. Duvall. 2016. "African Fire Cultures, Cattle Ranching, and Colonial Landscape Transformations in the Neotropics." *Geographical Review* 106 (2): 294–311.

Toniolo, A. 2004. "The Role of Land Tenure in the Occurrence of Accidental Fires in the Amazon Region: Case Studies from the National Forest of Tapajos, Para, Brasil." PhD dissertation, Indiana University, Bloomington.

Uriarte, M., M. Pinedo-Vasquez, R. S. DeFries, K. Fernandes, V. Gutierrez-Velez, W. E. Baethgen, and C. Padoch. 2012. "Depopulation of Rural Landscapes Exacerbates Fire Activity in the Western Amazon." *Proceedings of the National Academy of Sciences* 109 (52): 21546–21550.

Usup, A., Y. Hashimoto, H. Takahashi, and H. Hayasaka. 2004. "Combustion and Thermal Characteristics of Peat Fire in Tropical Peatland in Central Kalimantan, Indonesia." *Tropics* 14 (1): 1–19.

Vegas-Vilarrúbia, T., V. Rull, E. Montoya, and E. Safont. 2011. "Quaternary Palaeoecology and Nature Conservation: A General Review with

Examples from the Neotropics." *Quaternary Science Reviews* 30 (19–20): 2361–2388.

Viana, R. V. R. 2013. "Diálogos possíveis entre saberes científicos e locais associados ao Capim-Dourado e ao buriti na região do Jalapão, TO." Master's thesis, Biosciences Institute, Universidade de São Paulo.

Walter, B. M. T., and J. F. Ribeiro. 2010. "Diversidade fitofisionômica e o papel do fogo no bioma cerrado." In *Efeitos do regime de fogo sobre a estrutura de comunidades de cerrado: Projeto Fogo*, edited by Heloisa S. Miranda, 59–76. Brasilia: Instituto Brasileiro do Meio Ambiente (IBAMA).

Wantzen, K. M., A. Siqueira, C. N. d. Cunha, and M. d. F. Pereira de Sá. 2006. "Stream-Valley Systems of the Brazilian Cerrado: Impact Assessment and Conservation Scheme." *Aquatic Conservation: Marine and Freshwater Ecosystems* 16 (7): 713–732.

Welch, James R., Eduardo S. Brondízio, Scott S. Hetrick, and Carlos E. A. Coimbra Jr. 2013. "Indigenous Burning as Conservation Practice: Neotropical Savanna Recovery amid Agribusiness Deforestation in Central Brazil." *PLOS ONE* 8 (12): e81226.

Yeager, C. P., A. J. Marshall, C. M. Stickler, and C. A. Chapman. 2003. "Effects of Fires on Peat Swamp and Lowland Dipterocarp Forests in Kalimantan, Indonesia." *Tropical Biodiversity* 8: 121–138.

Yule, C. 2010. "Loss of Biodiversity and Ecosystem Functioning in Indo-Malayan Peat Swamp Forests." *Biodiversity and Conservation* 19 (2): 393–409.

CHAPTER 5

Fire Use among Swidden Farmers in Central Amazonia

Reflections on Practice and Conservation Policies

ANGELA MAY STEWARD

The practice of swidden agriculture, or shifting cultivation, has received increasing attention from scientists and policy makers working with forest conservation issues in Amazonia (Fearnside 2005, 2008; Padoch and Pinedo-Vasquez 2010; Ribeiro et al. 2013; van Vliet et al. 2012). While the topic has long been an area of scientific inquiry, especially within the human-environment sciences (Ribeiro et al. 2013) recently it came into the spotlight in two specific contexts: the increasing frequency of large-scale or "mega" fires in the region and the development of altered fire regimes (Hardesty et al. 2005). Numerous studies have attempted to elucidate the contributions of swidden agriculture as ignition sources (e.g., Carmenta et al. 2013).[1] Shifting cultivation and the role of smallholders in regional deforestation trends is also widely discussed in the context of debates on global climate change (e.g., Fearnside 2008, 2011).

Deforestation is responsible for 15 to 20 percent of global carbon emissions and 70 percent of Brazil's emissions (Alencar et al. 2010; do Valle 2010). Deforestation is second only to the energy sector in its impacts. Attention has thus turned toward maintaining tropical forest vegetation cover as a way to prevent future climate change. Activities aimed toward reforestation attempt to improve the capacity of existing forested areas to absorb carbon.

Conservation initiatives under the category of REDD (reducing emissions from deforestation and [forest] degradation) are efforts "to create a financial value for the carbon stored in forests, offering incentives for developing countries to reduce emissions from forested lands and invest in low-carbon paths to sustainable development" (UNDP 2016). REDD+ goes beyond merely addressing deforestation and forest degradation to include sustainable management of forests and enhancement of forest carbon stocks. On the ground these initiatives often include agreements between government agencies, nongovernmental organizations (NGOs), and forest communities that offer benefits (monetary or in the form of health and social services) for communities in exchange for their agreements to continue to engage in or, in some cases, shift toward sustainable forms of forest use that are thought to guarantee their conservation.

Brazil, with the world's largest carbon stocks on the planet, is primed as the key player in programs such as REDD and REDD+ (Santilli 2010). Shifting cultivation, which involves farmers clearing vegetation cover and burning biomass, has come under scrutiny, particularly in protected areas poised to receive REDD and REDD+ benefits (Padoch and Pinedo-Vasquez 2010). Proponents of agroecological approaches to smallholder agriculture (often also proponents of REDD+ initiatives) concur that the use of fire in farming systems is ecologically destructive for both forests and soils, and propose alternative methods that aim to implement permanent agricultural systems without the use of fire (Cardoso 2010). Farmers are often trained in agroecological techniques as part of REDD/REDD+ projects and are encouraged (or required) to discontinue swidden practices in return for benefits offered by these programs.[2]

Proponents of REDD, REDD+, and similar initiatives often do not distinguish between the various ways in which smallholders in Amazonia practice swidden agriculture. Furthermore, few studies have been conducted on how Amazonian smallholders use fire as a tool in agricultural landscapes (Carmenta et al. 2011; Carmenta et al. 2013).[3] Understanding how fire is applied within specific social and environmental contexts is essential to understanding the potential and actual impacts of swidden agriculture on forest cover, soils, and forest regeneration.

The objective of this chapter is to describe and compare swidden agriculture in upland (*terra firme*) and floodplain (*várzea*) sites located in two sustainable development reserves in the Middle Solimões region of the state of Amazonas, Brazil. Both are beneficiaries of REDD+ programs. This chapter sheds light on the ways in which farmers use fire in the landscape and discusses producers' perceptions of fire's role in land use systems. Because some farmers participating in this study have also experimented with agroecological alternatives—particularly, implementing agroforestry plots without using fire—I also include their reactions to these experiences.

Background and Context
Study Area and Participant Farmers

Research for this chapter was conducted in the Mamirauá and Amanã Sustainable Development Reserves in the Middle Solimões region, Amazonas State, Brazil (Figure 5.1). The Mamirauá reserve is located at the confluence of the Solimões and Japurá Rivers, with its easternmost portion close to the city of Tefé. The Amanã reserve neighbors Mamirauá, extending eastward. Together with Jaú National Park, these three conservation areas form one of the largest blocks of protected tropical forest on Earth, covering 6.5 million hectares (Queiroz 2005). The Mamirauá reserve is unique in being one of the only conservation areas on the planet that protects floodplain forests, where a great number of aquatic environments are encountered because of floodplain dynamism (Queiroz 2005). The Amanã reserve encompasses forested areas in upland, floodplain, and *paleovárzea* environments;[4] areas of flooded black-water forests (*igápo*) are also found, contributing to this local pattern of biodiversity (Irion et al. 2011; Valsecchi and Amaral 2009).

Seasonal flooding patterns have great impact on the ecology of the region and local resource management practices. Floodplain areas of the Middle Solimões region are subjected to dramatic seasonal changes in water levels, with river waters rising from 10 to 12 meters during the flood season due to upriver precipitation patterns and annual off-melting in Andean regions (Queiroz and Peralta 2006; Ramalho et al. 2010). Agriculture on the floodplain is organized around these patterns, being confined to seasonal periods of low water (*vazante*) and drought or dry periods (*seca*). Upland areas in Amanã

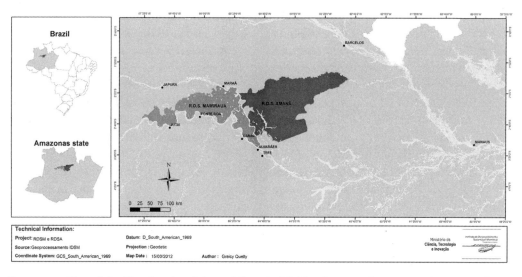

FIGURE 5.1. Map of the Mamirauá and Amanã Sustainable Development Reserves in Amazonas State, Amazonas State, Brazil (map by Greicy Quelly de Araújo Rodrigues).

escape annual flooding, and intermediate zones classified as paleovárzea are flooded during exceptional years. Here, agricultural activities are developed throughout the year, guided by seasonal rainfall patterns. Areas planted with more permanent crops, such as fruit and timber species, are not at risk of being damaged seasonally by rising waters.

In 2011, 11,532 residents were distributed among 207 communities in the Mamirauá reserve, with 3,860 people distributed among eighty-six communities in the Amanã reserve (Instituto de Desenvolvimento Sustentável Mamirauá 2011). Reserve residents belong to the rural Amazonian peasantry of Brazil and are locally referred to as *ribeirinhos* (riverine peoples). Ribeirinhos are rural dwellers of mixed cultural backgrounds (Indigenous, European, and African) who emerged as a social group during Portuguese colonization. Various authors discuss the group's attributes of adaptability, flexibility, and mobility as key to their autonomy and persistence despite the region's environmental and economic dynamism (Harris 2000; Lima-Ayres 1992; Steward 2007).

Characteristic of the ribeirinho economic system is its diversity. Families engage in a myriad of productive activities, including swidden agriculture, fishing, hunting, and timber and nontimber forest extraction (Adams, Murrieta et al. 2006). More recently, scholars have discussed the role of nonrural incomes and wages in Amazonian domestic economies (Lima 2009; Peralta and Lima 2010; Steward 2007). Socioeconomic surveys conducted in the Mamirauá and Amanã reserves in 2010 and 2011 have demonstrated that rural production activities accounted for 37 percent of total income generated, while direct monetary income (wages, commerce, and services) accounted for 19 percent, and social benefits (pensions and government cash transfer programs) for 44 percent of total household income. The most important income-generating production activities are fishing and agriculture, with agriculture being more important in upland areas (IDSM 2011; Peralta and Lima 2010). Farmers interviewed for the current study engage in agriculture for both subsistence and marketing purposes. As is characteristic of Central

Amazonia in general, smallholder farmer systems are based on swidden agriculture, with manioc (*Manihot esculenta* L.) being the predominant crop. While nonrural incomes increasingly make up a larger portion of domestic economies, men and women identify closely with rural work, describing themselves as farmers or fishers depending upon their main occupation (Adams et al. 2006).

Residing in conservation areas classified as sustainable use zones, Amanã and Mamirauá residents use and manage natural resources in accordance with the norms of this reserve category as defined by the Brazilian National Conservation Area System (Sistema Nacional de Unidade de Conservação). The two reserves are within the state of Amazonas and are managed by the Amazonas State Center for Conservation Areas (Centro Estadual de Unidades de Conservação de Amazonas). Natural resource use is guided by formalized management plans approved by the reserves' advisory boards, which include reserve residents, members of civil society, and government representatives. In addition to establishing guidelines and norms for resource use, the plans establish "use zones," designating certain areas for permanent protection and others for sustainable use (Queiroz and Peralta 2006). A revised management plan for the Mamirauá reserve was approved during an advisory council meeting in August 2014; the Amanã reserve's plan is being elaborated and should be completed by the end of 2017.

The Mamirauá Institute for Sustainable Development was created in 1999 as a civil society organization to undertake research with support and oversight from the Brazilian Ministry of Science, Technology, and Innovation (Ministério da Ciência, Tecnologia e Inovação). The institute promotes scientific research on biodiversity conservation by means of participatory management and promotion of sustainable use of natural resources in Amazonia (IDSM 2014a), with an emphasis on the Middle Solimões region. While in the past the Mamirauá Institute was directly responsible for managing both reserves, since 2010 it has provided support for management in the form of research on natural resource management, conservation, and biodiversity. Through its management programs, the Mamirauá Institute provides assistance to communities with sustainable fish capture using participatory approaches, community forest management, community-based tourism, smallholder agriculture, and actions aimed at improving the quality of life for reserve residents (IDSM 2014b).

Environmental Projects and Polices Impacting Fire Use

Agroforestry Initiatives of the Mamirauá Institute

Beginning in 2010, the Mamirauá Institute's Family Agriculture Program (now called the Agricultural Ecosystems Management Program [Programa de Manejo de Agroecossistemas]) began promoting agroforestry initiatives within both the Mamirauá and Amanã reserves, focusing primarily on upland areas in the Amanã reserve. Agroforestry initiatives were part of the program's agricultural extension efforts, which included educational activities such as workshops, courses, and farm visits conducted on a monthly basis by program technicians. Through these activities, the program sought to introduce agricultural techniques and technologies promoting the sustainable management of agricultural ecosystems and their resources within both reserves. Agroforestry initiatives specifically sought to disseminate novel forms of agricultural management (different from local practices) based on agroecology principles.[5] These initiatives began with an extensive course administered in the Amanã reserve involving approximately forty participant farmers over its duration. The course was organized into five themed modules: human-forest interactions, fire uses and impacts, nutrient cycling, native seed use, conservation, and the importance of maintaining productive diversity. The course integrated theory and practice. A program highlight was the implementation of agroforestry demonstration areas as a joint effort involving farmers, technicians, and researchers. The agroforestry model used to establish these areas was proposed by a technician from the Amazonian Social-Participatory Certification Association (Associação de Certificação Socio-Participativo da Amazônia) from Acre State. The technician was invited by the acting coordinator of the Agricultural Ecosystems Management Program to teach the course. Central to this method introduced by the Mamirauá Institute was the establishment of agricultural areas without the use of fire and the integration of existing forest biomass into the system. The model calls for the cultivation of fruit and forest (timber and nontimber) species in the same area. These species develop at different rates and thereby allow farmers to harvest different products throughout the year. They are also associated with different stages of forest succession, so that production can be distributed over time (from one to twenty-five years). The following steps are considered necessary to implement agroforestry demonstration areas per the method adopted by the Mamirauá Institute:

1. Identification of forest fallow to be managed and delimitation of an area of approximately 30 x 30 m on farmers' property.
2. Elimination of weedy and bottom-stratum species of forest fallow; spreading and distribution of remains across the ground to be incorporated into the organic matter of the soil.
3. Planting of banana seedlings following a spacing pattern of 3 x 3 m; planting of *cará* (taro) randomly in the spaces between bananas.
4. Felling of large hardwood species in the direction of cultivated bananas and placing trunks on top of planted bananas; covering with leaves and branches to form furrows of organic matter.
5. Planting of pineapples on the edge of each furrow, 1 m apart; planting of a diversity of annual crops between the furrows; planting seeds of woody fruit species alongside these crops in high densities.

Three agroforesty demonstration areas were established on farmers' properties in 2011 and 2012. Since 2012, five additional areas were developed using the same method. Although researchers and technicians involved in these initiatives have since changed their viewpoints, course proponents initially believed that this alternative model was more "environmentally correct" compared to traditional swidden agricultural practices. Course instructors emphasized the negative effects of applying fire to soils, explaining that nutrients are quickly washed away postburn, weakening soils in productive areas. Despite studies documenting high levels of agrobiodiversity in farmers' fields and home gardens in the Amanã and Mamirauá reserves (Schmidt 2003), as well as their accounts of the various products they sell during the year, course developers also operated according to the notion that annual fields were essentially monocultures of manioc. The developers considered their proposed model a sound alternative because they believed it would promote productive diversity. As an ultimate goal, advocates hoped that by disseminating these agroforestry practices, farmers would abstain from using mature forests to establish manioc fields, and that by emphasizing the development of agroforestry systems, producers would also diversify their production. Crop diversity was expected to improve diets and allow families to sell numerous products in the regional market, thus relieving them from an assumed economic dependency on manioc products.

Today, agroforestry initiatives of the Agricultural Ecosystems Management Program have been integrated into the larger, institutional project BioREC (Biodiversity Conservation and Sustainable Use in Conservation Areas [Conservação e Uso Sustentável da Biodiversidade em Unidades de Conservação]). In addition to the Agricultural Ecosystems Management Program and the related Amazonian Agriculture Research Group, the project involves a forest ecology research group, a community forestry management program, and environmental educators from Mamirauá Institute. The primary objective of the project is to support participatory natural resource management activities in the Mamirauá and Amanã reserves via research and dissemination of knowledge regarding sustainable agriculture (including cultivation and pasture systems), timber and nontimber forest management, environmental education, and environmental protection and monitoring (IDSM 2014b).

The BioREC project receives exclusive support from the Amazon Fund, established by the Brazilian federal government and administered by the Brazilian National Bank for Economic and Social Development (Banco Nacional de Desenvolvimento Econômico e Social). Its donors are first world governments, national and international businesses, multilateral institutions, NGOs, and individuals. The Amazon Fund (2014) secures financing to support projects aiming to combat and prevent deforestation and to promote conservation and sustainable use of Amazonian forests. The fund runs parallel to national efforts to reduce greenhouse gas emissions, including the National Action Plan to Prevent and Control Deforestation in the Amazon (Plano de Ação para Prevenção e Controle do Desmatamento na Amazônia Legal) and the National Plan on Climate Change (Plano Nacional sobre Mudança do Clima), which aim to reduce emissions in the energy and transport sectors and from deforestation (Amazon Fund 2014). The Amazon Fund itself is considered an important large-scale REDD+ pilot program (Viana 2010). In this way, the BioREC project and others supported by the Amazon Fund can also be categorized as REDD+ initiatives. In the Amanã and Mamirauá reserves, BioREC specifically seeks to reduce emissions emanating from deforestation related to agricultural activities and illegal timber extraction. It further aims to improve the capacity of current forests to absorb carbon by promoting restoration of degraded pasture and timber fallows.

Forest Conservation Allowance Program (Programa Bolsa Floresta)

In addition to the Mamirauá Institute's BioREC project, the communities in both reserves are integrated into the Forest Conservation Allowance Program (Programa Bolsa Floresta), administered by the Sustainable Amazonas Foundation (Fundação Amazonas Sustentável), which manages the environmental products and services of state-managed conservation areas. The program is characterized as a REDD+ initiative and was based on the broader concept of compensating forest-dwelling communities for maintaining forest cover through sustainable use (Viana 2008). The Forest Conservation Allowance Program was developed in 2007 and targets traditional populations living in conservation areas as well as Indigenous peoples within recognized territories in Amazonas State. Today the program is active in fifteen conservation areas in Amazonas State and has more than 35,000 beneficiaries. This includes 1,937 families in the Mamirauá reserve and 753 families in the Amanã reserve, representing the majority of reserve families (FAS 2014).

Program participants agree to 0 percent deforestation. In exchange for attending workshops covering good land use practices, sustainable forestry, and climate change (Viana 2009), they receive financial benefits through four types of payments: (1) the Family Forest Conservation Allowance (Bolsa Floresta Familiar), a cash payment of R$50 (US$21) paid to female heads of households; (2) the Forest Conservation Allowance for Community Associations (Bolsa Floresta Associação), which provides 10 percent of the total value of Family Forest Conservation Allowance stipends to the reserve association; (3) the Forest Conservation Income Allowance (Bolsa Floresta Renda), which grants roughly R$4,000 (US$1,714) to each community per year depending on its size (in this case, communities present projects to the Sustainable Amazonas Foundation, requesting materials to support productive activities involving fisheries, extraction of nontimber forest products, agroforestry, handicraft production, management of native honeybees, and sustainable timber management); and (4) the Forest Conservation Social Allowance (Bolsa Floresta Social), which grants up to R$4,000 (US$1,714) to communities for education, health, communication, and transportation services (for this last category, communities present proposals for projects such as the purchase of boats and outboard motors to serve as ambulances, and funds for physical improvements to schools and health posts).

Although the program prohibits opening agricultural fields in mature forest areas and support is generally not provided for swidden agriculture, clearing fallows in secondary forest is permitted. According to program coordinators, opening forested areas is acceptable in the case of newcomers who need to establish agricultural and dwelling sites (Fernanda Martins, personal communication).[6] Clearing of floodplain forests to establish residences and farms is also permitted when people migrate due to erosion or collapsing riverbanks, which commonly occur due to regional flood dynamics. In these cases, Sustainable Amazonas Foundation coordinators advise against the use of fire to establish new fields, preferring agroforestry approaches. However, coordinators and community leaders both say that in-field supervision of land use activities is nonexistent, and therefore program restrictions actually serve as recommendations.

Research Approach

Data on local agricultural systems and practices presented in this chapter are based on fieldwork conducted between February 2012 and November 2014. Data regarding swidden systems in upland and floodplain systems, including fire use and farmers' perceptions of fire utility, were gathered through farm visits and semistructured interviews conducted by the author and a research fellow. In interviews with male and female heads of households, we discussed topics such as agricultural techniques, history of land use on farmers' lands, and changes in practices resulting from broader policy or economic changes. Farm visits involved shadowing farmers during preparation of agricultural areas, planting of fields, and harvesting. The findings presented in this chapter also benefit from information presented in miscellaneous documents as well as monthly activity reports produced by researchers and technicians from the Agricultural Ecosystems Management Program, Research Group

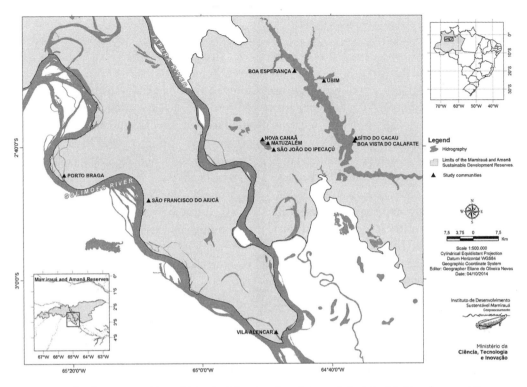

FIGURE 5.2. Study communities within the Mamirauá and Amanã Sustainable Development Reserves, Amazonas State, Brazil (map by Greicy Quelly de Araújo Rodrigues).

on Amazonian Agriculture, Mamirauá Institute, Tefé, Amazonas State. Figure 5.2 shows interview locations in both reserves.

Research activities also pertained to a larger project, "Ethnography of Rural Production Systems in the Middle Solimões Region," that was undertaken by the Mamirauá Institute's Amazonian Agriculture, Biodiversity, and Sustainable Management Research Group. The project was designed to better understand local agricultural practices and their contributions to household economies in the Middle Solimões region. The fire component of the project was added after observing differences between farmers' use of burning to manage agricultural areas. This apparent diversity stood out in contrast to assumptions held by Mamirauá Institute researchers and technicians that all burning practices were uniform among residents regardless of their cultural backgrounds, life experiences, and the environments in which they worked. Despite the general scarcity of information on burning practices, implicit in the Agricultural Ecosystems Management Program's agroforestry initiatives was the effort to reduce fire use by reserve farmers. Essentially, the program sought to curb a burning practice that it knew little about.

The proposal to better understand local agricultural systems only after agroforestry initiatives seeking to replace components of these systems were already underway brings up questions regarding effective and just approaches to applied research and extension. Initiatives seeking to improve or replace local practices should ideally be based on previous understanding of these systems. The discussion presented here is thus a critical reflection of the premises upon which the Agricultural

Ecosystems Management Program's agroforestry initiatives were developed.

It should further be noted that I contributed to the Agricultural Ecosystems Management Program's agroforestry actions as a program coordinator from February 2012 to March 2016, having inherited objectives strongly influenced by regional discourse on deforestation and common misunderstandings of local swidden agricultural practices.[7] The discussion presented in this chapter seeks to transcend these biases by presenting the complexities of these systems and their positive attributes. This reflection is part of an ongoing process of restructuring our research and extension approaches based on a more complete understanding of local systems and farmers' demands.

Characterization of Swidden Systems and Fire Use

Residents in the study area practice swidden fallow agriculture; this practice involves the general steps of clearing a forested area, burning felled vegetation, further clearing and sometimes reburning of the charred material, planting, weeding, and harvesting.[8] In the uplands, fields are generally planted for one to two years, after which they are left in fallow. Fallows are either abandoned and recolonized by forest species or enriched with fruit and timber species. In the second instance, fields are managed until they gradually form an orchard or enriched fallow, known as a *sítio* (Brookfield et al. 2002; Viana et al. 2016). The burning of felled vegetation serves as a quick and efficient way to enrich soils with nutrients essential for agricultural production. The fallow period, ranging from four to ten years depending on the soil type (Richers 2010), is equally important for the recuperation of forest vegetation and, when respected, serves to prevent soil degradation over time.

The agricultural landscape in the study area is a mosaic made up of swiddens of different ages (from one to three years), enriched fallows, unmanaged fallows, home gardens, and forested areas that may potentially be integrated into the swidden cycle. Previous studies conducted in the Amanã and Mamirauá reserves have demonstrated that ribeirinho farmers maintain high levels of specific and intraspecific diversity in land use areas (Lima et al. 2012; Pinedo-Vasquez et al. 1996; Schmidt 2003). Agrobiodiversity is high at the field, community, and regional levels. Average field size varies between floodplain and upland environments from .42 to .52 hectares and 1.54 to 2.10 hectares, respectively (Richers 2010).[9]

Manioc has long been the principal crop and staple carbohydrate for ribeirinho farmers and other traditional peoples across Amazonia (Adams et al. 2006; Lima et al. 2012). Manioc is planted in annual fields with a variety of other plants in smaller quantities. Common species include banana, sweet pepper, yam (*cará*), squash, pineapple, and papaya, among which intraspecific diversity is most notable in the first three. In addition, avocado, lime, peach palm, and açaí palm fruits are commonly planted alongside these faster growing crops. Fruit species consequently remain in the system as it transitions from an annual field to an enriched fallow or home garden (*quintal*). In floodplain zones, beaches (*praias*) and mudflats (*lamas*) appear annually with receding river levels. Farmers use these areas to plant short-cycle crops such as corn, beans, squash, melon, and watermelon. These garden plots are different from the manioc fields established on the highest stretches of river banks (*restingas*). In the remainder of this section, I describe in more detail the specific management steps involved in making manioc fields in both upland and floodplain areas within the study area. Productive calendars for both environments are presented in tables 5.1 and 5.2.

TABLE 5.1. Production calendar showing *roça* management activities in the *terra firme*, Amazonas State, Brazil

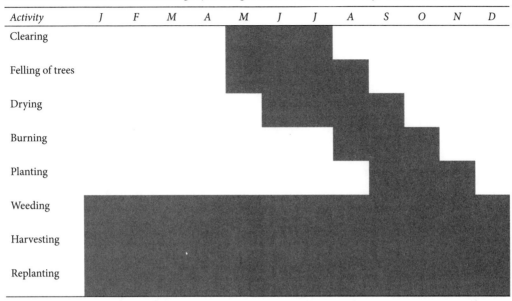

Field Preparation

To establish a manioc field, farmers must first identify the area to be cultivated. This decision is determined by several factors, including the land available to each family,[10] proximity to residential areas, and location of manioc processing areas. The closer fields are to the residence, the better for the family because it eases transport of goods and optimizes work conditions (Lima 2004; Lima-Ayres 1992; Steward and Lima 2014). In the floodplains, planting is restricted to the highest stretches of riverbanks, which are the last to be inundated by floodwaters. Furthermore, in the floodplains, where the dynamics of the environment create a diversity of soil types, farmers also choose where to plant in accordance with the best types of soil for specific varieties that are propagated from *manivas* (manioc stems) (Lima et al. 2012).

In the uplands, farmers also consider their preference for planting in forest fallows (*capoeiras*) or areas of mature forest, which require different types of labor. Forested areas require more labor to clear mature trees, while more recently cleared forests can be weeded less frequently because growth is less prevalent. Given its stage of succession, forest fallow seed banks are dominated by quicker growing grasses and weedy species. Thus, forest fallows are easier to clear initially but require more work to maintain. Families thereby base their decisions on where to place agricultural fields according to their capacities to mobilize different types of labor. Households with grown male children, for instance, may opt to place gardens in more mature forests, whereas those with young children or many female members may opt to work in forest fallows (Lima 2004; Viana et al. 2016).

Having located a suitable forest or fallow parcel, farmers start the process of clearing primary or secondary vegetation during the dry season, beginning as early as May or June in the uplands or July in the floodplains (Tables 5.1 and 5.2). The size of a field is determined by a family's consumption needs and commercial intents, as well as labor availability (discussed further below). Farmers first

TABLE 5.2. Production calendar showing *roça* management activities in the *várzea*, Mamirauá Sustainable Development Reserve, Amazonas State, Brazil

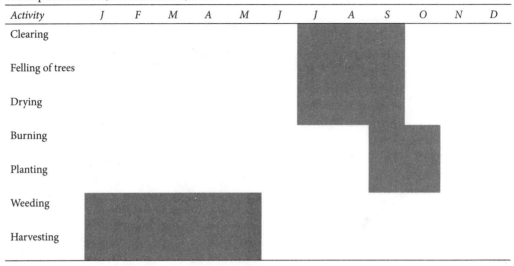

remove young trees and vines with machetes. The larger trees are then felled, traditionally with axes but increasingly with chainsaws, particularly in the uplands. Residents also use chainsaws to chop up previously felled material to facilitate subsequent burning.

On the other hand, families residing in the floodplains generally establish fields in areas of young forest fallow. Because agricultural areas are refreshed with alluvial deposits each year, it is possible to cultivate the same area in subsequent years. Farmers reported planting the same fields successively for four to fifteen years. Fields are abandoned when producers notice a decline in production (measured by a reduction in the size and thickness of manioc tubers) and are left in fallow for an average of 4.8 years (compared to 7.6 years in the uplands) (Richers 2010). Consequently, floodplain fallows present less pronounced forest regrowth compared to those in upland areas. This factor, in conjunction with smaller field size, reduces the time needed for the initial clearing phase. In both the floodplains and uplands, the area cleared for fields includes an uncultivated border area (*aceiro*) that permits sunlight to fall on the field and from which brush is removed to serve as a firebreak.

Regardless of the environment, farmers agree that preparing fields is the most arduous phase of the production cycle. They often mobilize extrafamilial laborers to meet work demands. In the study area, this was done in three ways: (1) through the organization of collective work groups (*ajuri*), in which families help prepare each other's manioc fields; (2) trading days (*troca de dia*), where a smaller number of farmers trade days of work; and (3) payment of a daily wage (*diária*) for agricultural work. In some communities, for example, it is common to pay a daily wage for clearing a forest parcel with a chainsaw. The use of extrafamilial labor was also observed during planting and harvesting, as well as in the processing of manioc tubers into flour, as described below (Steward and Lima 2014).

Burning Phases

After fields have been opened, felled debris is left to dry. In upland forests, drying periods last fifteen days to two months. Farmers reported that forest fallow in both upland and floodplain environments requires approximately fifteen to twenty days for material to dry. They emphasized, however, that drying times vary according to rainfall patterns. For instance, if it rains after fifteen days, one must wait an additional five days for the material to dry before burning. People expressed a general preoccupation with promoting a "good" or "complete" burn. We observed that in upland areas, felled materials were sometimes piled in long individual rows throughout the field. In other cases, farmers simply cut down (*baixar*) the general mass, chopping through vines and branches to prepare for ignition.

The act of burning is usually conducted by one or two farmers, generally men, although women in the study area also reported helping in this process. In cases of fields near community centers, large groups of residents participate in burns, which are considered exciting events. Burning begins during the hottest time of the day, from noon to 1:00 p.m. Fire is generally placed against the wind, and dried brush is ignited with a torch beginning on one edge of the field. Working simultaneously, farmers ignite the felled material at different points as they move forward to the other end of the field in two or three parallel lines (depending on the number of people working). In this way, as farmers advance, they continuously place themselves in front of the burning biomass. A floodplain farmer described a variation of this process, reporting simply igniting the "peaks" or highest points of the brush piles with matches. Fire then spreads from these points throughout the field, burning it entirely.

After this initial burn, farmers often apply a second round of burning in a practice referred to as *coivara*, a word of Indigenous origin that refers to the practice of piling branches and tree trunks—those not fully consumed in an initial burn—into small mounds and burning them once again (Pedroso Junior et al. 2008). Coivara can also include digging up and breaking apart stubborn tree stumps and large roots to burn them more thoroughly. Afterward, all of the piles of partially burned biomass scattered across the fields are burned again in situ. This piling and reburning practice is important for two reasons: it further clears piles of brush to make way for crops, allowing farmers to increase the size of planted areas; and it creates nutrient rich microsites for planting specific crops. In our interviews, farmers unanimously agreed that bananas are ideally planted where piled materials were reburned. Others mentioned that corn, yams, and squash also grow well in areas rich in charred materials.

Although some families practice coivara reburning in all of their fields as a preferred method for clearing vegetation, others do so with much less frequency. Most farmers reported that they employed coivara reburning only when fields did not burn thoroughly the first time. In the Baré community in the Amanã reserve, for instance, one researcher observed a farmer practicing coivara reburning in a field prepared seven months beforehand. Half of the field had been planted with manioc, while the other half, which "did not burn well," was still covered in scattered debris. In this case, the farmer decided to pile up the unburned debris and burn it again to make room for additional manioc plants. Working alone, the farmer poured rubbing alcohol into the center of each debris mound before igniting it (Fernanda Viana, personal communication; see figure 5.3).

In upland areas around Lake Amanã, farmers from the communities of Calafate, Ubim, and Boa Esperança explained that they apply fire only once after clearing forest parcels to lower debris piles and reduce smaller branches, vines, and weedy vegetation. The felled logs are left crisscrossed throughout the fields, unburned, and farm-

FIGURE 5.3. People practicing *coivara* in the community of Baré within the Amanã Sustainable Development Reserve, Brazil (photos by Fernanda Maria de Freitas Viana).

ers plant manioc and other crops between them. These upland farmers explained that this practice accelerates field preparation because they can move directly to planting after burning. Coivara (reburning) is considered too time-consuming and labor-intensive for fields larger than 1 hectare. Some people also stated that the slow decay of unburned tree trunks better enriched soils and prevented them from rapidly becoming "tired" (*cansadas*).

In the floodplains, fire is essential for establishing new field clearings from areas of forest fallow. After an area has been prepared and planted, farmers harvest the last batch of manioc tubers while simultaneously weeding and clearing. Vegetative debris is left in the field to be carried away by floodwaters. After the water recedes, these areas are free of weeds and enriched with a new layer of alluvial deposits. In the second and subsequent years, farmers weed any remaining growth as needed or move directly to planting manioc. In some cases, they collect manioc debris from the previous year and burn it in small piles, which are subsequently used to plant bananas and other crops that are considered particularly adapted to growing in soils rich with charcoal. Field areas in the floodplains are used from two years to as many as fifteen before being left to fallow.

After these burning phases, farmers plant almost immediately, waiting as few as five days for the embers to cool. During this time, farmers seek out seeds and other planting materials. Manioc, which is propagated by stems, is collected from older fields in uplands areas. Planting material is also collected from floodplain stocks that escaped annual floods, such as those conserved on floating platforms or elevated structures constructed in manioc fields. Planting is considered lighter work compared to field preparation, and men, women, and children of all ages participate. In the Amanã reserve, one commonly observes a division of labor whereby men dig small pits (*covas*) and women follow behind dropping in manioc cuttings and covering the pits (Steward and Lima 2014). Depending on the size of fields and the number of individuals mobilized to help, fields can be planted in one to five days.

Following planting, farmers manage the fields and care for their plants until harvest time. In the uplands, manioc fields are weeded as frequently as every two months if located in former fallows, or as little as once every six months if located in previously forested areas. In the floodplains, weeding begins about one month after planting. Fields are weeded at least three times before harvesting. Since weeding is done by both male and female farmers, it is common to see entire families engaging in this chore together. The frequency of weeding influences the development of manioc tubers, with unweeded areas yielding smaller, thinner ones.

In upland areas, fields are harvested in as few as nine months, but more commonly after one year of growth. While some manioc varieties mature faster, others are selected because they mature more slowly and therefore can be left in the soil longer. Ideally, farmers harvest manioc little by little, as needed for consumption and sale. In floodplain areas, farmers select and plant varieties that develop more quickly (about six months), allowing them to harvest in accordance with annual flood patterns (Lima et al. 2012). From April to June, depending on the speed of rising waters, farmers busily harvest manioc just ahead of floods.

While fields are being harvested, manioc tubers are transported to processing areas where families grate them, extract the bitter juice (*tucupi*), and toast the meal to produce flour (*farinha*), which is then stored for household consumption or separated for sale.

Fire Revisited: Farmers Perspectives on Fire Use

To better understand farmers' perceptions of fire and its utility, forty residents (twenty-one from upland areas and nineteen from floodplains) were interviewed specifically about agricultural burning. Only two of the forty residents said that they did not use fire at any point during the field preparation process. Being residents of the floodplains, these families maintained very small manioc fields that were easily cleared by hand. They then planted manioc directly in their field alongside the decaying weeds. One of these farmers described an alternative method of establishing a new field from forest fallow. Instead of burning, he used a chainsaw to chop debris into small pieces and then planted manioc within the decaying organic matter.

Two other floodplain residents who were interviewed said that they did not "always" use fire to prepare new areas. One reported that in some years, when vegetation is dominated by plants belonging to the *Heliconia* and *Musa* genera, it can be cleared by hand without burning. In floodplain fields, such as those on the beaches and mudflats that appear during the low-water season, crops are planted directly, without preparing the site with fire. In the uplands, fire is not applied when manioc stems are replanted during tuber harvesting. Such areas are usually replanted two or three times, depending on productivity, before being left to fallow.

The farmers who were interviewed reported a positive association between burning and plant development in agricultural areas.

They also noted that fire facilitates cultivation of manioc and stimulates tuberous growth. As previously mentioned, farmers commonly mention bananas as growing particularly well in coivara (reburn) sites. They also said that fire facilitates cultivation by softening soils and disintegrating hard, tough roots. Two farmers disagreed with claims that fire destroys soil health, explaining that fire does not enter the soil; rather, it burns surface brush and serves as fertilizer (*adubo*) for plants. One farmer (community of São João de Ipecaçú, Amanã reserve) explained in local terms what causes some garden areas to become degraded or "tired" (cansada):

> If we plant a field . . . in an area of forest fallow this year, we harvest it next year. If we let just one year go by before clearing again, the area doesn't yield nice manioc [*batata bonita*]. This is what we call terra cansada. Soils do not yield thick tubers when you plant in the same place every year. It's as if a woman had a child every year. There comes a time when the mother has no milk for the baby, and the woman becomes weak. It is the same thing with soils. You have to let the forest grow again for it to remain strong so that when you plant, things grow.

Farmers uniformly stated that fire use is the most efficient way to clear vegetation to establish agricultural areas. When considering not using fire, a few exclaimed that it "takes courage" to clear brush by hand, taking down the forest branch-by-branch or vine-by-vine. Others explained that it is possible to clear small areas without burning, but not large ones. Burning was also described as an effective means for controlling weeds (*ervas daninhas*) throughout the production cycle. Farmers noted that unburned areas require more frequent weeding to maintain fields until harvest time.

Even though most families still prefer to use fire to prepare agricultural areas, many had experimented with planting in unburned soils. Of the forty farmers we interviewed specifically regarding burning, three of them had participated in Agricultural Ecosystems Management Program agroforestry initiatives in the Amanã reserve. Two of them maintained demonstration areas on their properties, and the other participated in work groups to establish these areas. Three other families mentioned establishing new fields (or at least portions of fields) without fire after their burns did not take hold due to unexpected rainfall. One farmer described painstakingly clearing debris by hand instead of subjecting the area to a second round of burning. Another used a chainsaw to reduce remains into smaller pieces. Both of these farmers reported that their crops developed well, in one case even better than in previous years with burning. Those who had planted without burning (eight total) expressed the belief that it is possible to produce manioc without fire, and that crop development depends on how well one cares for the plants. However, they emphasized that this practice is not a viable option for larger fields, especially for small families with limited means of mobilizing extrafamilial labor. Some interviewees mentioned the large number of people (twenty to thirty) that helped prepare agroforestry experimental areas as an example of the high labor input needed to establish manioc fields without fire.

Farmers participating directly in agroforestry experiments as caretakers of demonstration areas offered interesting perspectives on the practice of preparing fields without fire. The first participant interviewed said she found the experience gratifying in the sense of learning a new technique and seeing that planting differently is possible. She confirmed that commonly cultivated crops such as sweet and bitter manioc, bananas, and açaí palm initially grew well in her family's experimental area. However, given that the area was established in forest fallow that had been intensively used in the past, the growth of grasses and other types of weeds was intense and not practical to maintain because her family had limited means of mobilizing labor. The plot eventually was integrated back into the community's swidden agricultural landscape. She believed that burning would have helped curb the growth of weeds.

In general, establishing fields without using fire was associated with having to expend more effort to control weeds. As one farmer said, "Now, there is one thing: it requires work. At weeding time, my wife simply cries!" (participant farmer, Amanã reserve, quoted in Steward et al. 2016).

Others found the agroforestry method compelling, especially for establishing enriched fallows. One farmer whose experimental area was established in April 2013 explained that skipping the fire phase allowed him to speed up the production process:

> This way we are not dependent on the sun and rain. We simply finish preparing the area and can plant immediately. There's no need to wait [for the brush] to dry, burn, and plant. Let's suppose that with a normal field, we would have had to wait until July to prepare a new field and then plant in August. This way, here we are in August and already have tall [sweet] manioc plants! (Steward et al. 2016)

In general, farmers said they enjoyed learning a new method of preparing agricultural areas. Of the eight initial experimental fields, six remain in the landscape as enriched fallows, with two of these standing out for their diversity. Farmers working in these two areas remain dedicated to managing and enriching them, having also reproduced the model on their own and within their communities. In all cases, however, farmers see the model as an alternative method for preparing enriched fallows, not as a replacement for establishing manioc fields, which they continue to prepare with fire due to its efficiency in clearing and suppressing weeds. Thus the model serves as a complement to traditional swidden agricultural practices in the region, but because of the extra work required in the preparation phase, its use is restricted to the establishment of relatively small enriched fallows.

Discussion
Diversity of Techniques and Practices

Regarding the use of fire in swidden agricultural practices in the study site, we can draw several conclusions. First, despite the agroforestry efforts of the Agricultural Ecosystems Management Program, and those of the Forest Conservation Allowance Program, this traditional practice remains largely integrated into families' management strategies. As evident in interviewees' descriptions of both systems, fire is used less frequently by floodplains farmers, who, given the dynamism of the environment, use a variety of techniques to prepare agricultural fields. At the same time, however, floodplains farmers agree that fire is a useful land management tool for establishing new agricultural areas.

Conservation and rural development advocates have criticized swidden agriculture as a rudimentary and simplistic practice involving only "slashing and burning" of forest vegetation (Padoch and Pinedo-Vasquez 2010). Contrary to this commonly held belief, the findings presented here demonstrate that farmers use precise and varied methods to apply fire, accumulating experience and confidence in the practice over time. As one resident of Cacau, Amanã reserve, explained: "This practice cannot be taken lightly. It can be quite dangerous for the unpracticed and so not everybody has the courage to conduct burns."

Fire use in local agricultural systems falls into three different categories: the initial burn, reburn (coivara), and restricted localized fires in floodplain landscapes to clear small brush piles.[11] We also observed three different approaches to applying fire to brush. These practices are combined in various ways depending on the needs of each farmer and the dynamics of each specific environment. Farmers emphasized that each family does things its own way, suggesting that more research may reveal greater diversity in practices. As a follow-up to this initial study, additional research is needed to understand the different impacts of fire on soils and the dynamics of forest regeneration following different burning regimes and management practices.

Perceptions of Fire Use and Impacts

When Agricultural Ecosystems Management Program researchers and technicians began the agroforestry course, they assumed that participants lacked previous experience planting without fire and persisted in using it because they lacked knowledge about alternatives. Through interviews with farmers after the course, we found that it is common for areas to be "poorly" or partially burned due to quick shifts in weather conditions, yet farmers planted them anyway. Thus, some farmers are skilled in cultivating areas without fire, but they choose to burn because they consider it practical. Their choice to use fire to prepare fields is not the result of not knowing how to farm differently.

Divergence in understandings of how to conserve soils and mitigate the effects of burning also deserve mention. From technicians' and researchers' perspectives, fire may provide immediate fertilizing effects but has the long-term effect of weakening soils by transforming biomass that eventually will be carried away from the agricultural site by rains. In contrast, farmers believe burning and creating charcoal feeds the soil and creates space in the forest for plant cultivation.

The overuse of soils (not fire) is what degrades them. This perspective was expressed by farmers in the analogy of a mother whose body is weak and whose milk dries up from having many children in quick succession. For farmers, letting areas lie fallow is an important means of caring for soils in productive and potentially productive areas, permitting their long-term use. Farmers' observations documented in this chapter are in line with a recent study reviewing the global literature on impacts of swidden agriculture on soils. Ribeiro and colleagues (2013, 720) determined that swidden systems "are not unsustainable per se in relation to soil system dynamics." Rather, demographic changes such as population growth and urbanization, along with climate and other environmental changes, can place pressure on societies practicing swidden agriculture. If such transitions spur land use changes—such as increases in the number of cultivation cycles and decreases in fallow periods—the resilience of these systems might be compromised (Ribeiro et al. 2013).

Evident from the comparison of the two systems is the reduced impact of swidden agriculture on floodplain forests compared to upland forests. This is due, in part, to the fact that floodplain farmers open smaller field areas and use them continuously. In the uplands, fields are larger, and their locations in the landscape change with much greater frequency. Upland farmers tend to open new field areas each year in either forests or forest fallow. We must understand, however, that swidden landscapes in this environment are also in constant change. Fields characterized by reduced vegetative growth compared to forests represent the first phase in the swidden cycle. Areas rotate from forest to fields, to fallow, and back to forests. Alternatively, they can be maintained for longer periods as home gardens or enriched fallows. In this case, forest regeneration processes and the maintenance of biodiversity are central to the functioning of the swidden system. As such, swidden practices in general are better described as resulting in long-term forest management, not forest clearing. This is, of course, different from other common land use practices cited in Amazonia, such as the establishment of pastures or plantations, where the goal is to permanently transform forests into intensive production spaces.

Manioc production and agriculture in general have enormous social importance in the region. As Lima (2004; Lima-Ayers 1992) demonstrates, the social reproduction of families is closely integrated with manioc production processes. For instance, the separation of young couples from their parents' or in-laws' home coincides with their establishment of independent manioc fields (Lima 2004). Further emphasizing the social significance of manioc fields in Amazonia is the celebration of fields' birthdays and ritual practices associated with manioc production. One such practice is the ceremonial planting of the heart of the field (*coração da roça*), described as a special area at the core of an agricultural field with a specific arrangement of plants. Because it is believed to give the field its vitality, the heart is planted first and harvested last (Lima et al. 2012). These practices go hand in hand with the observation that farmers actively integrate new manioc cuttings into their fields, trading old varieties for new ones, or adding new varieties while maintaining the old—practices that have resulted in agrobiodiversity creation in the Middle Solimões landscape. The importance of agrobiodiversity for the social reproduction of rural communities has been well established (Jarvis et al. 2013; Veteto and Skarbø 2009), as has the global importance of agrobiodiversity for provisioning important ecosystems services (Thrupp 2000). It has also been documented that smallholders, such as those from the Amanã and Mamirauá reserves, are largely responsible for maintaining this diversity (Brush 2008). Thus the impacts of swidden agricultural practices must be weighed against the threat of agrobiodiversity losses that could accrue from their continued prohibition.

Conclusions

Agroforestry initiatives promoted by the Agricultural Ecosystems Management Program have been reformulated, and technicians have changed their approaches based on their experiences with farmers. The alternative techniques promoted by technicians are now considered complements to farmers' local practices. The issue of fire is approached more neutrally, and technicians discuss the potential advantages of agricultural methods that do not involve burning without denigrating fire use. Likewise, as mentioned above, the regional coordinator for the Sustainable Amazonas Foundation who is responsible for overseeing the Forest Conservation Allowance Program in the Mamirauá and Amanã reserves emphasizes that the foundation does not entirely prohibit swidden agriculture and fire use, but rather recommends "good practices," stressing that farmers are not fined for unsanctioned land use nor are they directly monitored in the field. Furthermore, for many communities, the Forest Conservation Allowance Program has been an important resource for funding local projects. The program has also resulted in different forms of social organization within the reserve, the impacts of which deserve further attention. Despite shifts in extension and applied research approaches, the leniency of the Sustainable Amazonas Foundation, and the benefits of both agroforestry and forest allowance initiatives, the promotion of extension practices based on "improving" local swidden practices has reinforced negative stereotypes associated with swidden agriculture (i.e., practices are backward or non-ecological) among practitioners working on issues of conservation in the Middle Solimões region. Aside from being disrespectful, defaming local agricultural practices that are interwoven with the fabric of local society poses a threat to the important partnerships that evolved between conservation scientists, technicians, and residents during establishment of the reserves. In the Mamirauá and Amanã reserves, ribeirinho groups are largely responsible for forest preservation. Strengthening community alliances is crucial to the continued and future conservation of these areas (Carneiro da Cunha 2009; Queiroz and Peralta 2006). This clearly cannot be accomplished by vilifying important land use practices.

Based on our research group's previous experiences, it seems scientists and other conservation professionals are easily distracted by local farmers' relatively small impacts on forests and fail to recognize the most urgent threats to forests in Amazonas State and Amazonia in general. While the forests of Amazonas State remain largely conserved (only .4 percent deforested), southern regions have been subjected to widespread deforestation and are considered particularly vulnerable to future clear-cutting. As in other areas of Amazonia, the main drivers of deforestation in this region have been land grabbers claiming areas near highways for cattle pastures and agricultural plantations (Fearnside 2005; Viana 2008). Paving of major roads (BR-319, BR-230, and AM-174), already approved by the federal government, is projected to place even more pressure on this region (Viana 2008). Highways serve as vectors, giving access to previously remote areas of forest for illegal timber extraction and mining, clearing by cattle ranchers, and establishment of plantations by local farmers and landless migrant farmers who play a smaller yet still significant role in forest cover reduction (Fearnside 2008). Models simulating future trends predict that as much as 30 percent of Amazonas State could be deforested by 2050 (Soares-Filho et al. 2006).

Considering these regional patterns, scholars addressing the potential benefits of REDD projects argue that initiatives should first be directed toward conservation units in agricultural frontiers or near highways, which are most vulnerable to deforestation (Santilli 2010). In some cases, such as in the Juma Sustainable Development Reserve in the state of Amazonas, income generated by trading carbon credits on the voluntary market allowed for the creation and subsequent management of the conservation area. This reserve was strategically placed near major highways to help prevent further deforestation (Viana 2008). In the context of REDD programs, conservation areas and Indigenous lands in more remote areas, such as the Middle Solimões region, function as future carbon reserves. Advocates argue that interventions in such areas should not only mitigate impacts of residents' practices on forest cover, but also focus on securing tenure rights and improving social conditions for local communities so that traditional peoples and land use practices will persist (Santilli 2010). Natural resource management programs based on solid understandings of local land use practices can help ensure their sustainability in the face of eventual changes. This recommendation should be taken seriously in the case of swidden practices in the Mamirauá and Amanã reserves. The risks of suppressing local swidden practices outweigh any small potential benefits in reducing deforestation caused by traditional farmers' activities as practiced today.

Acknowledgments

First and foremost, I extend my gratitude to all the farmers of the Mamirauá and Amanã reserves who participated so eagerly in this study. I also thank the teams of the Agricultural Ecosystems Management Program and the Research Group in Amazonian Agriculture, Mamirauá Institute, for their collaboration during all phases of this project. I extend special gratitude to Luiza Câmpera for her help in data collection on local fire practices, and to Camille Rognant and Samis Vieira do Brito for their insights on the agroforestry program results. I acknowledge the Mamirauá Institute for its logistical support and the Brazilian Ministry of Science, Technology, and Innovation (Ministério de Ciência, Tecnologia e Inovação) for project financing.

Notes

1. Large fires are usually agricultural fires that get out of hand, most originating in large pastures or plantations. The issue of deforestation links back to increasing frequency of large fires because clearing causes changes in local climates and contributes to droughts that prolong the burning season (Bowman et al. 2011).
2. For the remainder of the text, REDD is used to refer to both REDD and REDD+ initiatives and projects.
3. In the same way, the carbon balance within swidden landscapes is also poorly understood (Brunn et al. 2009).
4. In the Amanã reserve, várzea areas are subject to annual flooding. Upland areas, in comparison, escape annual flooding because they are slightly higher in elevation. Paleovárzea areas are intermediate zones that occur in older geologic formations (consisting of old alluvial deposits). Parts of these areas are inundated during high flood years, but, for the most part, those in the study area escape annual floods (Irion et al. 2011). Local residents refer to areas of paleovárzea as terra firme, the term used for the remainder of this chapter.
5. According to its proponents, agroecology applies the concepts and principles of ecology to manage agricultural systems in a sustainable fashion. Activities are understood as sustainable when they do not interfere with the integrity of the environment. Patterns of agricultural production equally address the social, economic, and environmental goals of producers and their communities (Hart 1980).
6. New residents have kin or affinity ties with other community members.
7. As Padoch and Pinedo-Vasquez (2010) argue, principal elements of swidden systems—cutting and burning of the forest, comparatively low yields of staple crops in annual fields, and dynamism—stand in direct opposition to commonly held notions of sustainable forest use. Many well-intentioned scientists and extension agents distracted by these seemingly irrational and/or destructive practices fail to consider that these systems may also have positive attributes (i.e., biodiversity, resilience, and flexibility) and may contribute to the sustainable management of tropical landscapes.
8. The general characteristics of these systems in Amazonia and their potential impacts on forest cover and soils are well documented; see, for example, Pedroso Junior et al. 2008; Ribeiro Filho et al. 2013; and van Vliet et al. 2012 for comprehensive reviews.
9. By way of comparison, a single cattle ranch of a small farmer from the traditional north region of Mato Grosso covers an average of 872 hectares (Barbier 2013), whereas government-supported (SUDAM) projects for ranching in the 1970s and 1980s ranged from 18,126 hectares to 28,860 hectares.
10. Access to land in the region is governed through a system of usufruct rights where recognition of "ownership" is determined by the initial work a farmer invests in a given area. The area continues to belong to the farmer on the condition that he/she continues to invest work in the area. Abandoned areas become free for use by the community. Over time, "owners" may acquire an increasing number of forest fallow in various stages of succession, which they use to establish agricultural areas (Lima 2004; Lima-Ayres 1992).
11. Families use small fires to clear brush, debris, and household residues from home garden areas (quintais) during the summer months in both environments.

References Cited

Adams, Cristina, Rui Murrieta, Andrea Siqueira, Walter Neves, and Rosely Sanches. 2006. "O Pão da terra: Da invisibilidade da Mandioca na Amazônia." In *Sociedades caboclas amazônicas: Modernidade e invisibilidade*, edited by Cristina Adams, Rui Murrieta, and Walter Neves, 295–322. São Paulo: Annablume.

Adams, Cristina, Rui Murietta, and Walter Neves, eds. 2006. *Sociedades caboclas amazônicas: Modernidade e invisibilidade*. São Paulo: Annablume.

Alencar, Ane, Osvaldo S. Martins, Andre Nahur, Daniel Nepstad, Anrea Cattaneo, and Tracy Johns. 2010. *Brazil's Emerging Sectoral Framework for Reducing Emissions from Deforestation and Degradation and the Potential to Deliver Greenhouse Gas Emissions Reductions from Avoided Deforestation in the Amazon's Xingu River Basin*. Forest Carbon Portal.

Amazon Fund. 2014. "Frequently Asked Questions." Accessed October 1, 2014. http://www.amazonfund.gov.br/FundoAmazonia/fam/site_en/Topo/FAQ.

Barbier, Edward B. 2013. *Economics, Natural-Resource Scarcity and Development: Conventional and Alternative Views*. New York: Routledge.

Bowman, David M. J. S., Jennifer Balch, Paulo Artaxo, William J. Bond, Mark A. Cochrane, Carla M. D'Antonio, and Ruth DeFries 2011. "The Human Dimension of Fire Regimes on Earth." *Journal of Biogeography* 38 (12): 2223–2236.

Brookfield, Harold, Christine Padoch, Helen Parsons, and Michael Stocking. 2002. *Cultivating Biodiversity: Understanding, Analysing and Using Agricultural Diversity*. Rugby Warwickshire: ITDG.

Bruun, Thilde Bech, Andreas de Neergaard, Deborah Lawrence, and Alan D. Ziegler. 2009. "Environmental Consequences of the Demise in Swidden Cultivation in Southeast Asia: Carbon Storage and Soil Quality." *Human Ecology* 37 (3): 375–388.

Brush, Stephen B. 2008. *Farmers' Bounty: Locating Crop Diversity in the Contemporary World*. New Haven, CT: Yale University Press.

Cardoso, Thiago Mota. 2010. *O saber biodiverso: Práticas e conhecimentos na agricultura indígena do Baixo Rio Negro*. Manaus: Editora da Universidade Federal do Amazonas.

Carmenta, Rachel, Luke Parry, Alan Blackburn, Saskia Vermeylen, and Jos Barlow. 2011. "Understanding Human-Fire Interactions in Tropical Forest Regions: A Case for Interdisciplinary Research across the Natural and Social Sciences." *Ecology and Society* 16 (1): 53.

Carmenta, Rachel, Saskia Vermeylen, Luke Parry, and Jos Barlow. 2013. "Shifting Cultivation and Fire Policy: Insights from the Brazilian Amazon." *Human Ecology* 41 (4): 603–614.

Carneiro da Cunha, Manuela. 2009. *Cultura com Aspas*. São Paulo: Cosac and Naify.

do Valle, Raul Silva Telles. 2010. *Avoided Deforestation (REDD) and Indigenous Peoples: Experiences, Challenges and Opportunities in the Amazon Context*. São Paulo: Instituto Socioambiental, and Washington, DC: Forest Trends.

Fearnside, Philip M. 2005. "Deforestation in Brazilian Amazonia: History, Rates, and Consequences." *Conservation Biology* 19 (3): 680–688.

———. 2008. "The Roles and Movements of Actors in the Deforestation of Brazilian Amazonia." *Ecology and Society* 13 (1): 23.

———. 2011. "Brazil's Amazon Forest in Mitigating Global Warming: Unresolved Controversies." *Climate Policy* 12 (1): 70–81.

Fearnside, Philip M., and Reinaldo Imbrozio Barbosa. 2004. "Accelerating Deforestation in Brazilian Amazonia: Towards Answering Open Questions." *Environmental Conservation* 31 (1): 7–10.

Fundação Amazonas Sustentável [FAS]. 2014. "Programa Bolsa Floresta." Acessed October 1. http://fas-amazonas.org/pbf/.

Hardesty, Jeff, Ron Myers, and Wendy Fulks. 2005. "Fire, Ecosystems, and People: A Preliminary Assessment of Fire as a Global Conservation Issue." *George Wright Forum* 22 (4): 78–87.

Harris, Mark. 2000. *Life on the Amazon: The Anthropology of a Brazilian Peasant Village.* Oxford: Oxford University Press.

Hart, R. D. 1980. *Agrosistemas: Conceptos básicos.* Turrialba: CATIE.

IDSM (Instituto de Desenvolvimento Sustentável Mamirauá). 2011. "Banco de dados do levantamento sociodemográfico da Reserva de Desenvolvimento Sustentável Amanã, 2011." Tefé, Amazonas: Instituto de Desenvolvimento Sustentável Mamirauá.

———. 2014a. "Institucional/Objetivos." Accessed October 1, 2014. http://www.mamiraua.org.br/pt-br/institucional/objetivos/.

———. 2014b. "Programa de Manejo de Agroecossistemas." Accessed October 1, 2014. http://www.mamiraua.org.br/pt-br/manejo-e-desenvolvimento/programa-de-manejo-de-agroecossistemas/.

Irion, Georg, José A. S. N. de Mello, Jáder Morais, Maria T. F. Piedade, Wolfgang J. Junk, and Linda Garming. 2011. "Development of the Amazon Valley During the Middle to Late Quaternary: Sedimentological and Climatological Observations." In *Amazonian Floodplain Forests*, edited by Wolfgang J. Junk, Maria T. F. Piedade, Florian Wittmann, Jochen Schöngart, and Pia Parolin, 27–42. Netherlands: Springer.

Jarvis, Devra Ivy, Christine Padoch, and H. David Cooper. 2013. *Managing Biodiversity in Agricultural Ecosystems.* New York: Columbia University Press.

Lima, Deborah. 2004. "The Roça Legacy: Land Use and Kinship Dynamics in Nogueira, an Amazonian Community of the Middle Solimões Region." In *Some Other Amazonians: Perspectives on Modern Amazonia,* edited by Stephen Nugent and Mark Harris, 12–37. London: Institute for the Study of the Americas.

———. 2009. "The Domestic Economy in Mamirauá, Tefé, Amazonas State." In *Amazon Peasant Societies in a Changing Environment,* edited by Cristina Adams, Rui. S. Murrieta, Walter A. Neves, and Mark Harris, 131–156. Netherlands: Springer.

Lima, Deborah, Angela Steward, and Bárbara T. Richers. 2012. "Trocas, experimentações e preferências: Um estudo sobre a dinâmica da diversidade da mandioca no Médio Solimões, Amazonas." *Boletim do Museu Paraense Emílio Goeldi: Ciências Humanas* 7 (2): 371–396.

Lima-Ayres, Deborah de Magalhães. 1992. The Social Category Caboclo: History, Social Organization, Identity and Outsider's Social Classification of the Rural Population of an Amazonian Region (the Middle Solimões). PhD dissertation, University of Cambridge.

Padoch, Christine, and Miguel Pinedo-Vasquez. 2010. "Saving Slash-and-Burn to Save Biodiversity." *Biotropica* 42 (5): 153–174.

Pedroso Júnior, Nelson Novaes, Rui Sérgio Sereni Murrieta, and Cristina Adams. 2008. "The Slash-and-Burn Agriculture: A System in Transformation." *Boletim do Museu Paraense Emílio Goeldi. Ciências Humanas* 3 (2): 153–174.

Peralta, Nelissa, and Deborah de Magalhães Lima. 2010. "A Comprehensive Overview of the Domestic Economy in Mamirauá and Amanã in 2010." *Uakari* 9 (2): 33–62.

Pereira, Kayo, Julio Cesar, Bianca Ferreira Lima, Raimundo Silva dos Reis, and Elizabeth Ann Veasey. 2008. "Saber tradicional, agricultura e transformação da paisagem na Reserva de Desenvolvimento Sustentável Amanã, Amazonas." *Uakari* 2 (1): 9–26.

Pinedo-Vasquez, Miguel, Christine Padoch, and Inuma C. Jomber. 1996. *Identifying and Understanding Agricultural, Agroforestry and Forest Management Systems and Techniques Practiced in Mamirauá.* Tefé, Amazonas: Sociedade Civil Mamirauá.

Queiroz, Helder L. 2005. "A Reserva de Desenvolvimento Sustentável Mamirauá." *Estudos Avançados* 19 (54): 183–203.

Queiroz, Helder L., and Nelissa Peralta. 2006. "Reserva de Desenvolvimento Sustentável: Manejo Integrado dos Recursos Naturais e Gestão Participativa." *Dimensões humanas da biodiversidade: O desafio de novas relações sociedade-natureza no século XXI*, edited by Garay Becker and Bertha K. Becker, 447–476. Petropólis: Editora Vozes.

Ramalho, E. E., Joana Macedo, T. M. Vieira, João Valsecchi, J. Calvimontes, Mirim Marmontel, and Helder L. Queiroz. 2010. "Ciclo hidrológico no ambientes de Várzea da Reserva de Desenvolvimento Sustentável Mamirauá—Médio Rio Solimões, período de 1990 a 2008." *Uakari* 5 (1): 61–87.

Ribeiro Filho, A. A., Cristina Adams, and Rui S. Murrieta. 2013. "The Impacts of Shifting Cultivation on Tropical Soil: A Review." *Boletim do Museu Paraense Emílio Goeldi, Ciências Humanas* 8 (3): 693–727.

Richers, Bárbara. 2010. "Agricultura migratória na Várzea: Ameaça ou uso integrado?" *Uakari* 6 (1): 27–37.

Santilli, Márcio. 2010. "Indigenous Lands and the Climate Crisis." In *Avoided Deforestation (REDD) and Indigenous Peoples: Experiences, Challenges and Opportunities in the Amazon Context*, edited by Raul Silva Telles do Valle, 9–19. São Paulo: Instituto Socioambiental, and Washington, DC: Forest Trends.

Schmidt, Morgan J. 2003. "Farming and Patterns of Agrobiodiveristy on the Amazon Floodplain in the Vicinity of Mamiraua, Amazonas, Brazil." Master's thesis, University of Florida, Gainesville.

Soares-Filho, Britaldo Silveira, Daniel Curtis Nepstad, Lisa M. Curran, Gustavo Coutinho Cerqueira, Ricardo Alexandrino Garcia, Claudia Azevedo Ramos, Eliane Voll, Alice McDonald, Paul Lefebvre, and Peter Schlesinger. 2006. "Modelling Conservation in the Amazon Basin." *Nature* 440 (7083): 520–523.

Steward, Angela. 2007. "Changing Lives, Changing Fields: Diversity in Agriculture and Economic Strategies in Two Caboclo Communities in the Amazon Estuary." PhD dissertation, Graduate Center, City University of New York.

Steward, Angela, and Deborah Lima. 2014. "Interações na Toça: Por uma ecologia das práticas da produção de mandioca no Médio Solimões/AM." Paper presented at the Twenty-Ninth Meeting of the Brazilian Association of Anthropology, Federal University of Rio Grande do Norte, Natal, Brazil, August 3–6.

Steward, Angela May, Camille Rognant, and Samis Vieira do Brito. 2016. "Roça sem fogo: A visão de agricultores e técnicos sobre uma experiência de manejo na Reserva de Desenvolvimento Sustentável Amanã, Amazonas, Brasil." *Biodiversidade Brasileira* 6 (2): 71–87.

Thrupp, Lori Ann. 2000. "Linking Agricultural Biodiversity and Food Security: The Valuable Role of Agrobiodiversity for Sustainable Agriculture." *International Affairs* 76 (2): 283–297.

UNDP (United Nations Development Program). 2016. "Branching Out for a Green Economy, REDD+." Accessed March 15, 2017. http://www.unep.org/forests/REDD/tabid/7189/Default.aspx.

Valsecchi, João, and Paulo V. Amaral. 2009. "Perfil da caça e dos caçadores na Reserva de Desenvolvimento Sustentável Amanã, Amazonas, Brasil." *Uakari* 5 (2): 33–48.

van Vliet, Nathalie, Ole Mertz, Andreas Heinimann, Tobias Langanke, Unai Pascual, Birgit Schmook, and Cristina Adams. 2012. "Trends, Drivers and Impacts of Changes in Swidden Cultivation in Tropical Forest-Agriculture Frontiers: A Global Assessment." *Global Environmental Change* 22 (2): 418–429.

Veteto, James R., and Kristine Skarbø. 2009. "Sowing the Seeds: Anthropological Contributions to Agrobiodiversity Studies." *Culture and Agriculture* 31 (2): 73–87.

Viana, Fernanda Maria Freitas, Angela Steward, and Bárbara T. Richers. 2016. "Cultivo itinerante na Amazônia Central: Manejo tradicional e transformações da paisagem." *Novos Cadernos NAEA* 19: 93–122.

Viana, Virgílio M. 2008. "Bolsa Floresta (Forest Conservation Allowance): An Innovative Mechanism to Promote Health in Traditional Communities in the Amazon." *Estudos Avançados* 22 (64): 143–153.

———. 2009. "Financiando REDD mesclando o mercado com fundos do governo." International Institute for Environment and Development. Accessed March 15, 2017. http://pubs.iied.org/17053PIIED/?w=NR&p=101.

———. 2010. *Sustainable Development in Practice: Lessons Learned from Amazonas. Environmental Governance No. 3*. London: International Institute for Environment and Development. Accessed March 15, 2017. http://pubs.iied.org/pdfs/17508IIED.pdf.

CHAPTER 6

Restoration, Risk, and the (Non) Reintroduction of Coast Salish Fire Ecologies in Washington State

JOYCE K. LECOMPTE

In June of 2007, sixty tribal members, federal land managers, natural and social scientists, policy makers, and commercial harvest representatives traveled from throughout the United States and Canadian Pacific Northwest to attend the first ever Big Huckleberry Summit. The one day conference was held to share multiple perspectives about this berry, its cultural significance, and concerns about its well-being.

Big huckleberry is found in mid-elevation meadows, primarily throughout the Pacific Northwest (Figure 6.1). People have been traveling to the mountains to harvest the sweet blue-black berries for at least 6,000 years (Stenholm 1989). People go in the late summertime when the days are simultaneously hot and carry the cool undertones of the coming autumn on the breeze. Along with their human harvesters, deer and elk, black and grizzly bear, mountain beaver, marmot, other small mammals, and several species of resident and migratory birds also rely on the foliage and fruit of big huckleberry and the other plants that grow along with them.

Berry harvesters—particularly Native Americans for whom big huckleberry is a traditional food—have been concerned for some time about the effects of land management practices and competition from nontribal harvesters on their ability to gather enough huckleberries to meet their needs. Throughout the day, summit participants listened as Muckleshoot, Nisqually, Umatilla, and Warm Springs tribal members shared stories about their profound connections to big huckleberry and the importance of maintaining those connections for the cultural and physical well-being of their people. Before they were alienated from their traditional homelands and federal fire suppression policies were vigorously pursued, Native experts used to burn huckleberry habitat to reinvigorate the land and to discourage trees from invading the meadows (Barrett and Arno 1999; Deur 2002; French 1999; Lepofsky et al. 2005; Mack 2003; Main-Johnson 1999; Turner 1999). The elders who spoke that day emphasized that burning, in addition to ensuring a future supply of the berries, is a responsibility—part of a reciprocal obligation to take care of something that also takes care of them.

I co-organized the Big Huckleberry Summit with Warren KingGeorge, a Muckleshoot tribal member and oral historian for the tribe, and Laura Potash, a botanist for the

FIGURE 6.1. Big huckleberry (Swəda?χ) (*Vaccinium membranaceum* Douglas ex. Torr.) (photo by Joyce K. LeCompte).

Mount Baker–Snoqualmie National Forest. We had recently begun a collaboration to implement a huckleberry enhancement project on a portion of the Muckleshoot Tribe's ceded territories that are now part of the national forest. Our shared vision for the summit was to provide a venue for cross-cultural sharing of knowledge, communication of the importance of the berry within tribal communities to land managers and natural scientists, and development of a network of support for those working on projects similar to our own.

Throughout its range, the human-huckleberry nexus has been a node in a complex mesh of socioecological relations since time immemorial (Anzinger 2002; Deur 2002; Fisher 1997; Hunn and Selam 1990; Keefer 2007; Mack and Mcclure 2001; Naxaxalhts'i [McHalsie] 2007; Trusler and Johnson 2008; Turner, Deur, and Mellott 2011; Turney-High 1941). But the nature of that network has dramatically changed from the kinds that were maintained by Indigenous peoples prior to Euro-American colonization and the eventual development of a state apparatus that has had the effect of managing and controlling those relations. Dynamic processes of becoming are integral to the relational ecologies that compose westside forests like those in the Mount Baker–Snoqualmie National Forest, where our huckleberry enhancement project was unfolding. Here, the interaction of very long lived tree species, long- and short-term changes in climate, and infrequent yet major disturbance events makes it difficult to claim any sort of ecological equilibrium (Sprugel 1991). A central theme of the Big Huckleberry Summit was that both natural and anthropogenic fire in Pacific Northwest forests are agents of disturbance that have been integral to the development and maintenance of huckleberry habitats over time (Hemstrom and Franklin 1982; Agee 1993; Franklin and Dyrness 1988; Mah 2000). At the end of the day, as people summed up what they had learned, the then–district ranger for the Mount Baker-Snoqualmie National Forest stood and remarked, "One thing I have come to believe after today is that we must find ways to reintroduce fire on the landscape."

Over the eight years that have passed since the summit, I have attended numerous conferences devoted specifically to discussions of big huckleberry and to the management of traditional Indigenous plant foods. I have also been closely involved as a participant-observer in two collaborative huckleberry enhancement projects in the Mount Baker-Snoqualmie National Forest. Both projects proposed using mechanical thinning of trees, as opposed to burning, as a primary mode of disturbance to allow light into the understory. While one of the project's management plans included broadcast burning after thinning, this part of the project was not accomplished.

In the past eight years I have also interviewed dozens of tribal members, land managers, and natural scientists about their views on the reasons for, and social-ecological effects of, loss of huckleberry habitat in the Pacific Northwest. Given that mechanical removal of trees has been so central to the projects that I have been involved with, and fire so marginal to them, how do natural scientists, land managers, and tribal members view the relative merits of burning versus mechanical thinning of trees in big huckleberry habitat? From the perspectives of those directly involved with the two projects in the Mount Baker–Snoqualmie National Forest, are there particular benefits to the use of fire that are not met through mechanical means? What are the perceived environmental and social risks, and structural barriers that have thus far prevented the reintroduction of fire on the land? Conversely, what are the perceived risks to both humans and nonhumans of *not* reintroducing fire to the land?

The problem of whether and how big huckleberry is to be cared for is a matter of environmental justice. In the human realm at least, Native Americans bear a greater burden when it comes to the loss of accessible and productive meadows.

Paradoxes of Disturbance

In her analysis of the "blasted landscapes" within which human–matsutake mushroom relations can thrive, Anna Tsing (2014, 92) asks, "Which disturbance regimes are we willing to live with?" Like the mushrooms that ground Tsing's questions, huckleberry vitality is also disturbance-dependent. In their resistance to human efforts to cultivate them, huckleberries also share with the matsutake a certain recalcitrance. Echoing Tsing's (2014, 90) observations of human-matsutake relations, I too find that relations with huckleberries and how we might care for them "press us into multispecies ecologies in which control may be impossible." Deliberations over the kinds of disturbance that might be desirable or even possible to care for big huckleberry are inevitably political processes involving negotiation, diplomacy, and risk for both the humans and nonhumans involved.

Thus, as important as this very specific problem is on its own, examining deliberations over the role of fire in these projects also opens a space for asking broader questions about how we might live together—and well—in postcolonial spaces such as the US Pacific Northwest.[1] A critical examination of the fire-related deliberations that took place within these projects is also a historically specific window into processes of becoming in which the Forest Service, local tribes, fire, huckleberries, and numerous other bioculturally diverse actors that constitute this thing we call a forest are implicated. The forms that these evolving social-ecological collectives take are also dependent on the discursive and material relations that are brought forward from the past into the present, including Forest Service policies, political economies, colonial histories, and the materiality of the forest itself.

At the same time, the physical processes that constitute the material world upon which we depend seem largely indifferent to us, and there is much about the materiality of life that exceeds our ability to control or predict. There are many earth and life processes over which we have little control that do go on without any intervention from humans at all and will continue to go on without us long after we have gone (Clark 2011). As much as the human story is deeply entangled with the use of fire as technology and source of meaning (Pyne 1997, 1998; Wrangham 2009), fire on Earth does go on with or without human assistance (Pyne 2001; Vale 2002).

At a *landscape* scale, fire regimes on the west side of the Cascade Range are characterized by infrequent but catastrophic fires (Agee 1993; Hemstrom and Franklin 1982 [but cf. Walsh et al. 2015]). Coast Salish ancestors most certainly directly experienced those probably terrifying fires, as well as igniting smaller controlled burns to meet their needs. As Tsing (2014, 92) reminds us, " . . . humans are both vectors and victims of disturbance." Thus it should not be surprising that these ancestors seem to have had a deep appreciation for both the destructive and productive capacities of fire (Ballard 1929; Snyder 1968).

The Indigenous fire ecologies of Coast Salish ancestors certainly played a substantial role in shaping the land at a *human* scale (LeCompte-Mastenbrook 2015). Anthropogenic fire regimes have the potential to alter natural ones by shortening fire-return intervals and the timing of burns, which in turn has the potential to substantively alter species composition and habitat structure (Anderson and Rosenthal 2015; Bowman et al. 2011; Sullivan et al. 2015). Prior to the devastating effects of the first smallpox epidemic in the area in the late eighteenth century, an estimated 13,000 Coast Salish people lived in some 140 permanent villages along the saltwater, rivers, lakes, prairies, and uplands of the Puget Sound basin (Boyd 1999; Smith 1940). The cumulative effects of their fire regimes may have been considerable.

Alien diseases and the processes of colonialism that interfered with and eventually extinguished Indigenous fire ecologies have had cascading effects that continue to impact the physical, emotional, and spiritual well-being of Native communities today (Norgaard 2014). How might the reintroduction of a form of Indigenous fire ecology ameliorate the cultural disturbances wrought by colonialism? Anna Tsing's (2014, 92) follow-up question to her first seems particularly apt when it comes both to catastrophic fire and the cascading effects of colonialism: "Given the realities of disturbance we do not like, how shall we live?"

Although their histories are quite different, both Tsing's and my case studies focus on Forest Service lands that share a legacy of the ecological, political, and economic effects of intensive industrial logging. Yet some species, such as matsutake mushrooms and big huckleberry, thrive even in the ruins of these "blasted landscapes" (Tsing 2014). That big huckleberry was one of the first plants to colonize the moonscape that was Mount St. Helens after the 1980 eruption is a testament to its incredible resilience after disturbance (Yang 2006; Yang et al. 2008). Tsing encourages us to consider what forms of disturbance are inimical to life, and which offer opportunities for multiple species to flourish. But as I describe below, the answer to those questions depends on whom one is asking.

Research Sites

The huckleberry enhancement projects that I describe here involved members and staff of the Muckleshoot and Tulalip tribes, two of ten federally recognized Coast Salish treaty tribes. Portions of their ceded hunting, fishing, and gathering territories are now part of the Mount Baker–Snoqualmie National Forest (Figure 6.2).[2] The ancestors of the Muckleshoot and Tulalip people were party to the so-called Stevens Treaties, which include the following passage about reserved rights to resources:

> The right of taking fish at all usual and accustomed grounds and stations, is further secured to said Indians in common with all citizens of the Territory, and of erecting temporary houses for the purpose of curing, together with the privilege of hunting, [and] gathering roots and berries, and pasturing their horses on open and unclaimed lands. (United States 1855; United States 1854)

Coast Salish and other federally recognized tribes in Washington State are involved with virtually every agency at the local, state, and federal level when it comes to the management of natural resources. The tribes have vigorously employed legal and rhetorical tactics, and have engaged in civil disobedience to defend their treaty-reserved rights to fish, hunt, and gather (American Friends Service Committee 1970; Cohen 1986; Wilkinson 2000). Indeed, it has been noted that the perennial battle to exercise their treaty rights has become a defining characteristic of Coast Salish identity (Harmon 1998).

The most notable outcome of these efforts was the 1974 Boldt Decision, which affirmed the tribes' right to 50 percent of the salmon fishery and made them co-managers with the state of Washington of the fishery. Today the tribes see themselves as co-managers not just of the fishery (which from a legal standpoint now includes shellfish), but also of all natural resources related to treaty-reserved rights. Over the past fifty years, many tribal governments have developed formidable natural resources management apparatuses of their own, composed of tribal members, scientists (typically nontribal), policy analysts, and attorneys. The considerable political influence of Washington State treaty tribes can be attributed not only to the 1974 Boldt Decision, but also to the economic success of many tribal casinos and other economic endeavors. As Muckleshoot tribal historian Warren KingGeorge said to me, "That's why this village has a voice. You know, it's always been there, it's just that with the help of our enterprises? Our financial success? Our voice is even louder now."

For these reasons, the tribes will be here into the future, continuing to advocate for the well-being of resources and places that matter the most to them. In that way at least, those of us who care about the fate of this place and the nonhumans who also dwell here are tied to the fates of the tribes.

The 1.7 million acre Mount Baker–Snoqualmie National Forest stretches from the lowlands of the western flanks of the Cascade Range to its crest, extending approximately 150 miles from the US-Canadian border south to the boundary of Mount Rainier National Park (Figure 6.2). The land grows magnificent coniferous trees, and since the early twentieth century, the primary focus of the Forest Service has been timber extraction. However, despite its long history of extractive industrial logging, through much of its history under federal management the national forest has also been a place to "consume" the aesthetics of wild nature. Like the

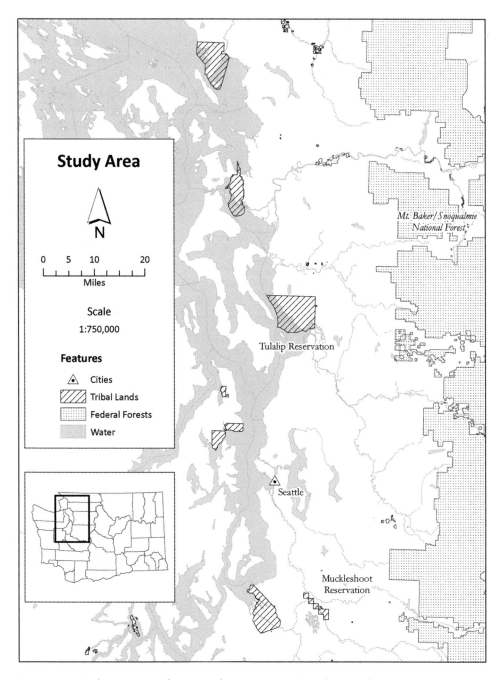

FIGURE 6.2. Study area in northwest Washington State (photo by David R. Carlson).

region of which it is a part, the Mount Baker–Snoqualmie National Forest is geologically and geographically complex, situated in the Cascade volcanic chain, which is overlain on the national forest's northern section by the North Cascades subcontinent. This portion of the national forest, in particular, is what gives Mount Baker–Snoqualmie the honor of having the most rugged, steep, and inaccessible terrain found in the national forest system.

Drawn to its dramatic landscapes and lush forests, people from the greater Seattle metropolitan area (mostly white middle- and upper-class urbanites) have long valued the forest for its recreational opportunities and a place for respite from city life (Beckey 2003; Klingle 2007). As I discuss in greater detail below, beginning in the late 1980s, continuing with the advent of the Northwest Forest Plan in 1994, and intensifying since then with the increasing probability of litigation in all but the narrowest parameters of acceptable timber harvest, industrial logging in the forest is almost nonexistent. With its proximity to the most populous area in Washington State, Mount Baker–Snoqualmie is now most valued for and managed as a place of recreation and biodiversity conservation. Each year it is estimated that more than two million people visit the national forest. The vast majority reside no more than 50 miles away and visit for the purpose of recreation (USDA Forest Service 2005). Given that the Puget Sound metropolitan area is one of the fastest-growing regions in the United States (Soper 2014), this trend is likely to intensify in the future.[3]

Fire's Allies

The district ranger's utterance at the end of the Big Huckleberry Summit regarding the need to reintroduce fire to the land reflects a broadly shared, though by no means unanimous or unequivocal, perspective. All of the tribal members and most of the Forest Service staff whom I have spoken with recognize not only the potential direct benefits of intentional burning to the flourishing of huckleberry, but also the broader ecological and cultural significance of doing so.

In terms of ecological benefits, all of the tribal members and many of the federal land managers I have spoken with say that fire cleans and invigorates the land in a way that only fire can. Along with increasing the availability of light, fire releases nutrients bound in the fuels that burn, making them available to the huckleberry plants. Although a burn may kill their tops, such an event reinvigorates the plants by encouraging the sprouting of new shoots from surviving roots. Burning may also reduce the likelihood of disease and insect infestation. Other kinds of vegetation also grow back lushly after a fire, increasing the quality and quantity of forage for deer and elk. All of these are benefits that are understood to be difficult to achieve without the help of fire. In fact, mechanical disturbance in the absence of fire may have the opposite effect. Accumulated dead wood suppresses the growth of vegetation and also makes it difficult for both human and nonhumans to move through the forest.

Russell Moses, a Tulalip tribal member who worked for the Mount Baker–Snoqualmie National Forest for seventeen years, views fire as essential for the regeneration of the forest as a whole. "I liken fire to what beavers do down here," he says. "[Fires] just completely change the whole kind of microsite. Enough that it releases a lot of woody debris." He describes how during the heyday of logging in the forest, if slash accumulated in upper-elevation landscapes and was not burned, after a few years the wood had formed such a dense, compacted mat that nothing would grow through it. He has observed that wood decomposes much more slowly in the mountains than in the lowlands, so these effects can last for a very long time.

To incorporate burning into the care of big huckleberry is to also leave open a space for the possibility of cultural revitalization—and for all that flows from it to flourish. Warren KingGeorge says that Coast Salish ancestors used fire to "clear the table" in preparation for the next generation to enjoy: "Our ancestors used the fire to cleanse ... a hillside of the old gifts when the life has been used up.... And so this plate is cleaned. Wash the plate, get it ready for the next meal."

One Forest Service wildlife biologist who has worked closely with the Tulalip on a huckleberry enhancement project in an area known as Pyrola Meadows described in an interview the connections between cultural benefits to tribal members and to the forest as a whole:

> I think it was a tool that was used traditionally ... so I think one of the concerns is just the reconnection of this tool with tribal persons.... I think that just to keep all the tools that you have available to you. If you don't use it, then you sort of lose the knowledge.... And so the mechanical means of working with the huckleberry is probably one of the things that perhaps was not as readily available in previous times as it is now. So it's being able to adapt to what tools are available now. But not lose some. Where fire may have some benefits or have some consequences in how the burn turns out that we aren't aware of with doing things mechanically.

Restoration and Risk

The collaborative projects that have formed the basis of my observations come at an interesting time and place in terms of the history of American federal-tribal politics related to natural resources management. While the political and economic power of Washington State's treaty tribes is ascending, the opposite is true in many ways for the Forest Service. This creates a situation where the power dynamics between the Forest Service and tribes are less unequal than they have been since the federal government and the ancestors of contemporary Coast Salish tribal members entered into a trust relationship under the Stevens Treaties 150 years ago. It is within this context of changing power relations that the potential for decolonizing Forest Service land management ideology and practice lies. The reintroduction of fire on the land to tend to big huckleberry in a manner reminiscent of Coast Salish ancestors would be one measure of such a shift. In this section I illuminate the broader context within which these collaborations take place, reflecting on what is at stake for both the tribes and the Forest Service when it comes to reintroducing fire to care for the huckleberries. What do those involved in these projects believe is being restored? What is at risk? And how does that shape what it is possible to accomplish?

While some might argue that tribal governments and, by extension, tribal members risk becoming bureaucratized through processes of natural resources co-management (see, for example, Nadasdy 2003), it has been my experience that tribal members who engage with outside agencies and their staff are perfectly capable of speaking the bureaucratic and scientific language of natural resources management and policy without becoming culturally unmoored. In speaking with King-George about the problematic concept of 'co-management,' for example, he told me that the tribes use the term because (as it is enshrined in the Boldt Decision) it is one that nontribal agencies are able to understand; however, *co-beneficiary* is the term that he has settled on as an alternative that is more consistent with a Coast Salish worldview:

> The term "management," it's like we're dictating where the plants are going to grow, and how much of it is going to grow, and then we determine who is going to get half of what. I think that's a terrible attitude to have, especially when these things are considered gifts. We're beneficiaries of these gifts, and that's how it should be approached.

This perspective is a reminder that while tribal sovereignty is growing in the Pacific

Northwest, it is not necessarily an end in itself. Rather, tribal sovereignty is a means to ensure the vitality of resources, which in turn support the restoration of tribal members' physical and emotional health, and the flourishing of their relationships to the ancestors and to the land, whose well-being is their responsibility (Krohn and Segrest 2010; LeCompte-Mastenbrook 2015). Like all of us, contemporary Coast Salish people are the products of histories of becoming—subject to all the human and nonhuman transformations of the world to which we are the heirs. But it seems that the truth of this informs the present to a much greater extent in Coast Salish communities than in the settler communities that co-inhabit this place. Ancestral signatories to the Stevens Treaties and the deep-time ancestors who tended the land never seem to be far from the minds of their descendants.

For many Coast Salish tribal members, with that recognition, along with the awareness that their ancestors fought to preserve their descendants' rights to these resources, comes a responsibility to care for them. What is at stake now is the sense that continued cultural erosion could conceivably put an end not only to exercising their sovereign rights as enshrined in federal law, but also recognizing the responsibilities inherent in these relationships. While the physical health benefits associated with the consumption of huckleberries are certainly important to tribal members, creating opportunities to harvest huckleberries is perceived as ameliorating the risk of cultural loss through the restoration of reciprocal relationships.

The Mount Baker–Snoqualmie National Forest, where the huckleberry enhancement projects are taking place, falls within the purview of those who are required to manage the roughly 24 million acres of public land in Washington, Oregon, and northern California in accord with the goals and rationale of the Northwest Forest Plan, which was signed into law in 1994 after years of unarguably excessive clearcut logging, primarily on the west side of the Cascade Range (Dietrich 1992). The plan's stated vision is to restore and protect the health of late successional forest habitats relied upon by two federally listed threatened species (the northern spotted owl and the marbled murrelet), while at the same time ensuring the economic health of timber-dependent communities and a continuing stream of social benefits to the broader public (USDA and USDI 1994). This vision—and the ecosystem management approach that informs the Northwest Forest Plan's standards and guidelines—essentially relies on a triple bottom line sustainability model that, in theory, is attendant to balancing social, environmental, and economic considerations in forest management planning and implementation.

To this end, public forests within the Northwest Forest Plan area were divided into seven land allocation types based on different goals to achieve "desired future conditions," along with guidelines outlining the means to achieve those goals (Table 6.1). For instance, 30 percent of the forested lands managed by the Forest Service and Bureau of Land Management was set aside as "late-successional reserves" to be managed for old-growth structural conditions favored by the northern spotted owl and marbled murrelet. The majority of timber was to be cut from the 16 percent designated as "matrix" lands, which were in essence the areas that had not been allocated to one of the seven designated types. In addition to matrix lands, the plan allows for limited timber harvesting in late-successional reserves if the harvest will accelerate old-growth conditions[4] or help prevent catastrophic fires.

In the current political and economic atmosphere, and with the Northwest Forest Plan in place, it is quite difficult to accomplish anything in the Mount Baker–Snoqualmie National Forest. As with the federal budget

TABLE 6.1. Northwest Forest Plan land allocations

Land allocation	Description/strategy	Allowable harvest	Plan area	MBSNF
Congressionally reserved areas	Wilderness, national monuments, parks, wildlife areas, etc.	No timber harvest	30%	47%
Late-successional reserves	Maintenance of functional old-growth ecosystems	West side forests: restoration thinning in stands < 80 yr	30%	36%
Adaptive management areas	Experimental forests designed to integrate and achieve ecological, economic, and other social and community objectives. Ten total in plan area.	Timber harvest/salvage varies depending on focus of AMA.	6%	1%
Managed late-successional areas	Protection of rare or threatened species. Includes mapped owl pair areas or unmapped habitat buffers.	Some timber harvest	1%	*
Administratively withdrawn	Recreational areas, viewsheds, backcountry, etc.	No timber harvest	6%	5%
Riparian reserves	Protection of water quality and aquatic species. Includes all riparian areas whether fish-bearing or not.	None within riparian buffer	11%	**
Matrix	Lands not included in other categories. Expectation is that majority of timber harvest will occur on matrix lands.	Timber harvest allowed; must leave 15% green trees/acre.	16%	9%
Other (not classified)				1%

* Included in late-successional reserves.
** Included in matrix.

in general, the Forest Service budget has undergone serious cuts since policies promoting neoliberalization were put in place and began to be implemented during the Reagan administration. Funds for all of the national forests are allocated based on projected annual timber harvests, which have shrunk in part as a result of the Northwest Forest Plan, but also as a result of the difficulty of doing any cutting on the forest at all. I have been told repeatedly that this is due to the ever-present likelihood of any timber sale being challenged by environmental NGOs. In one interview I conducted, the vegetation manager for the Mount Baker–Snoqualmie National Forest explained the broader repercussions, and thus risks, of potential lawsuits from his perspective:

> It just . . . takes a tremendous amount of time and energy away from anything else you do, and therefore money as well. But I think the biggest risk is that, if you end up losing, then you're basically set back in everything you've done up to that point. That effort is basically for nothing. So if you are dealing with limited resources and people available, and money, you want to put them where you will have the highest probability of success. So for example, rather than regenerating matrix stands in old-growth

now, we thin LSRs [late-stage reserves], which was a minor component of the Northwest Forest Plan. It was recognized as a possibility, but on some forests that is the primary activity they do in timber right now. Because there's a much lower risk and a higher probability. It's not so much a risk of failure as the probability of success.

According to University of Washington forest ecologist Jerry Franklin, one of the main architects of the Northwest Forest Plan, the plan as adopted was not what he and others originally designed. The plan as envisioned was intended to be flexible and adaptive, and subject to revision based on what was learned about the forest through the monitoring protocols that were supposed to be an integral part of the plan's standards and guidelines. For instance, the distribution and extent of land allocations were designed to vary across the landscape over time, but "everyone else wanted to lock [them] in place—land managers, the timber industry, enviros."

From Franklin's perspective, the main parties involved wanted certainty from a system that is inevitably dynamic and subject to change—something that the main stakeholders did not want to acknowledge. In addition, because the Forest Service wanted to avoid the possibility of conflict—particularly litigation—and get on with management, they continued to concede more and more to what Franklin calls "the grassroots enviros." The Forest Service vegetation manager I interviewed described this in the context of how past litigation on one forest influenced the decision-making processes on all of the forests mandated to follow the Northwest Forest Plan. "People become much more circumspect due to litigation in the past," he explained. "[It] not only sets you back, but also guides your thinking for what is an acceptable project."

After the Northwest Forest Plan was signed into law, salvage logging, the logging of old-growth stands, and, eventually, logging of all mature trees in the matrix were taken off the table—all of which, according to Franklin, happened without any public process at all.

Franklin says that in the case of the Forest Service, the fear of litigation can be traced to the structure of the agency's budget. Federal forests and the people who work for them are evaluated and rewarded based on their accomplishments, not the risks they might be willing to take. As the forest vegetation manager explained to me, forests get no credit for projects that are not completed. A timber sale that is laid out but then delayed or dropped is not considered an accomplishment. The response of the Forest Service has been to reduce management, with the exception of tree thinning (a practice that cannot continue indefinitely). Thus while the Northwest Forest Plan projected that average annual timber harvest would drop from an estimated 1.5 billion board feet per year to about .5 billion under the new management strategies, in practice, timber harvests on many national forests are increasingly rare, including Mount Baker–Snoqualmie. For example, in the early 1980s about 385 million board feet was being harvested every year from the Mount Baker–Snoqualmie. The current annual harvest budget is approximately 17 million board feet. Most of this harvest is for the purpose of improving wildlife habitat through the creation of forest openings, or accelerating old-growth forest characteristics through thinning (though the effectiveness of this strategy is debated).

An abstract understanding of the effects of current policy and plans and annual allowable timber harvests in the Mount Baker–Snoqualmie does little to shed light on how the current status of the Forest Service affects staff members, who often seem stressed and embattled (McLain 2000). Franklin characterizes Forest Service conservatism as being related to increasing external scrutiny, the structure of the agency's budget, and the general inability of large agencies to respond adaptively to change. Processes of neoliberalization and attendant shrinking budgets, however, are also systemic factors affecting the Forest Service's sensitivity to public scrutiny, and thus also how it functions.

Clearly, one of the highest priorities for managers of the Mount Baker–Snoqualmie and other national forests within the Northwest Forest Plan area is to repair the damage done by excessive logging in the past—to restore old-growth conditions and suitable habitat for endangered species—despite having limited funds to do so. The success of these efforts is one measure used to evaluate the performance and accomplishments of Mount Baker–Snoqualmie staff. These efforts may also be a way for the Forest Service to regain legitimacy in the eyes of the broader public. As other forest studies scholars have observed, projects involving community forestry and collaborative management are particularly appealing, allowing land management agencies to not only satisfy the expectations of the governing bodies that oversee them, but to also extend their access to revenue and perhaps restore a sense of trust in stewardship of public lands (Charnley and Poe 2007; McCarthy 2005, 2006). Considered in this context, while the threat of litigation is on the one hand perceived as an impediment to certain kinds of disturbance as envisioned in the initial Northwest Forest Plan, it is also a factor motivating broader public involvement when it comes to Forest Service decision-making. As several Forest Service staff explained to me, public engagement in decision-making processes is one way to mitigate some of the risk of litigation. As the vegetation manager for the Mount Baker–Snoqualmie National Forest noted:

> Looking at how we do things now, the whole concept of collaboration with all the different groups in a lot of ways is designed and intended to reduce some of the risk, and to allow us to develop projects that are beneficial. So that we can all discuss the benefits and the risks of that project. And that maybe we can do things that we the Forest Service wouldn't have thought of doing ourselves. Because being fairly risk averse, we might have shied away from it. But you might get greater support on a project [identified by] the broader public, and once you have that, then you know, we would see that as reducing the risk.

I am not trying to imply that Forest Service collaboration with the broader public is simply a means of gaining public approval for its own internally predefined goals. Several Forest Service staff members described increased engagement with the broader public and with the tribes as having been deeply meaningful to them personally, and also to having improved the way the Forest Service functions. As one staff member noted, understanding other people's points of view about what they think the Forest Service should be doing, and what it is currently doing, is important. Speaking specifically of his involvement with the huckleberry enhancement projects, the Mount Baker–Snoqualmie vegetation manager described his understanding of some of the benefits to the Forest Service of that collaboration:

> Hearing those other views—like maybe some of our regulations are there to be just regulations and they don't provide any real benefit to anyone.... It's just part of the whole broadening our perspectives, and kind of stepping back from our jobs—being within our jobs and also stepping back far enough that we can really take a more objective view, whereas we tend to get focused on our day to day "this is how we've always done it," so we just keep doing it that way. A real benefit to me was just that perspec-

tive of stepping back and looking at, well, why do we do it this way? I think it's good that people question why we do this, because if I can't give an answer, then maybe we shouldn't be doing it that way. . . . It's not just this one committee. The whole concept [of collaboration] applies at a broader scale.

In addition to a general trend toward greater public involvement in making decisions about management of Forest Service lands, the agency, as a federal entity, also has a trust responsibility to federally recognized tribes. Because the tribes are legally wards of the state, the federal government plays the paternalistic role of trustee of their treaty-reserved rights to resources. But within this specific context I would argue that the trust relationship is less a factor than the capacity of tribes to effectively litigate on the one hand, and to bring much needed resources to the agency on the other. As one of the Forest Service staffers who has been closely involved with these projects explained:

> I think that the Forest Service is really stressed. Perhaps even more so when resources are tight, the role of partnerships becomes very attractive. . . . The tribes [are also] taking a stronger leadership role in how they will approach the Forest [Service] with their interests. And they have the resources now to have additional staff to support [these] interactions.

The trust relationship does, however, provide the legal and ethical foundation for the argument that tribes often make to federal land management agencies: that they are not just another "constituency" equal to other public or stakeholder groups. I return to this point below, but it is worth mentioning here given that the different public groups that the Forest Service is beginning to engage with more broadly are heterogeneous in terms of political influence and the extent to which their ideas about forest disturbance align with those of the tribes.

It is within this broader context that, at least from the perspectives of human actors involved in controversies over huckleberries and fire's roles in their well-being, treaty tribes now, more than at any time in the history of Euro-American colonialism, are increasingly empowered to shape the discourse and practice of natural resources co-management. The treaty tribes of the Pacific Northwest are the ones who have raised the status of big huckleberry as a matter of concern, and in doing so they have encouraged other humans (i.e., Forest Service staff) to begin to act on behalf of these plants' well-being.

Of course, a prescribed fire that burns out of control—which would be considered a failure on the part of Forest Service staff—could potentially affect the public's perception of the Forest Service's legitimacy and efficacy, and therefore is a risk. The primary role of the eight-person full-time, year-round Mount Baker–Snoqualmie fire staff is to suppress fires, not to start them—ostensibly reducing the risk to resources, life, and property. In a typical year, close to fifty wildfires ignite in the Mount Baker–Snoqualmie, approximately half of which are anthropogenic and half caused by lightning. The response of the fire staff is predicated on their predictions of fire behavior, its potential negative ecological and economic impacts, and the risk it poses to fire crews. The potential benefits of fire are not taken into account in this calculus.

Huckleberry habitat, however, has been most negatively impacted by the prioritization of certain species and habitat types over others in the Northwest Forest Plan—particularly old-growth forests over early-succession forests ones. Ironically, as fire suppression was increasingly integrated into national forest management practices, clearcut logging to some extent took the place of fire as an agent of disturbance that benefited big huckleberry habitat (Anzinger 2002). Under the Northwest Forest Plan, both forms of disturbance are suppressed, and their combined effects have led to the loss of huckleberry habitat becoming a major concern. As shown in Table 6.1, the proportion of land allocations in the Mount Baker–Snoqualmie where disturbance is allowed under the Northwest Forest Plan is quite small. Nearly half of Mount Baker–Snoqualmie lands are designated wilderness, over a third are allocated to late successional reserve or managed late successional areas, and about 10 percent are either matrix or adaptive management areas where timber harvesting and other anthropogenic forest disturbance is theoretically less constrained. This minimal disturbance negatively affects not just big huckleberry, but many kinds of culturally important plants, which often thrive on disturbance (Mah 2000). Most importantly for big huckleberry, however, despite estimates that there is a moderate to high probability of big huckleberry occurring on about one-third of Mount Baker–Snoqualmie managed lands, in reality less than 4 percent of this potential habitat is included in the so-called matrix lands where anthropogenic disturbance would be ostensibly less controversial (Lesher et al. 2015).

Thus the ability to actually act on tribal matters of concern (on the part of tribes and Forest Service staff collaborators) is substantially constrained by forest policy as it is currently practiced. These concerns are made explicit in a 2010 report documenting the effects of the Northwest Forest Plan, fifteen years after its implementation, on federally recognized tribes. Nearly half of the twenty-two federally recognized tribes in Oregon and Washington whose members were interviewed for the study felt that the Northwest Forest Plan had negatively impacted treaty-reserved gathering rights, either through effects on the resources themselves or loss of access due to road closures. Tribal members who participated in the study specifically mentioned cedar bark and big huckleberry as resources of concern (Lynn and MacKendrick 2011).

Although the Northwest Forest Plan recognizes potential conflicts between implementation and its effects on gathering rights, and is claimed to provide a higher level of protection for trust resources than the previous forest plans that it amends, the plan's language sets up a false dichotomy between its stated goals and the relationships of local tribal members with the forest. The plan explicitly states that treaty-reserved rights will not be restricted unless it is determined that a restriction is necessary for the preservation of a particular species at issue, that its preservation cannot be accomplished without the restriction, and the restriction does not discriminate against Native American activities (USDA and USDI 1994).

Such an understanding does little, however, to acknowledge the impacts of the *absence* of Native Americans' active presence on the health of the forest. This contradiction is also highlighted in a report examining the effects of the Northwest Forest Plan on treaty tribes ten years after its implementation. The study highlights many of the challenges and contradictions inherent in the conceptualization and implementation of the plan, where tradeoffs between protection of and access to some resources and practices over others are not desirable but seem inevitable.

While the ten-year Northwest Forest Plan tribal monitoring report gives the impression that forest health and the condition of aquatic and riparian habitat had improved, about half of those interviewed stated that changes related to the plan had made access to resources and exercise of treaty rights worse. Another common theme was the generally negative effect of the Northwest Forest Plan on project implementation related to tribal trust resources (Stuart and Martine 2005).

It seems that huckleberry enhancement projects are most likely to gain the financial and staff support required to implement them when they are small and mesh well with other goals and objectives of the Northwest Forest Plan. So-called synergies are much easier to justify, both ecologically and economically. For example, there is a sense that thinning projects emphasizing benefits to wildlife (as opposed to emphasizing benefits to humans) are less likely to meet with resistance from environmental NGOs, and more likely to receive federal funding, particularly if the projects are small. "We are, in a sense, the wildlife," Russell Moses once told me, speaking of Pyrola Meadows, one of the huckleberry enhancement projects that the Tulalip tribe has implemented with the Forest Service. This project is being conducted on only 32 hectares or so. Each year for three years, Moses and Jason Gobin, a Tulalip tribal member and the head of the tribe's Forestry Department, brought young people to the site to help thin a few acres of forest canopy and move the brush to where it could be chipped or, ideally, burned. "If I went up to treat 80 acres at once," says Gobin, "oh no, that would never happen. But if we do 3 acres, wait another year, do 3 acres, 10 acres, 4 acres, you know we can just sit there below the radar of everybody else and get the work done."

Another unintended consequence of changes in disturbance regimes is a certain amount of de-skilling of the Mount Baker–Snoqualmie fire staff because of the shift of orientation from one that included both prescribed burning and fire suppression, to one almost entirely devoted to fire suppression. For example, prescribed burning of logging debris was an integral part of clearcut logging, and the national forest fire crews were typically responsible for this. Brush and debris, or BD, crews were an integral part of the process, and they developed substantial knowledge about working with fire in the particular environments that they were burning. A Forest Service staff member familiar with this practice made this point when discussing perceptions about potential "economies of scale" that might come with burning huckleberry meadows, as opposed to treating them strictly with mechanical methods:

> It's also an economy of familiarity. If we're doing the same things over and . . . again? We're gonna get much better at it. And the box we're gonna walk out and start to assess is gonna be much smaller /'cause we're gonna know where we're gonna go with it, what we're gonna try, and when we're gonna try it. [When it's] a new thing, it takes a lot more time than if it's the twentieth one we've done in a five-year period.

Not surprisingly, under these conditions it is actually quite difficult to accomplish any projects that would benefit the production of big huckleberry, whether fire is involved or not. Thus far, huckleberry enhancement in the Mount Baker–Snoqualmie National Forest has not been considered a high enough priority to explicitly include in any forest-wide planning documents or the annual budget, so Forest Service staff and tribal members look for grants or other funding sources, typically creating yet more delays.

The National Environmental Policy Act (NEPA) mandates that an environmental assessment be conducted to evaluate the potential impacts to natural and cultural resources of every action taken on the forest. Even for very small projects, it costs many thousands of dollars for Forest Service botanists, ecologists, wildlife biologists, hydrologists, silviculturists, fire experts, and so on to conduct their analyses, make recommendations, and propose adjustments to the initial plan. Tribal members sometimes join Forest Service staff for the interdisciplinary field trips that are an integral part of the

NEPA process, and are included in the discussions about what the plan should involve, but that is not a given. I once asked Warren KingGeorge about this after an interdisciplinary NEPA field trip where I'd noticed that he was mostly ignored. He said it didn't bother him. "I felt that the trip wasn't about me," he said. "It wasn't about the tribe. The trip was about the Forest Service getting familiar, so instead of just looking at their tables, and their data, and their pies, and their ratios, this allowed them—these players in the management game—to add some context. And I was okay with it. I mean that's part of the game."

Jason Gobin says he feels that the NEPA process is one of those areas where Forest Service bureaucracy interferes with the urgent task of just getting out on the land and getting the work done. He has been instrumental in writing the "prescription," as they are called in Forest Service parlance, for treatment at Pyrola Meadows, but has little patience for the minutiae and bickering that can take place in the process: "Rather than doing, people just spend a lot of time thinking about doing it. And everybody's coming to the same conclusion, but they're coming from a different direction. You know, and it's just like, you guys are wasting time here debating this. Let's just do it! What's the harm in doing it?"

The possibility exists that the fire expert involved in the interdisciplinary process can preclude the use of fire in the project at any point in the planning or implementation process. Burn plans are complex documents that require even more time to develop, and thus cost more money, than a project not involving prescribed fire. Even a burn plan for a very small project, such as the one that was developed for Pyrola Meadows, can take several days to develop, not including time spent in the field. The Mount Baker–Snoqualmie fire staff officer explained to me that burn plans are highly prescriptive and designed to minimize risk and take every unanticipated event into consideration. "There have been fires caused by prescribed fires," he says. "Burn plans are intended to lower the risk of that."

The assistant fire staff officer added that the planning and the burning alone are expensive, but then there is also the question of where the funds will come from to implement a contingency plan should the fire escape:

> That's the risk. . . . Okay, we have the money to pay for this project, but then what about the 'what if'? Then who's gonna pay for that bill? Who's gonna be responsible for that bill?

It is thus not too surprising that unless there are good relations with stakeholders and a strong desire on the part of the Forest Service to support a burn, it is easy to make excuses for not including prescribed fire in a project plan.

This is particularly true when there is no unequivocal consensus about the benefits of prescribed fire to huckleberry productivity. Although, as previously mentioned, tribal members and many Forest Service staff who have worked with them perceive ecological, social, and cultural benefits to burning, other staff members are skeptical about reintroducing fire, particularly given the associated risks. One of the fire management officers who works on the south end of the Mount Baker–Snoqualmie agrees that there is cause for concern that huckleberry meadows are filling in now that cutting in these upper-elevation landscapes is no longer permitted, and that something needs to be done about it. But he does not necessarily feel that prescribed fire is the solution because there are "too many variables" that affect the way fire behaves. He expressed doubt that burning would actually yield the results that people are hoping for. His sense is that fires in these particular landscapes tend to be very severe and kill all the vegetation (including huckleberries). From his observations, it can take up to twenty years for a huckleberry meadow to recover from one of these burns, if it recovers at all. He suggested that in the past, Native people who burned the forest did so because that was the only tool available, and he suggested that mechanical thinning is a better approach that minimizes risk. Thinning 5 to 15 acres would be manageable from the standpoint of available labor and would also be effective in terms of opening up spaces for huckleberry production and elk habitat. From his perspective, projects of this scale would also be less likely to meet with resistance or concern from environmental groups.

This was also the perspective of Jan Henderson and Robin Lesher, recently retired forest ecologists who devoted their careers to understanding the structure, composition, and processes of forested communities in the Mount Baker–Snoqualmie. Between the two of them, they have more than sixty years of experience studying forest ecology there. Both agree that what big huckleberry needs is light, and that the plants are not particular about how that actually happens. They do wonder, however, if fire might be beneficial over some longer interval, such as the estimated natural fire-return interval of 150 to 500 years described by forest and fire ecologists for upper-elevation huckleberry habitats (Agee 1993; Franklin and Dyrness 1988; Hemstrom and Franklin 1982). They say that such infrequent fire may be necessary for nutrient cycling and moderating insects and disease.

What actually matters, though, about the nature and extent of forest disturbance caused by historical burning by local tribes, and to whom? One reason that questions about Native peoples' use of fire have come to matter is exemplified in books like Thomas Vale's 2002 edited volume, *Fire, Native Peoples and the Natural Landscape*. Focusing on case studies from across the United States, the contributors attempted to settle a long-standing (and ongoing) debate about whether the landscape observed by the first Euro-Americans to arrive was a "pristine" or "humanized" landscape.[5] In his introduction to the volume, Vale (2002, 7) characterizes the argument as often devolving to a polarized binary between natural scientists arguing for the pristine model, and social scientists (particularly anthropologists) arguing for the humanized model, with much of the dialogue dominated by "arm–waving, [and] careless generalizations" (6). On one level, the debate as characterized by Vale revolves around the extent to which precolonial Native American fire regimes affected the structure and composition of North American forests at a landscape scale. From Vale's (2002, 8) perspective, what is most at stake in these debates are questions about "the legitimacy of hands-off protection [of nonhuman nature] as a policy" versus "the need for manipulating nature as a universal policy for the management of nature preserves." Paleoecologists Cathy Whitlock and Margaret Knox (2002) extend this argument in their chapter on fire regimes in the Pacific Northwest, characterizing the argument for long-term anthropogenic influence on the landscape as being driven by economic interests—particularly salvage logging.

Over the course of my fieldwork, I too have observed and been troubled by what can only be described as the outright appropriation of Indigenous histories of resource management to serve the kind of political-economic agenda described by Whitlock and Knox (2002). What both the contributors to Vale's edited volume and those who would appropriate Indigenous fire histories to suit their own political economic interests fail to acknowledge, however, is the well-being of the actual living descendants of the people around whom their arguments revolve.

In the first place, following John Locke's notion of property as articulated in his "Two Treatises of Government" (cited in Arneil 1996), by demonstrating that they managed resources by burning, Indigenous people can argue that they too "improved the land," potentially bolstering claims to rights to land and resources—albeit by appealing to a Western political-economic framework. Similarly, the inability to continue traditional land management practices, or even utilize traditional resources that rely on disturbance to be viable, can be seen as a threat to tribal sovereignty (Norgaard 2014). Indigenous fire, then, is not simply a chemical reaction or a component of biological processes that can be separated from human meanings (Fowler 2013). Indigenous burning is also a "cultural keystone process" (Turner and Garibaldi 2004; Norgaard 2014), the absence of which will inevitably have cascading social effects. It is not simply another agent of disturbance: within that agency is attached a suite of social relations. Jason Gobin and I had the following dialogue about these ideas:

Jason: And so, there was a lot of that knowledge that was lost. Especially way back when with the disease, and then you had the schools and everything. So there's a lot of the knowledge that basically was just gone. You know, the people who were specialized in doing that thing were just gone, and weren't able to keep going and teaching everybody. So I think it's important for us to do the work, to try and just get up there, and reconnect with that work, and reconnect with the land up there.
Joyce: Yeah. And as that happens, you know, that remembering happens, you observe the way fire behaves, and the way it feels at the time of—you know, when a fire is good. Like this feels like a good burn.
Jason: Yeah.
Joyce: And then it starts to become a part of you.
Jason: Yeah. And it's just . . . you just have to observe it. But the problem with fire is, that we have only limited opportunity to do it.

Conclusion

With seemingly little to go on in regard to understanding the ecological effects of Indigenous burning practices on huckleberries specifically and ecosystem composition and structure more generally, forest ecologists may underestimate the impacts of anthropogenic burning on forest composition and development. The integration of "best available science" into NEPA analysis is a basic requirement of that process. If little is known about Indigenous burning in the past, then it is difficult to incorporate it as best available science in the present. And if some version of Indigenous land management is not incorporated into the planning process, that absence becomes a self-perpetuating cycle. For example, there is little information about the effects of small-scale, repeated burns on huckleberries and their habitats. These types of burns are not only potentially important for huckleberry production, but also ecologically in terms of species composition more generally. The point is that the historical role of small-scale fires, along with their potential effects both past and present, and ultimately what huckleberries might need or want are open questions. The significance of small-scale fires and their ecological effects should not simply be dismissed out of hand because of knowledge gaps in the present.

Just as important as the ecological implications are the social implications of our lack of knowledge. Also important is recognizing and acknowledging that Indigenous people were here, and that through their practices of burning and tending, they developed their own kinds of attachments and relations to the land that shaped their sense of themselves; in itself this can be seen as a form of restoration. It might even be considered a kind of restorative justice. On the other hand, lack of recognition of Indigenous fire ecologies by ecologists and land managers can be seen as yet another instance of neocolonial "epistemological imperialism" (Perley and Heatherington 2011).

It is time to begin the work of collectively constructing a more inclusive understanding of both past and present human-fire relations in westside forests in the Cascade Range. Part of that is recognizing that these forests and their fire ecologies have involved people from the time they developed their current patterns of process, structure, and composition—some seven thousand years ago (Brubaker 1988; Dunwiddie 1986; Gavin et al. 2003; Leopold and Boyd 1999; Lertzman et al. 2002). Given that we find ourselves in a time when all fires in the Pacific Northwest have at least some anthropogenic influences, it is an especially salient time to do so.

Notes

1. I recognize that some readers will disagree with my use of the term postcolonial to refer to a still-colonized landscape. I find the term useful as conceptualized in postcolonial studies that highlight the ongoing effects of historical events as well as contemporary colonial practices— particularly settler colonialism.
2. Along with the Muckleshoot Indian Tribe and the Tulalip Tribes, the Lummi, Nooksack, Samish, Upper Skagit, Sauk-Suiattle, Stillaguamish, and Puyallup tribes also claim portions of the forest as part of their ceded territories, though not all of these tribes have treaty-reserved rights.
3. See Breslow 2011 for an overview of the literature on relationships between urbanity and the politics of environmental preservation and conservation.
4. Forest thinning theoretically accelerates old-growth conditions by more quickly moving forests through the canopy closure and stem exclusion phases of stand development.
5. See Denevan 2011 for an overview of the history and contours of the debate.

References Cited

Agee, James K. 1993. *Fire Ecology of Pacific Northwest Forests*. Washington, DC: Island.

American Friends Service Committee. 1970. *Uncommon Controversy: Fishing Rights of the Muckleshoot, Puyallup, and Nisqually Indians*. Seattle and London: University of Washington Press.

Anderson, M. Kat, and Jeffrey Rosenthal. 2015. "An Ethnobiological Approach to Reconstructing Indigenous Fire Regimes in the Foothill Chaparral of the Western Sierra Nevada." *Journal of Ethnobiology* 35 (1): 4–36.

Anzinger, Dawn. 2002. "Big Huckleberry (*Vaccinium membranaceum* Dougl.) Ecology and Forest Succession, Mount Hood National Forest and Warm Springs Indian Reservation, Oregon." Master's thesis, Oregon State University, Corvallis.

Arneil, Barbara. 1996. "The Wild Indian's Venison: Locke's Theory of Property and English Colonialism in America." *Political Studies* 44: 60–74.

Ballard, Arthur C. 1929. "Mythology of Southern Puget Sound." *University of Washington Publications in Anthropology* 3 (2): 31–150.

Barrett, Stephen, and Stephen Arno. 1999. "Indian Fires in the Northern Rockies." In *Indians, Fire and the Land in the Pacific Northwest*, edited by Robert Boyd, 50–64. Corvallis: Oregon State University Press.

Beckey, Fred. 2003. *Range of Glaciers: The Exploration and Survey of the Northern Cascade Range*. Portland: Oregon Historical Society Press.

Bowman, David M. J. S., Jennifer Balch, Paulo Artaxo, William J. Bond, Mark A. Cochrane, Carla M. D'antonio, and Ruth Defries. 2011. "The Human Dimension of Fire Regimes on Earth." *Journal of Biogeography* 38 (12): 2223–2236.

Boyd, Robert T. 1999. *The Coming of the Spirit of Pestilence : Introduced Infectious Diseases and Population Decline among Northwest Coast Indians, 1774–1874.* Vancouver, BC, and Seattle: University of British Columbia Press and University of Washington Press.

Breslow, Sara Jo. 2011. *Salmon Habitat Restoration, Farmland Preservation and Environmental Drama in the Skagit River Valley.* Seattle: University of Washington Press.

Brubaker, Linda B. 1988. "Vegetation History and Anticipating Future Climate Change." In *Ecosystem Management for Parks and Wilderness*, edited by James K. Agee and Darryll R. Johnson, 41–62. Seattle: University of Washington Press.

Charnley, Susan, and Melissa R. Poe. 2007. "Community Forestry in Theory and Practice: Where Are We Now?" *Annual Review of Anthropology* 36 (1): 301–336.

Clark, Nigel. 2011. *Inhuman Nature: Sociable Life on a Dynamic Planet.* Los Angeles: Sage.

Cohen, Fay. 1986. *Treaties on Trial: The Continuing Controversy over Northwest Indian Fishing Rights.* A report prepared for the American Friends Service Committee. Seattle: University of Washington Press.

Denevan, William M. 2011. "The Pristine Myth Revisited." *Geographical Review* 101 (4): 576–591.

Deur, Douglas. 2002. "Huckleberry Mountain Traditional Use Study." Final Report. National Park Service and USDA Forest Service, Rogue River National Forest, WA. http://soda.sou.edu/Data/Library1/030212b1.pdf.

Dietrich, William. 1992. *The Final Forest : Big Trees, Forks, and the Pacific Northwest.* Seattle : University of Washington Press.

Dunwiddie, Peter W. 1986. "A 6000-Year Record of Forest History on Mount Rainier, Washington." *Ecology* 67 (1): 58–68.

Fisher, Andrew H. 1997. "The 1932 Handshake Agreement: Yakama Indian Treaty Rights and Forest Service Policy in the Pacific Northwest." *Western Historical Quarterly* 28 (Summer): 187–217.

Fowler, Cynthia. 2013. *Ignition Stories: Indigenous Fire Ecology in the Indo-Australian Monsoon Zone.* Durham, NC: Carolina Academic Press.

Franklin, Jerry F., and C. T. Dyrness. 1988. *Natural Vegetation of Oregon and Washington.* Corvallis: Oregon State University Press.

French, David. 1999. "Aboriginal Control of Huckleberry Yield in the Northwest." In *Indians, Fire and the Land in the Pacific Northwest*, edited by Robert Boyd, 31–35. Corvallis: Oregon State University Press.

Gavin, Daniel G., Linda B. Brubaker, and Kenneth P. Lertzman. 2003. "Holocene Fire History of a Coastal Temperate Rainforest Based on Soil Charcoal Radiocarbon Dates." 84 (1): 186–201.

Harmon, Alexandra. 1998. *Indians in the Making: Ethnic Relations and Indian Identities around Puget Sound.* Seattle: University of Washington Press.

Hemstrom, Miles A., and Jerry F. Franklin. 1982. "Fire and Other Disturbances of the Forests in Mount Rainier National Park." *Quaternary Research* 18: 32–51.

Hunn, Eugene S., and James Selam. 1990. *Nch'i-Wana: "The Big River"—Mid-Columbia Indians and Their Land.* Seattle: University of Washington Press.

Keefer, Michael. 2007. "The Kootenay Huckleberry Case Study." Paper presented to the Thirtieth Annual Meeting of the Society of Ethnobiology, University of California, Berkeley, March 29–31.

Klingle, Matthew. 2007. *Emerald City: An Environmental History of Seattle.* New Haven: Yale University Press.

Krohn, Elise, and Valerie Segrest. 2010. *Feeding the People, Feeding the Spirit: Revitalizing Northwest Coastal Indian Food Culture*. Bellingham, WA: Northwest Indian College.

LeCompte-Mastenbrook, Joyce. 2015. "Restoring Coast Salish Foods and Landscapes: A More-than-Human Politics of Place, History and Becoming." PhD dissertation, University of Washington, Seattle.

Leopold, Estella, and Robert Boyd. 1999. "An Ecological History of Old Prairie Areas in Southwestern Washington." In *Indians, Fire and the Land*, edited by Robert Boyd, 139–160. Corvallis: Oregon State University Press.

Lepofsky, Dana, Douglas Hallett, Ken Lertzman, Rolf Mathewes, Albert [Sonny] McHalsie, and Kevin Washbrook. 2005. "Documenting Pre-contact Plant Management on the Northwest Coast: An Example of Prescribed Burning in the Central and Upper Fraser Valley, British Columbia." In *Keeping It Living: Traditions of Plant Use and Cultivation on the Northwest Coast of North America*, edited by Douglas Deur and Nancy Turner, 218–239. Vancouver: University of British Columbia Press.

Lertzman, Ken P., Daniel G. Gavin, Douglas J. Hallett, Linda B. Brubaker, Dana S. Lepofsky, and Rolf W. Mathewes. 2002. "Long-Term Fire Regime Estimated from Soil Charcoal in Coastal Temperate Rainforests." *Conservation Ecology* 6 (2): 5. http://www.consecol.org/vol6/iss2/art5/.

Lesher, Robin D., Jan A. Henderson, and Chris Ringo. 2015. "Distribution of Big Huckleberry in the Mount Baker–Snoqualmie National Forest." In *Distribution and Recreational Harvest of Mountain Huckleberry "Swədaʔχ" in the Mount Baker–Snoqualmie National Forest*, edited by Libby Halpin Nelson. Marysville, WA: Tulalip Tribes Natural Resources Department.

Lynn, Kathy, and Katie MacKendrick. 2011. "Strengthening the Federal-Tribal Relationship: A Report on Monitoring Consultation under the Northwest Forest Plan in Oregon and Washington." Report prepared for the Tribal Relations Office of the USDA Forest Service Region 6 and the US Department of the Interior Bureau of Land Management, Eugene, OR.

Mack, Cheryl A. 2003. "A Burning Issue: American Indian Fire Use on the Mount Rainier Forest Reserve." *Fire Management Today* 63 (2): 20–24.

Mack, Cheryl A., and Rick H. Mcclure. 2001. "Vaccinium Processing in the Washington Cascades." *Journal of Ethnobiology* 22: 35–60.

Mah, Shirley. 2000. "Relationship Between Vital Attributes of Ktunaxa Plants and Natural Disturbance Regimes in Southeastern British Columbia." Master's thesis, University of British Columbia, Vancouver.

Main-Johnson, Leslie. 1999. "Aboriginal Burning for Vegetation Management in Northwest British Columbia." In *Indians, Fire and the Land in the Pacific Northwest*, edited by Robert Boyd. Corvallis: Oregon State University Press.

McCarthy, James. 2005. "Devolution in the Woods: Community Forestry as Hybrid Neoliberalism." *Environment and Planning A* 37 (6): 995–1014.

———. 2006. "Neoliberalism and the Politics of Alternatives: Community Forestry in British Columbia and the United States." *Annals of the Association of American Geographers* 96 (1): 84–104.

McLain, Rebecca Jean. 2000. "Controlling the Forest Understory: Wild Mushroom Politics in Central Oregon." PhD dissertation, University of Washington, Seattle.

Nadasdy, Paul. 2003. *Hunters and Bureaucrats: Power, Knowledge, and Aboriginal State Relations in the Southwest Yukon*. Vancouver, Toronto: UBC.

Naxaxalhts'i [McHalsie], Albert (Sonny). 2007. "We Have to Take Care of Everything That Belongs to Us." In *Be of Good Mind: Essays on the Coast Salish*, edited by Bruce Granville Miller,

82–130. Vancouver: University of British Columbia Press.

Norgaard, Kari Marie. 2014. "The Politics of Fire and the Social Impacts of Fire Exclusion on the Klamath." *Humboldt Journal of Social Relations* 36: 77–101.

Perley, Bernard C., and Tracey Heatherington. 2011. "Epistemic Imperialism and the Anthropology of Mischief: Techno-Science, Expert Knowledge and Unruly Subjects." Session Abstract: Knowledge and Value in a Globalising World—Disentangling Dichotomies, Querying Unities. Accessed February 14, 2015. http://www.anthropologywa.org/iuaes_aas_asaanz_conference2011/0002.html.

———. 2015. "Epistemic Imperialism and the Anthropology of Mischief: Techno-Science, Expert Knowledge and Unruly Subjects." Abstract for the IUAES, ASS, ASAANZ Conference, 2011. Accessed February 14. http://www.anthropologywa.org/iuaes_aas_asaanz_conference2011/0002.html.

Pyne, Stephen J. 1997. *World Fire : The Culture of Fire on Earth*. Seattle: University of Washington Press.

———. 1998. "Forged in Fire: History, Land and Anthropogenic Fire." In *Advances in Historical Ecology*, edited by William Balee, 62–103. New York: Columbia University Press.

———. 2001. *Fire: A Brief History*. Seattle: University of Washington Press.

Smith, Marian W. 1940. *The Puyallup-Nisqually*. New York: Columbia University Press.

Snyder, Warren A. 1968. *Southern Puget Sound Salish: Text, Place Names and Dictionary*. Vol. 9. Sacramento: Sacramento Anthropological Society.

Soper, Taylor. 2014. "Here Are the 10 Fastest-Growing Regions of the U.S. (Seattle Is One of Them)." *Geekwire* (March 27). Accessed March 15, 2017. http://www.geekwire.com/2014/seattle-population/.

Sprugel, Douglas G. 1991. "Disturbance, Equilibrium, and Environmental Variability: What Is 'Natural' Vegetation in a Changing Environment?" *Biological Conservation* 58 (1): 1–18.

Stenholm, Nancy. 1989. "The Botanical Assemblage of Layser Cave, Site 45LE223." Report prepared for the Randle Ranger District, Gifford Pinchot National Forest, USDA Forest Service, Randle, WA.

Stuart, Claudia, and Kristen Martine, technical editors. 2005. "Northwest Forest Plan—the First Ten Years (1994–2003): Effectiveness of the Federal-Tribal Relationship." Tech. Paper R6-RPM-TP-02-2006. USDA Forest Service, Pacific Northwest Region, Portland, OR.

Sullivan, Alan P., Jean N. Berkebile, Kathleen M. Forste, and Ryan M. Washam. 2015. "Disturbing Developments: An Archaeobotanical Perspective on Pinyon-Juniper Woodland Fire Ecology, Economic Resource Production, and Ecosystem History." *Journal of Ethnobiology* 35 (1): 37–59.

Trusler, Scott, and Leslie Johnson. 2008. "'Berry Patch' as a Kind of Place—the Ethnoecology of Black Huckleberry in Northwestern Canada." *Human Ecology* 36 (4): 553–568.

Tsing, Anna Lowenhaupt. 2014. "Blasted Landscapes (and the Gentle Arts of Mushroom Picking)." In *The Multispecies Salon*, edited by S. Eben Kirksey, 87–109. Durham, NC: Duke University Press.

Turner, Nancy. 1999. "'Time to Burn': Traditional Uses of Fire to Enhance Resource Production by Aboriginal Peoples in British Columbia." In *Indians, Fire and the Land in the Pacific Northwest*, edited by Robert Boyd, 185–218. Corvallis: Oregon State University Press.

Turner, Nancy, and Ann Garibaldi. 2004. "The Nature of Culture and Keystones." *Ecology and Society* 9 (3): r2. http://www.ecologyandsociety.org/vol9/iss3/resp2/.

Turner, Nancy J., Douglas Deur, and Carla Rae Mellott. 2011. "'Up on the Mountain': Ethnobotanical Importance of Montane Sites in Pacific

Coastal North America." *Journal of Ethnobiology* 31 (1): 4–43.

Turney-High, Harry Holbert. 1941. *Ethnography of the Kutenai*. Menasha, WI: American Anthropological Association.

United States. 1854. "Treaty between the United States and the Nisqually and Others Bands of Indians." No. 43293021.

———. 1855. "Treaty between the United States and the Dwamish, Suquamish, and Other Allied and Subordinate Tribes of Indians in Washington Territory : January 22, 1855, Ratified April 11, 1859."

USDA Forest Service. 2005. "National Visitor Use Monitoring Report: Mount Baker–Snoqualmie National Forest." Portland, OR: USDA Forest Service, Region 6.

USDA and USDI. 1994. Record of Decision on Management of Habitat for Late-Successional and Old-Growth Forest Related Species within the Range of the Northern Spotted Owl (Northwest Forest Plan). US Department of Agriculture, Forest Service, and US Department of Interior, Bureau of Land Management.

Vale, Thomas R., ed. 2002. *Fire, Native Peoples, and the Natural Landscape*. Washington, DC: Island Press.

Walsh, Megan K., Jennifer R. Marlon, Simon J. Goring, Kendrick J. Brown, and Daniel G. Gavin. 2015. "A Regional Perspective on Holocene Fire-Climate-Human Interactions in the Pacific Northwest of North America." *Annals of the Association of American Geographers* 105 (6): 1–23.

Whitlock, Cathy, and Margaret Knox. 2002. "Prehistoric Burning in the Pacific Northwest: Human versus Climatic Influences." In *Fire, Native Peoples, and the Natural Landscape*, edited by Thomas R. Vale. Washington, DC: Island Press.

Wilkinson, Charles. 2000. *Messages from Frank's Landing: A Story of Salmon, Treaties, and the Indian Way*. Seattle: University of Washington Press.

Wrangham, Richard W. 2009. *Catching Fire : How Cooking Made Us Human*. New York: Basic Books.

Yang, Suann. 2006. Helens Seed Dispersal, Community Context and Population Genetics." PhD dissertation, Washington State University, Pullman.

Yang, S[uann]., J. G. Bishop, and M. S. Webster. 2008. "Colonization Genetics of an Animal-Dispersed Plant (*Vaccinium membranaceum*) at Mount St. Helens, Washington." *Molecular Ecology* 17 (3): 731–740.

CHAPTER 7

The Critical Role of Firefighters' Place-Based Environmental Knowledge in Responding to Novel Fire Regimes in Hawai'i

LISA GOLLIN AND CLAY TRAUERNICHT

THE "WICKED" PROBLEM OF WILDFIRES IN HAWAI'I

Human-caused wildfire has become a frequent disturbance across the main, inhabited islands of the Hawaiian archipelago. Over the past decade, an average of more than 1,000 ignitions burn more than 8,000 hectares each year statewide, which as a proportion of total land area is comparable to and in some years exceeds the proportion burned in the western United States (Trauernicht, Pickett et al. 2015). In addition to providing the source for 99 percent of all ignitions in Hawai'i, human activities have dramatically increased the flammability of its landscapes through land use practices that have promoted the spread and establishment of fire-prone non-native grasses. Combined with strong orographic rainfall patterns (rain shadows) and episodic droughts such as those associated with El Niño events, these anthropogenic drivers create potential for frequent year-round, often destructive wildfires. Fire regimes have changed dramatically in terms of recurrence, intensity, seasonality, and the negative effects on Hawaii's native, largely fire-sensitive ecosystems. Importantly, the increasing flammability of Hawaii's landscapes places a disproportionate burden on the fire response professionals who work to contain and suppress theses fires in extraordinarily complex and challenging topographic and environmental conditions, and with very little access to the tools available for fire response in the US mainland.

Contemporary wildland fire management in Hawai'i is focused primarily on suppression efforts. Novel fire regimes wrought by dramatic ecological and social changes, described below, have largely negative impacts on natural resources and local communities. With prevention and fire risk reduction efforts largely incommensurate relative to the prevalence and extent of wildland fire statewide (Trauernicht, Pickett et al. 2015), the responsibility for protecting resources falls squarely on the shoulders of local fire responders. As such, and out of necessity to do their jobs safely and effectively, this group of practitioners has developed a deep understanding of how the drivers of fire—vegetation (fuels), climate and weather, topography, and land use—affect fire behavior and constrain tactical response. This knowledge is especially critical given the high degree to which these drivers vary over space and time in Hawai'i and the lack of fire science available to inform suppression efforts relative to well-studied mainland ecosystems (Weise et al. 2010; Trauernicht, Pickett et al. 2015).

Wildfire response jurisdictions in Hawai'i are split among multiple county, state, and federal agencies, many of which maintain mutual aid agreements for joint response and resource sharing. Hawaii's four county fire departments (Hawai'i, Kaua'i, Maui, and the City and County of Honolulu) provide critical initial response for the majority of wildfires in the state, yet they are primarily trained and equipped for structure fires. The Hawai'i Division of Forestry and Wildlife is trained and prepared specifically for wildland fire response and is directly responsible for fire suppression on state lands (more than 640,000 hectares or 41 percent of state land area). However, the Division of Forestry and Wildlife employs no full-time firefighters, acting instead as a reserve-style resource with personnel being pulled from their primary forestry duties and mobilized for fire response, typically after county agencies are first alerted. The only full-time wildland fire suppression programs in the state are with federal agencies at Hawai'i Volcanoes National Park and the US Army garrisons on Hawai'i and O'ahu Islands. These federal programs also often assist county and state suppression efforts both with personnel and air support (helicopter bucket drops). Assistance often requires, however, that federal resources be directly threatened during a fire incident. Despite frequent interagency cooperation, suppression efforts are also constrained by Hawaii's geography, which prevents the mobilization and sharing of heavy equipment such as water trucks, brush trucks, and bulldozers among islands during wildfires.

The focus on fire suppression in Hawai'i stems in part from the relative sensitivity of native vegetation to the effects of fire (D'Antonio and Vitousek 1992; LaRosa et al. 2008).

Wildland fires were relatively infrequent in Hawai'i prior to human settlement, limited to areas of volcanic activity and rare lightning strikes (Burney et al. 1995; Smith and Tunison 1992). Humans, however, have forever altered the fire regime on the islands, beginning with the arrival of the first Polynesian settlers an estimated 800 (Wilmshurst et al. 2011) to 1,000 years BP (Kirch 1986). The use of fire for agriculture and land clearance by Pacific Island peoples has been linked to the expansion of fire-adapted vegetation types in New Zealand (McWethy et al. 2010), Fiji (King 2004), and Micronesia (Dodson and Intoh 1999). Historical accounts describe intentional landscape burning by Hawaiians to access and manipulate key resources such as the indigenous *pili* grass (*Heteropogon contortus* [L.] P.Beauv. ex Roem. & Schult.), for thatching, and other plants for food and fodder (McEldowney 1979; Kirch 1982). The extent of forest conversion by Hawaiians was likely small relative to post-European impacts (Maly and Wilcox 2000) and may also be attributed to factors other than fire, such as seed and seedling predation by introduced rats (Athens 2008).

The most important changes to Hawaii's landscape with respect to fire have occurred within the past century. The rise of the plantation and ranching industries in the late nineteenth and early twentieth centuries led to large-scale clearance of dry, mesic, and wet native forest as well as the introduction of a suite of non-native pasture grasses. Ironically, following the disastrous 30,000 acre Hamakua fire of 1901 on Hawai'i Island, it was recognition of the value of forests for water provisioning to the plantations that prompted development of a forest warden program and Forest Reserve System by the territorial government (Commissioner of Agriculture and Forestry 1903). Although these changes in vegetation and fire use for sugarcane harvesting likely increased wildfire occurrence,

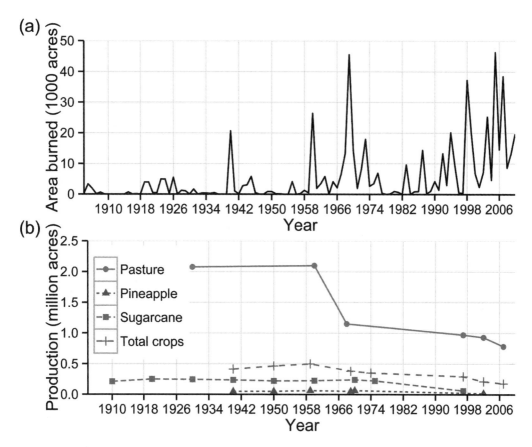

FIGURE 7.1. Trends in the spatial extent of wildfire occurrence and agricultural production in Hawai'i over the past century. The top graph illustrates the increase in annual area burned statewide for Hawai'i from 1904 to 2012 coinciding with the decrease in area under production for ranching and plantation agriculture depicted in the bottom graph.

plantations and ranches provided infrastructure and resources (people, roads, water, and heavy equipment) and actively assisted government agencies in fire notification and suppression (DOFAW 2010). In terms of actively managed lands, however, these industries have declined by over 60 percent from peak production in the 1950s–1960s, at which time plantations and ranches occupied approximately 5 percent and 50 percent of Hawaii's total land area, respectively (Schmitt 1977; Trauernicht, Pickett et al. 2015; Figure 7.1). Agricultural abandonment not only results in the expansion of fire-prone non-native grasslands and shrublands, or derived savannas (see Veldman and Putz 2011), but also significantly reduces the resources historically provided by agricultural operations for fire suppression.

Widespread expansion of non-native derived savannas—combined with increasing human-caused ignitions in the wildland-urban interface, strong rain shadow effects, and episodic droughts—has resulted in a fourfold increase in area burned statewide over the past century (Trauernicht, Pickett et al. 2015; Figure 7.1). In terms of area burned, wildfire is most prevalent in dry, leeward areas of Hawai'i and during the drier summer months. Interannual increases in fire occurrence, however, are strongly driven by drought events, such as those that occur during El Niño episodes (Chu et al. 2002; Dolling et al. 2005), during which large fires can occur in climatically wet areas; namely, windward regions and rainforests. With significant drying trends projected for Hawai'i, the relationship between drought and increased fire activity indicates that fire occurrence and impacts will worsen with climate change.

The pattern of agricultural abandonment in Hawaii's wildland-urban interface effectively creates landscape-scale wicks of fire-prone grasslands that connect developed areas with high-ignition densities to upland forested areas critical to watershed functioning and native species conservation. The impacts of wildfire on valued resources are especially acute in Hawai'i given the sensitivity of native ecosystems to fire disturbance (LaRosa et al. 2008) and postfire invasion by fire-adapted non-native species (D'Antonio and Vitousek 1992), as well as the small land area and tight linkages between terrestrial and marine ecosystems. Unlike wildland-urban interface issues in the continental United States, where fire risk to people and infrastructure increases as development pushes into fire-prone ecosystems, fire risk in many of Hawaii's residential areas has increased as the surrounding landscapes change with agricultural abandonment and the expansion of fire-prone vegetation. Further, the rezoning and division of agricultural lands for new developments complicates access for fire response, and no legislation exists at the state level to ensure that the design of new residential areas reduces wildfire risk.

Hawaii's contemporary fire regime also differs from that of the US mainland in that wildfires occur year-round and are associated with completely novel ecosystems and fuel types. Hawaii's derived savannas are typically dominated by non-native grasses such as *Pennisetum clandestinum* Hoscht. ex Chiov. (*kikuyu* grass), *Pennisetum setaceum* (Forssk.) Chiov. (fountain grass), *Melinus minutiflora* P. Beauv. (molasses grass), and *Megathyrsus maximus* (Jacq.) B. K. Simon and S. W. L. Jacobs (Guinea grass), which attain extremely high fuel loads (e.g., 15 to 30 metric tons/hectare (Beavers et al. 1999; Castillo et al. 2003; Ellsworth et al. 2014) and ignite and burn at high intensities even under low fire-risk weather conditions (moderate relative humidity) (C. Trauernicht, personal observation; Figure 7.2). Weather and climate conditions related to fire occurrence, such as rainfall and wind, are incredibly variable over short temporal and spatial scales in Hawai'i, yet these gradients are poorly captured by available weather stations (e.g., Giambelluca et al. 2013; Weise et al. 2010). Importantly, novel fuel types and high environmental variability mean that many of the available tools for fire risk assessment and prediction, including the National Fire Danger Rating System and fire behavior models such as BEHAVE and FARSITE, were developed for mainland US ecosystems and perform poorly in Hawai'i (Beavers et al. 1999; Benoit et al. 2009; Weise et al. 2010). In addition, fire response agencies in Hawai'i lack access to the impressive array of interagency support available on the mainland because the major federal land management and fire suppression agencies (the US Forest Service and Bureau of Land Management) do not own land in the state. Another critical factor that must be calculated into operational response is protection of cultural and natural resources. Hawai'i has the dubious distinction of being the "endangered species capital of the world" (DOFAW 2015). Also threatened are the cultural resources found throughout the islands, including aboveground and subsurface archaeological sites, burials, and culturally significant areas for gatherings, rituals, and a suite of other

FIGURE 7.2. An experimental fire in guinea grass illustrating high fire intensity (4–5 meter flame lengths) despite relatively benign fire weather (70 percent relative humidity, 5 mph winds) (photo by Clay Trauernicht, 2013).

cultural practices. These factors have forced firefighters to develop sophisticated place-based understandings of local environmental conditions and fire behavior to do their jobs safely and effectively.

The objective of this chapter is to present how firefighters are responding to the "wicked" problem—in that, it is multifaceted, unique, and not readily solvable (Xiang 2013)—of wildfires in Hawai'i through adaptive, place-based knowledge. Based on the first author's prior ethnographic research and community engagement with *paniola* or *paniolo* (Hawaiian cowboys), taro farmers, salt collectors, fishers, plant-gatherers, and other cultural practitioners, and the second author's extension work with firefighters, we made a few assumptions. The first is that, like cultural practitioners, wildland firefighters possess an intimate knowledge of their natural surroundings, especially in terms of fuels, weather, topography, and fire behavior, and they are able to recognize variability and change. For example, native practitioners and professional firefighters similarly stress the growing unpredictability of managing and utilizing resources due to extreme weather events and climate change. The second is that firefighters respond to environmental fluctuations and change in Hawaii's diverse ecosystems with place-based knowledge that is an amalgam of personal experience, the experiences of native Hawaiians, and newly introduced resource management understandings and practices. Consequently, there exists an informal (oral) knowledge network—an oral history or narrative—linking the state's firefighters.

For several decades, anthropologists and ethnobiologists have been studying fire's evolutionary and cultural significance and how (usually Indigenous and Native) peoples use fire to manage resources and influence fire ecology (Welch et al., chapter 2, this volume). For more than forty years, social scientists in the United States have investigated the human dimensions of wildland fires with regard to applied topics such as public perceptions of, public education on, and community engagement in fuels management, controlled burns, postfire recovery, homeowner mitigation measures, and wildfire planning and policy, to name a few (Toman et al. 2013). Anthropologists (as well as ethnobiologists, geographers, sociologists, and other social scientists) have explored the applied significance of indigenous anthropogenic fire regimes and environmental stewardship. Applied research has been undertaken to promote biocultural diversity and restoration, and improve public and private sector land management, fire prevention, and mitigation programs. A few studies demonstrate how the reestablishment of Native American burning practices can help to foster tribal integrity and ecological restoration. For instance, certain fire management practices encourage the growth of culturally valued taxa, such as fungi and plants with edible berries, and wildlife that also provides a range of ecosystem services (Anderson 2006; Anderson and Barbour 2003). In particular, anthropologists have elucidated how fire use and control are deeply embedded in cultural identity, notably the Kodi of eastern Indonesia (Fowler 2013; and chapter 9, this volume) and the Xavante in the *cerrado* biome of Brazil (Welch 2014). At the same time, these authors explore the extent to which Indigenous groups may be culpable for deforestation and global warming, or conversely, whether they are in fact enhancing biodiversity and/or producing negligible survival emissions. These investigations are critical to reframing international conservation narratives that sometimes cast Indigenous groups as responsible for "overspending" global carbon budgets, leading to calls for suppression of anthropogenic burning.

Our research is also informed by integrative and adaptive knowledge and resilience in socioecological systems (Berkes and Folke 1998; Berkes et al. 2002), as well as the integration of Indigenous fire use schema into conventional resource management programs. Recognizing that federal fire management relies on national narratives at the expense of place-based sustainability science, Ray and her colleagues (2012) have proposed a solution that draws from the traditional ecological knowledge of Athabascan forest users in Alaska. Lake and Long (2014) describe successful collaborations between national forests and native tribes. Ryan and his colleagues (2012) have provided guidance to federal fire agencies on how to better understand, avoid, and mitigate potential detrimental effects of fire suppression and postfire rehabilitation activities on cultural and archaeological resources; the authors describe the effects of fire on tangible biocultural and physical resources such as plants and animals (foraged, hunted, or herded), ceramics, lithic artifacts, petroglyphs and pictographs, subsurface deposits, and more. Welch (2012) takes a community and landscape approach to also consider the value of intangible cultural resources to Indigenous peoples. Conceptual, oral, and behavioral beliefs and practices such as origin stories and histories tied to specific places or landscape features are interdependent and key to successful heritage management. All of these authors present several ways to actively involve Native people in the development of collaborative management plans.

Little attention has been paid to the environmental observations and strategies of firefighters. A small body of social science literature exists on wildland firefighters, such as Desmond's (2007) ethnographic exploration of how wildland firefighters understand and habituate to risk and death, Klein's (2000) analysis of decision-making in high-risk professions (e.g., Klein 2000), Desmond's (2007) exploration of the cultural construction of masculinity and gendered dimensions of wildland firefighting, and Eriksen's (2014) argument that wildfire management is a means through which conventional gender roles and power relations are perpetuated. With this chapter, we hope to contribute to the dialogue on fire ecology, management, climate change, and the value of firefighters' place-based knowledge in heterogeneous and fragile island ecosystems, and also point out the practical value of incorporating firefighter place-based knowledge into applied programs such as firefighter education, "best practices" and training guides, resource allocation, operational procedures, natural and cultural resource protection and management, planning, and policy.

Study Overview and Methods

The findings presented in this chapter grew out of a larger project conducted at the University of Hawai'i at Mānoa's Department of Natural Resources and Environmental Management. The wildland fire extension specialist, Clay Trauernicht (coauthor of this chapter) was principal investigator of the project Challenges to Rapid Wildfire Containment in Hawai'i. The project sought to identify factors limiting rapid initial attack and suppression by fire response agencies in order to better inform resource allocation and investment decisions to improve the capacity of wildfire suppression. The goal of the social science component of the project was to draw on the expert knowledge of Hawaii's fire and emergency response community via in-depth interviews with incident commanders (ICs) and chiefs and fire managers from the Hawai'i Division of Forestry and Wildlife, as well as from county and federal fire departments. The study results were synthesized and presented in a report with recommendations to multiple stakeholders and policymakers (Trauernicht, Gollin et al. 2015).

The social science research component of the Challenges project was performed by the senior author of this chapter, Lisa Gollin. Interviews were conducted from December 2014 to June 2015. Judgment, or expert sampling (following Bernard 2011), was used to identify and invite fire leaders and decision-makers to participate in the project. Fifteen ICs were interviewed on the islands of O'ahu, Hawai'i, Maui, and Kaua'i. The Maui participants also discussed fire response on the islands of Moloka'i and Lāna'i, which are within their response district. All but three of the ICs are originally from Hawai'i. The two who were originally from the continental United States had spent most of their firefighting careers in Hawai'i. Participants included individuals from various government fire response agencies, as well as one fire science educator with the Hawai'i Community College Fire Science Program who was formerly an IC with the National Park Service. Five of the participants were retirees: three from county fire departments and two from the Division of Forestry and Wildlife.

Participants were interviewed individually or with one other participant at their offices, fire stations, or homes. Semistructured, structured, and open-ended questions were asked about basic biographical information and firefighting experience; the most critical information needed upon receiving the call to a fire; initial attack and containment procedures and issues; wildland fire behavior characteristics in different fuel types, weather conditions, and terrain; agency coordination in each participant's jurisdiction and areas of mutual aid; availability of personnel and equipment; electronic resources (e.g., a Likert ranking of online, real-time, and software modeling programs); and recommendations for improving resource allocation and best practices.

Based on the assumptions noted above, as well as themes that emerged from pilot interviews, Gollin added a new section to the questionnaire. The questions explored "rules of thumb" or "size-up"[1] assessments (the latter term a common firefighting term we heard throughout the interviews) based on experiential indicators of fire behavior rather than standard tools; sensory cues and observations of the natural surroundings, such as odors, sounds, animal behavior, cloud formations, and observations of environmental or climate change over the course of the participant's career; and how managers factor protection of Hawaiʻi's biocultural diversity into fire mitigation. Participants with firefighting experience on the mainland were asked about how fire ecology and suppression in Hawaiʻi differ from that on the mainland.

Results
Location and Oral Transmission of Knowledge

The findings presented in this chapter are primarily drawn from the Environmental Observations and Place-Based Knowledge section of the interview protocol. Nevertheless, we note that environmental observations and place-based knowledge were addressed throughout the interview, starting with the first question on initial attack: "When you receive a call to respond to a wildfire, what are the immediate questions you need answered in deciding the actions to be taken?" Thirteen of the participants (87 percent) stated that location was the first question to be answered to coordinate a successful response, and it was cited as most critical for two reasons: first, to determine if the fire is in an area of primary response or automatic aid (secondary response); and second, but of no less importance, to guide operational strategies informed by what a few participants referred to as "the slide projector of past fires in the area"—prior site-specific knowledge of vegetation, weather, terrain, and fire behavior. The following statement from a Maui participant is a typical response: "We don't even have to ask what the fuel type is if we know the location. Like if it's at Lahaina Luna School, I know it's in cane grass [*Arundo donax* L.] and that it's going to hit the big trees eventually."

Another theme that arose from the interviews was the institutional memory of Hawaiʻi's fire response community. Interisland and interagency collaboration on fires as well as cross-agency trainings, meetings, conferences, and more provide opportunities for firefighters to share iconic fire stories and suppression strategies.[2] There is an intergenerational knowledge network, with stories and tactics being passed down from the handful of senior commanders who have trained most firefighters on the islands. For example, the phrase "your slide projector," referring to the experiential catalog one brings to a fire, was traceable to a senior forestry commander who, by several accounts, was one of the most influential fire educators on the islands. Throughout the interviews, participants often cited the sources of their knowledge, often mentioning two recently retired forestry division ICs, one on Oʻahu and the other on Hawaiʻi Island, both of whom were interviewed for this study. Participants across agencies and islands described these two men variously as the "brain trusts" or "icons" of Hawaiʻi wildland firefighting; in the words of a Kauaʻi fire chief discussing the Oʻahu retiree, "You know that [guy] is the man with the white hat."[3]

Place-Based Knowledge: Environmental Cues, Senses, and Surroundings

The place-based knowledge of Hawaii's firefighters is not a defined body of knowledge, but rather the result of diverse influences. By extension, while standard firefighting jargon permeates the language, the lexicon of Hawaii's firefighters also includes regional and local terms (which we attempt to sort out and explain throughout the chapter). Wildland firefighting place-based knowledge derives from a mix of Native Hawaiian ecological knowledge and cultural practices, ranching and farming resource management practices,[4] an individual's genealogical connection and/or familiarity with an area and its microenvironments, "talk story"(Hawaiian pidgin for exchanging stories and casual conversation) with other fire responders, and an individual's sensory acumen or "perceptual expertise" (Klein 2000, 45). Participants identified a number of environmental cues used in decision-making and to assess fire danger. The incident stories and examples shared for the Challenges project reflect this layering of personal and collective knowledge. Figure 7.3 presents the environmental cues that firefighters use, listed in order from most to least mentioned. Participants generally discussed cues from the fire science framework of the three principal environmental elements that affect wildland fire behavior: fuels, weather, and topography. They also spoke of concomitant phenomena such as smoke behavior, and sensory cues such as odors or the sound of fire; a few participants said they include animal behaviors in their sizeup assessments.

Relative humidity and wind are the two primary weather parameters influencing fire spread and determining whether containment will be successful. Wind feeds the fire, determining intensity and spread. Rain (or the wind dying down) starves it but feeds fuel loading on longer time scales that affect future fires. In Hawai'i, relative humidity, as with wind, fluctuates rapidly in time and space, especially under the influence of the islands' patchy and intermittent precipitation events and rugged terrain. Participants emphasized that relative humidity varies considerably, even in nearby locations such as neighboring ridgetops.[5] Topography is the most stable element of the fire behavior environmental triangle, but the rugged and varied terrain in many parts of Hawai'i places real physical constraints on the ability of firefighters to suppress fire safely and effectively. Importantly, the steepness and exposure of terrain differentially and dramatically affect fire activity. Fires move quickly up steep slopes and burn more intensely on drier, south-facing slopes, and terrain has complex effects on wind direction.[6] Fuel types are also relatively stable, but fire behavior is unpredictable given the other environmental variables in Hawaii's diverse and changing biogeographic setting. (Novel fuel types are considered in more detail in the next section.)

The environmental cues identified by participants and depicted in Figure 7.3 are not discrete categories. Environmental factors are interrelated and dynamic. An example is the relationship between wind, smoke, and fire activity; a firefighter may use the lean of a smoke column to determine wind direction and therefore where the fire is heading. As discussed in the introduction, many of the standard tools (e.g., fire modeling software, fire danger rating systems) were not designed for island geography, weather systems, and microclimates. As the fire science educator and former National Park Service IC remarked in his interview, "All of those [online, real-time, fire behavior modeling] programs you mentioned don't really fit here. It's a lot more complex and it's just something we deal with."[7]

The following observations provided by participants shed light on the critical role of place-based knowledge and sensory information from the natural surroundings in making decisions about fire management and suppression.

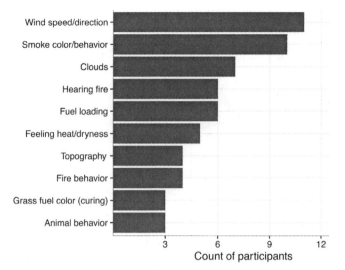

FIGURE 7.3. Environmental cues used by firefighters in Hawai'i to determine fire danger and guide decision making.

Wind

Wind (direction, speed, and microclimatic variation) was identified as the most critical variable in determining the course of a fire and operational response by most participants and was discussed more than any other environmental element. Eleven participants shared stories about wind characteristics and response tactics. Seven discussed cloud formations in relation to wind. In the words of a protection forester who runs the fire management program for Hawai'i Island:

> When you see thunder clouds forming, have fast moving clouds, or lenticular clouds, then you know you have the possibility of strong winds. Feeling the air drying out in your nose and mouth. The wind and terrain will determine which direction the wind is going to go. Wind more than terrain.

Particularly salient was understanding the diurnal and seasonal wind changes characteristic of an area. The chief of the Hawai'i Island fire department provided this example:

> Growing up here I know that . . . the Nīnole Loop at night, the wind always shifts from mountain to the ocean coming back up. Starts at about sunset. The wind goes [up from the] ocean and then at night mauka [mountain] to makai [ocean]. We were relieved by a crew. I remember mentioning that the winds shift at night. That worked out for the first day. But the second crew didn't pay attention and they left an apparatus out on the roadway. The wind shifted. [The fire] turned back and destroyed the vehicle.

Figure 7.4 is a schematic of the Maui Island wind system drawn by the forest management supervisor for the islands of Maui, Lāna'i, and Moloka'i. It displays how the prevailing trade winds from the northeast shoot through the saddle of central Maui between the west Maui Mountains and Haleakalā on the east creating, in his words, "an eddy effect . . . like water in a stream funnels between rocks." He explained that the swirling winds form along the higher elevations and affect fire behavior in the Kula Forest Reserve. While his home was untouched by an August storm a short distance from Kula, the reserve sustained major damage in one gulch where he estimated winds to be 90 to 100 mile per hour. "It mowed down a 400 acre swath. It snapped the large trees to the ground. You can look up on the ridge and see there is no longer a tree line in that area. That's a microclimate!"

The Critical Role of Firefighters' Place-Based Environmental Knowledge

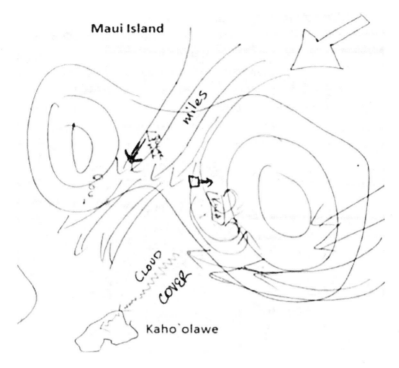

FIGURE 7.4. Illustration of the Maui wind system (drawing by Lance DeSilva, 2015).

The forest management supervisor emphasized that knowledge of wind speed and direction is critical to protecting the habitat restoration exclosure (an area enclosed to keep unwanted flora and fauna out) of the endangered *nēnē* (Hawaiian goose, *Branta sandvicensis* Vigors) from fire on the ridges of the western mountains where firefighters typically battle light, flashy grasses with strong winds. The Maui forestry IC explained that winds blowing on the west side are of minimal consequence "because the fire will back into the strong wind and it won't move too much."

If a fire starts east (mauka) of the nēnē pen, he said the trade winds are likely to steer the fire toward the fenced exclosure. He further described how the trade winds that shoot through the saddle of central Maui help to form the "Hawaiian cloud" that used to form rainstorms on the eastern island of Kahoʻolawe (referring to the *nāulu* rains and cloud bridge that once existed between Maui and Kahoʻolawe).[8]

The fire science professor and former federal IC shares the following story with his students. As a "burn boss,"[9] he would always spend a few days in an area ahead of a prescribed burn to become familiar with it and strategize. For a burn in the coastal area of west Hawaiʻi, he consulted with older fishers at Kawaihae Harbor who pay close attention to the lunar cycle and other natural cycles and weather patterns to determine which fish are running and the best time and places to catch them. They told the IC that within a couple days of the new moon, the winds die down at night, making it the best time for fishing, and they suggested he refer to the Hawaiian fishing calendar. He did, and planned the burn accordingly. On the day of the prescribed burn, the wind was howling ferociously, even into the evening. The IC began to doubt his decision. At 8:00 p.m. (as per the fishing calendar) the wind died down, allowing them to execute a "95 percent perfect burn." He tells his classes, "We didn't have to waste any of [the] taxpayers' money. The take-home message? Local knowledge is important!"

One retired firefighter who worked for the Hawai'i Fire Department and served as president of the Hawai'i Wildfire Management Organization discussed wind speed and fire behavior on volcanic terrain. When fire burns toward a *puʻu*, or cinder cone, the winds "will be squirrely as they blow over [it] because they won't keep on going the same direction. They will accelerate as they approach it and then . . . wrap around the sides and do all kinds of swirly things."

Relative Humidity

Four of the project participants discussed humidity in relation to sensory perception and knowledge of place. The educator and former National Park Service IC with years of experience fighting fires on the mainland shared the differences between relative humidity and fire activity in the continental United States versus Hawai'i. Whereas fire danger thresholds for relative humidity range from 15–30 percent on the mainland (NOAA n.d.), he discovered that Hawai'i can have very high intensity fires at much higher relative humidity:

> We can have fires here at 95 percent humidity. But at 95 percent humidity, they are not doing anything. What I found is that 60 percent [relative humidity] is the cut-off. . . . You start to see increased fire activity and more difficulty with the alien grasses.

The same participant had become so attuned to relative humidity, he said, "I can tell how many days it's been since there's been rain." Although the air may always feel moist to someone not accustomed to Hawaii's wet weather, the IC can feel what is unseasonably dry under these humid conditions. He recounted taking technicians from the mainland to service one of the remote automatic weather stations. "I told them, you need to check that again because I think that's not the right humidity. They checked again and their sensor was off. I was more sensitive than their sensor."

Smoke

Reading smoke—its form, velocity, color, and odor in relation to fire characteristics—is a standard subject taught in fire science programs and training modules. Eight of the participants discussed a color code for smoke particularly in relation to Hawaii's fuel sources. A chief with the Kaua'i Fire Department said, "The color of smoke is either black, brown, gray, or white depending on what's burning." He went on to explain that dark or black smoke indicates that heavy fuels such as cars, tires, or forest hardwoods are burning, while white smoke indicates lighter, flashy green fuels. The fire chief of the US Army's Pōhakuloa Training Area on Hawai'i Island uses smoke color and plume size to help evaluate whether he has the resources on hand to fight the fire, and if the fire will be an extended attack: "White, flowing smoke that means you have a convection pattern going in the sky and it's already creating its own weather. When you are at that point, it's beyond IA [Initial Attack]. If you see an anvil, it's beyond IA."

The fire chief developed his own rule of thumb to ascertain if his company has the resources available to put out the fire or if they need to request assistance. When he is a mile away from the fire, he holds up his thumb to the smoke column. If it is wider than his thumb, the fire is beyond an acre and beyond the 300 gallon water-holding capacity of a brush truck. "It's usually ten to fifteen minutes before I can get another brush truck. That's when we . . . call for more water, resources. We call it an 'all hands' fire."

Firefighters also note the direction the smoke column is leaning to ascertain wind direction. A Maui battalion chief described another smoke characteristic: "If there are explosive puffs, [the wind and fire are] moving quickly. If it's lazy billowing, slowly."

A few of the firefighters mentioned that where they lack access maps, sometimes the smoke plume is the only way to locate the fire from a distance.

The retired fire management officer from O'ahu who worked for the Division of Forestry and Wildlife (and who is considered one of Hawaii's wildland firefighting icons) is an old hand at smoke reading. He described how his knowledge of fire and smoke behavior came not only from formal fire science education, but also from watching the harvesting of sugarcane using backfire operations when he was growing up. He noticed that dispersed smoke aggregates in a column "because the fire is drawing a lot of oxygen, so everything is going to be drawn in the direction of the main fire. When the fires meet, it's all done."

Backfiring is a suppression strategy in which fire is intentionally set along the inner edge of a fire line to consume fuels and reduce the intensity and/or slow the progression of an approaching wildfire. The retired firefighter recalled a fire early in his career, one of his first backfire operations. Another firefighter was sitting at the bulldozer line in the safety zone where the backfire would be lit, casually smoking a cigar. When the cigar smoke suddenly changed direction toward the main fire approaching from the distance, "we knew it was the perfect time to set the backfire."

Olfactory cues are used by structural and wildland firefighters to identify what kind of fuel the fire is burning and to locate its source. It is common in structural fires for stations to receive "odor-of-smoke" calls. Firefighters familiar with this type of fire are able to recognize the difference between, for example, the distinctive sharp, pungent odor of an electrical fire and the burnt food smell of a stove fire (Sheridan 2014). Many wildfire ICs are able to identify the predominant fuel type by its smell, which is particularly useful for predicting fire behavior and protecting endangered native flora and fauna when visibility is obscured by clouds, inclement weather, dense smoke, or nightfall. The Hawai'i Island Pōhakuloa Training Area fire chief has become astute at differentiating vegetation types by smell, such as the mixture of nutty, woody, and "sweet maple bacon" of burning *kiawe* (mesquite, *Prosopis pallida* [Humb. & Bonpl. Ex Willd.] Kunth). He explained:

> Upon arrival, you can usually smell if something is wet and burning as opposed to if it's dry and burning. How accurate that is can potentially tell you what is burning. . . . You can tell the different types of fuels that are burning. You can tell if what's burning is in [native] mamane [*Sophora chrysophylla* (Salisb.) Seem.], naio [false sandalwood, *Myoporum sandwicense* A.Gray] or koa [*Acacia koa* A.Gray]. Mamane and naio are very oily and have a very pungent smell when they burn. Whereas if it's a hardwood like 'ōhi'a [*Metrosideros* spp.] or koa, it has a more woody smell—a more pleasant smell.

Terrain/Topography

Four of the firefighters that were interviewed discussed topography in relation to fire behavior and access issues. The corrugated folds of the island chain's tall mountains contribute to the landscape-scale "wicking" system, in which uncontained fire in lowland shrub and grasslands can quickly become an upland forest fire. A retired O'ahu county fire captain described how fire accelerates as it burns uphill, putting firefighters in danger and community assets—homes, structures, native forest—in the path of the fire: "You are always watching out for uphill situations or what we would call box canyons where the valley comes to a tight V which would almost in essence create a chimney-type effect."

The Big Island (the nickname for Hawai'i Island) is the youngest, most mountainous, and only volcanically active island in the archipelago. Active lava flows ignite wildfires along the flanks of the Kīlauea Volcano,[10] but inactive flows across the island pose a unique challenge to firefighters who have to navigate the porous and rough terrain of old lava fields. Interviewees from the Big Island explained that long dead lava fields serve a useful function because they can sometimes be used as natural fire breaks where vegetation is sparse. The fire science educator and former federal IC pointed out this contradistinction, "For most of the island you can't cut line because of the lava. You don't have soil. On the mainland, that's the Holy Grail."[11]

He uses an alternate technique to cutting line by hand that firefighters refer to as "black lining," which refers to the process of igniting pockets of unburned fuels between the fire line and the main fire.

The firefighters from the Big Island have a simple topography typology for lava. The spiky broken shards of lava known in Hawaiian as 'a'ā is treacherous to walk on, but bulldozers—the primary tool for cutting line in this environment—can be used to create fire breaks. However, bulldozers cannot be used on *pāhoehoe*—a smooth, unbroken, billowing, elephant skin–like lava. It is too tough and, most critically, there are lava tubes, cave-like subsurface channels, some of which are large enough to, in the words of the Pōhakuloa Training Area fire chief, "swallow equipment":

> The lava tubes in ancient times were used as areas of refuge. . . . Any time you get into the pāhoehoe lava, there's a good chance there's lava tubes in the area. You look for skylights [formed when a section of lava tube roof collapses] and other indicators. Normally, lava tubes run from mauka to makai rather than north to south. So, if you find a skylight, you have a pretty good idea of where the tube might be. [The lava] is going to follow the path of least resistance. We've had dozers fall into tubes. Now we try and put crews in front of the dozer so that we have eyes ahead.

Firefighters use many perceptual cues to guide their responses. Sensory cues provide critical information, particularly when visibility is poor. For example, the sound of a fire can indicate where the fire head is and how the fire is rolling. Six interview participants discussed such auditory cues. The federal fire chief described the sound and feel of a fire that is growing in ferocity:

> When a fire is getting ready to run, you got the wind going. If you're developing a line in front of a fire and all of sudden you feel a strong wind drawing across your face and the fire is behind you, that's not a good sign. Because it means that the fire is sucking that in, and it means that it's rolling. You hear people referring sometimes to . . . the fire as sounding like a freight train. Your audio cues will tell you how the fire is going. When you hit the fire with water, you're going to hit the seat of it, and it will sound like a hiss. You are going to hear that. If you have a duff fire and you hit it with a straight stream, you're going to hear a really low rumbling in the ground.

Two of the ICs (one from Hawai'i Island, the other Kaua'i) noted that animal evacuation patterns, such as that of wild pigs in the mountains or mice, rats, and even cane spiders in lowland fields, can serve as forewarning that the fire front is heading in their direction. The Hawai'i Island IC described how the heat of the fire drives everything in front of it: "If you're cutting line and all of a sudden you have all these pigs running past you, you know the fire is coming your way."

Another participant from Hawai'i Island and a retiree of the county fire department used 'io, the endemic Hawaiian hawk (*Buteo solitarius* Peale), as a guide. He said that when he sees hawks catching thermal updrafts, soaring high in the air, "The fire is likely to be able to allow a big column of smoke and heat to rise because it's an unstable weather pattern. There is no trade wind inversion overhead. So the fire is likely to get a good head of steam going."[12]

ECOLOGICAL SUCCESSION: GRASSES, SHRUBS, AND FIRE BEHAVIOR

The spread of fire-prone invasive grasses and, increasingly, pyrophytic environments was the consensus topic of concern for project participants. Division of Forestry and Wildlife ICs—responsible for managing and protecting native ecosystems and cultural resources on state lands—were more likely than county firefighters to discuss forest conversion due to fire. An IC with the O'ahu Island Division of Forestry and Wildlife commented, "If you have an area that is burned forest, then it comes back as grass. It very rarely recovers 100 percent to what it was before. It's irreversible."

Many others echoed that statement. A Maui forestry IC discussed ecological succession in Honua'ula. When he began his career, land formerly cultivated in sugarcane and pineapple at the lower elevations had come to be dominated by cane grass and other non-native species. There were stands of large eucalyptus trees climbing up the mountains. Native forest was on the ridgetops. A fire burned in Honua'ula on his second day on the job. The responders were able to put it out before it reached the native forest. It burned through the cane grass and stopped. Then it rained, and it took only a week for the grasses to grow back. The next year there was another fire. This one advanced up into the eucalyptus trees but did not touch the native forest. Grasses then grew in place of the eucalyptus.

> Everything burns: grass, timber, natives—or really, grass, grass, native. And now the natives are being replaced by grasses. It's a new vegetation succession. You can still see eucalyptus trees, but it's sparse. And it's an annual, or really every other year fire regime. It used to be dark with natives above the timberline. Now when you look at it, it's light [with grasses] all the way up the hill.

Counterintuitively, some grasses can be as dangerous when they appear green and alive as when they appear brown and dead (see molasses and kikuyu below). For the Hawaiian Islands, there are no curing guides, which use the color of prominent grass species to help firefighters assess the percentage of dead material and decreasing live fuel moisture to determine potential fire risk, intensity, and spread.[13] Several participants do use a heuristic code. The retired Big Island IC (another of Hawaii's wildland firefighting icons and repositories of knowledge) used the curing colors of fountain grass (green-yellow-golden-brown-gray) as an indicator of when to expect more fires: "We always used to joke about grasses. I used to drive along this corridor from Kona to Waimea where it's covered with fountain grass. I used to always tell the boys when I would see it gray, you better start packing your bags because we're going out on a fire soon."

Table 7.1 presents fuel types and associated fire behaviors of the most problematic non-native grasses and, for comparison, the native grass pili and native fern uluhe (*Dicranopteris linearis* [Burm.f.] Underw.). Species are listed from most to least challenging for fire response based on project participants' responses to free-listing exercises and follow-up questions. The results underscore what is novel about Hawaii's changing fire regimes. For example, firefighters often use previously burned areas as a safety zone (or, as they call it, "the black") from which to fight fire. Grasses generally burn quickly and "clean," leaving no unconsumed fuel to allow the fire to burn back over the same ground. In Hawai'i, burned grasslands do not reliably provide a safe spot for firefighters.[14]

Some of the non-native grasses listed in Table 7.1 result in what firefighters refer to as a "dirty burn" in which the fire does not consume all the fuels in an area, leaving partially burned spots susceptible to reigniting. Certain grasses can reignite quickly (within hours or days), even when they have high moisture content (appear largely green) in high humidity or precipitation. One of the ICs with many years of experience on the mainland and in Europe quickly learned that in Hawai'i, because of the constantly shifting winds, and aspect and slope of its upland terrain, "If you drive into the black to fight the fire, soon you will have the fire back around you."

Guinea grass is considered the most difficult to suppress. It has a high fuel load, grows tall, and has exceptionally high flame lengths. A Big Island forestry IC described Guinea grass as "a whole other animal," and he approaches a Guinea grass fire from "a different angle . . . paralleling it because the radiant heat is so high." When dry, it burns fast but can still lead to a dirty burn. One retired Oʻahu fire captain noted that the Guinea grass often reburns:

> A lot of times the clump in the center might not burn. It's ironic, you take the ash from the fire and that goes on the ground and then we shoot water on it and we're basically fertilizing it to grow again. . . . Sometimes Guinea grass doesn't burn completely. I think that has to do with the wind, how fast it's moving through. The drier parts are usually the ends of the grass versus the clump which is down by the moisture. The moisture goes from the ground and lessens as you get to the top of the grass, so there are situations where the clumps will still be green and still present after the fire. So, they can burn again or dry out again. Ideally, you'd attack these fires from the . . . black, but that's not always possible.

Fountain grass and buffel grass (*Cenchrus ciliarus* L.) are tussock grasses and, as one firefighter noted, they "can burn themselves out except if it's windy." According to a Maui participant, buffel grass may throw embers but does not tend to create spot fires because "if it does throw embers, it's most likely not going to ignite again."

Fountain grass, a popular landscaping ornamental, now covers vast expanses of west Hawai'i Island and has created a fire regime alien to the islands' dryland ecosystems (Blackmore and Vitousek 2000). Unlike buffel grass, it has high fuel loading. According to the federal IC working in the high-altitude plateau of the Pōhakuloa Training Area, where fountain grass is ubiquitous,[15] it has a "dualphasic burn": the top burns in the morning, then dries up and heats the bottom; when the wind switches directions in the afternoon, as it typically does, it reignites the grass that burned in the morning. "You get a hot fire that kind of kills everything in the ground," he said. "Then up where we are, you get a leaching of phosphorous which allows nothing to come back."

Cane grass resembles sugarcane and can grow up to 20 feet tall. A number of participants noted that cane grass holds moisture, meaning "it never fully burns out" and can throw embers that are carried by the wind and easily ignite elsewhere and reburn through the black. Molasses grass grows in mats and is found mainly in disturbed, dry to mesic open areas. Firefighters say it can create a high-intensity fire that is easily wind driven. Molasses grass no more than a foot high can put out a 15 foot flame and, as the fire science educator noted, it "has a resin in it that makes it burn hotter when it's green than when it's dry. It's really sticky stuff, like molasses."

Kikuyu is an important range and pasture grass. In Hawai'i, as in other parts of the tropics, it is also an inexpensive and easy to maintain golf course turf grass. The firefighters said kikuyu grass is slow burning, but the dense thatch of subsurface stolons and stems makes kikuyu fires difficult to extinguish: "For mop-up [postfire cleanup], it's the biggest headache." They pointed out that a lush green carpet of wet kikuyu grass can be deceptive. Although it may appear to be defensible space—a buffer zone against fire—kikuyu can end up in a dirty burn. A retiree from the Big Island explained:

> Kikuyu grass . . . will easily go 6 feet down and into the cracks of the rocks. They are deep-seated fires and they can run horizontally. In a kikuyu pasture the roots can easily go so far underground and spread underneath a 20 foot firebreak. . . . It can burn underground for a couple of days and pop up on the other side and the fire is off and running again.

Frequently mentioned non-native trees and shrubs were kiawe, *koa haole* (*Leucaena leucocephalum* [Lam.] de Wit), '*opiuma* (*Pithecellobium dulce* [Roxb.] Benth.), and eucalyptus (*Eucalyptus* spp. L'Hér.). Koa haole and kiawe, both introduced for cattle forage, are widespread throughout the islands. Each part of a koa haole tree has its own burn characteristics: the trunk and branches burn slowly but often survive after a fire moves quickly through a stand. An O'ahu retiree said such fires can move "so fast that the plant will still be alive and in three weeks you'll notice new branches sprouting. The leaves will burn off quickly unless it's dead but the main stem will burn for a while. It throws a lot of embers. The seeds hold a lot of heat and then land and start fires."

Kiawe, or mesquite, is a culturally valued fuelwood for *imu* (underground ovens) used to cook '*aha'aina* (Hawaiian feasts) foods such as pig, taro, and sweet potatoes.[16] It is also widely favored for barbecues. Firefighters discussed how difficult it is to fully extinguish pitoven fires, which can smolder underground and reignite weeks or months later and start or restart wildland fires. A few interviewees had seen fellow firefighters severely injured by falling into ash pits. They also provided other examples of the many potential problems characteristic of kiawe. For example, the large amount of leaf and bean pod detritus under the tree is highly flammable, can burn for a long time, and create spot fires. Also, the wood is so dense that tree cutters with chainsaws have a difficult time felling kiawe in the path of fire. According to one O'ahu firefighter, "It's almost impossible for us to put out every kiawe stump. It's just not a wise use of the time."

Kiawe stumps are always in danger of burning, however, and potentially rekindling the fire. Two of the Maui participants discussed how 'opiuma, introduced as a shade tree in pastures, burns hot and easily rekindles, and was increasingly becoming a fire hazard in leeward areas of Maui and on Moloka'i. Eucalyptus, many species of which are now found on all the major Hawaiian islands, takes over fallow fields and native habitats, resprouts prolifically after burning, and, as noted by a Big Island fire chief, has its own fire regime: "We've seen crowning type fires that we've never had before."

For the most part, project participants did not bring up native plants. Where grasses are concerned, exotic grasses pose the greatest fire threat. Native grasses such as pili, unlike

FIGURE 7.5. Wildfire burning through *uluhe* fern (*Dicranopteris linearis* [Burm.F.] Underw.) in native forest on Oʻahu, 2015 (courtesy of the Hawaiʻi Division of Forestry and Wildlife).

alien grasses, burn "slow and clean." In fact, the Big Island IC who established that fire risk rises when relative humidity falls below 60 percent in alien grasslands had determined that the cutoff is 40 percent for pili grass, suggesting that areas where pili grass remains the dominant fuel have a lower potential for burning. There is one notable exception. The native fern uluhe (Figure 7.5) has taken forestry division firefighters by surprise. Uluhe dominates many upland forests, draping the mountains in dense, thick foliage that often provides continuous vegetative ground cover up to 10 feet deep between emergent trees.

An Oʻahu forestry IC explained that under certain conditions, "it can burn as aggressively as Guinea grass or fountain grass. . . . The top layer is green, maybe one foot, but the bottom layer is brown. In the Koʻolau (mountain range) the bottom layer is wet and moist, but in between, the middle layer is all dead." He shared a video on his phone of an uluhe fire in central Oʻahu where the fire was so fierce, it created its own wind:

See how the smoke is going backwards. It's not even a wind-driven fire. It's burning that good without the wind at its back driving it. You can imagine how bad it would've been with winds. That was at six at night in January! The temperatures are cooler. It's not two in the afternoon. The humidity is recovering.

One of the retirees considered part of Hawaii's wildfire fighting "brain trust" described a suppression strategy that he came up with to manage uluhe fires:

We would construct a hand line by going direct, meaning constructing the line as close as possible to the fire's edge. That way you are eliminating the dry dead fuel from being burned and the potential for reburn is significantly reduced.

Environmental Change over Participants' Careers

In the words of a thirty-year forestry division veteran, "Firefighters are deeply in-tuned

TABLE 7.1. Novel fuel types and corresponding fire behavior and characteristics described by project participants.

	Species	Fire Behavior Descriptions
Non-native grasses	Guinea grass (*Megathyrsus maximus*)	Large flame height; High intensity; High rate of spread; Large fuel loads; Continuous fuel loads; Tall fuel bed; Embers/spotting; Low visibility; Requires large fuel breaks; Reburns 'through black'; Dangerous; Smolders/continues burning
	Fountain grass (*Cenchrus setaceus*)	High intensity; High rate of spread; Burns clean; Requires more water to extinguish; Reburns 'through black';
	Buffel grass (*Cenchrus ciliarus*)	Low intensity; Low flame height; High rate of spread; Flashy; burns clean; Easy to extinguish; Cures rapidly
	Molasses grass (*Melinis minutiflora*)	Large flame height; High intensity; Large fuel loads; Burns clean
	Cane grass (*Arundo donax*)	Low rate of spread; High fuel moisture; Embers/spotting; Reburns 'through black'
	Kikuyu grass (*Pennisetum clandestinum*)	Lower rate of spread; Large fuel loads; Smolders/continues burning; Difficult to extinguish; Matting/burns into ground; Difficult mop up
Native species	Uluhe (*Dicranopteris linearis*)	High rate of spread; Reburns 'through black', cures rapidly
	Pili grass (*Heteropogon contortus*)	Low flame height; Low intensity; Low rate of spread; Easy to extinguish; Requires lower RH (<40%) to ignite
Non-native trees & shrubs	Eucalyptus (*Eucalyptus* spp.)	Crown fires; Increased fire risk; Increased fuels (downed woody debris)
	'Opiuma (*Pithocellobium dulce*)	High intensity; Difficult to extinguish
	Kiawe (*Prosopis pallida*)	High intensity; Smolders/continues burning; Difficult to extinguish; Large quantities of surface fuels; Difficult to cut
	Haole koa (*Leucaena leucocephalum*)	High intensity; Ignites easily; Dangerous; Cures rapidly; Embers/spotting

to the environment, and must be in order to do our job." Figure 7.7 presents fire-related environmental changes observed over the course of participants' careers. The most common observation was the expansion of fire-prone grasses into native dry and mesic woodlands and the resulting emergent fire regimes.

The "brain trust" IC provided an example from Puʻuanahulu, where "fountain grass has burned off the native seed sources" and created grasslands where there were once native ʻōhiʻa shrubs (possibly *Metrosideros polymorpha* Guadich. or *Eugenia* sp. L.; mamane; and aʻaliʻi [*Dodonaea* spp. Mill.]). He remarked that younger firefighters believe the area to be natural grassland, being unaware that it was formerly a native and endemic shrubland. Some grass species, especially fountain grass, are capable of colonizing barren lava flows. On the Big Island, firefighters count on younger lava fields as natural fire breaks and to establish safety zones. *Kīpuka* (clear places or green oases within a lava bed) generally do not ignite if composed of pili and other natives. A Big Island forestry IC predicted that by the end of his career, as fountain grass colonizes lava substrate, "we won't be able to use lava flows as barriers."

Figure 7.6. Fountain grass (*Pennisetum setaceum* (Forssk.) Chiov.) colonizing lava substrate. Non-native grasslands and shrublands cover approximately 25 percent of land area in Hawai'i (photo by Lisa Gollin, 2015).

Several interviewees talked about droughts becoming longer, and how leeward-windward areas of the islands are uncharacteristically wetter or drier and more fire prone due to unprecedented fuel buildup from torrential rains followed by protracted hot and dry spells. Following are some of the statements made by ICs from each of the major Hawaiian islands where the interviews were conducted:

- "What used to be west O'ahu's fire problem is now central and west O'ahu's problem and in some regards even out to the Kahuku area of the North Shore."
- "Now more and more, you're getting a lot of heavy rains and long periods of drought."
- "The island is getting a lot drier. The average rainfall might be the same, but you are getting it all at one time or not at all. It's not well distributed."
- "In the last years it's been more wet than what we've seen in a long while. There are areas that are green that should be brown now."
- "I've seen places burn before I never thought would ever burn. I notice less trades and increases in Kona(s) [warm, southerly winds]."

The firefighters on the Big Island discussed the seven-year drought that generated unprecedented fires in the rainforests. One IC discussed the 2002 Kupukupu fire (named after the fern *Nephrolepis cordifolia* [L.] K.Presl), the primary fuel carrying the fire). He explained that this winddriven forest fire was different from past fires because when the lava flowed into the rainforest, the understory vegetation ignited rather than trees. He further described how unusually dry conditions and erratic winds had dried the ferns, which ignited and drew the fire into the forest at such high temperatures that the radiant heat made it impossible for firefighters to work on the ground:

- "I had to use bucket work." He said they battled flames 150 feet high at the fire's leading edge, which he characterized as "extreme fire behavior."

Seven interviewees discussed how rapid development in wildland-urban interface areas is changing the face of wildland firefighting.[17] For example, a Maui forestry IC recalled a conversation with a senior forestry IC about converting the Division of Forestry and Wildlife from a wildland agency to a "WUI [wildland-urban interface] agency because we have to work not in, but around, structures more and more."

The Big Island fire science educator offered three reasons fires are getting bigger across the islands. The first is climate change. The second is that firefighters have become more successful at suppressing fires, which naturally leads to fuel buildup. "The third," he said, "is that we have more people building homes out in the woods." This expansion into the wildland-urban interface means more human-caused ignitions. In the past, he recalled, they would have let a fire burn, "but now that there are homes there, we have to go put the fire out."

Somewhat paradoxically—given that Hawaii's ranching and plantation past is largely responsible for the contemporary widespread conversion of native ecosystems and environmental disturbance, and resultant higher-intensity fire regimes—several participants lamented the loss of the former agricultural infrastructure because they offered several presuppression and suppression services (see the section below on ungulates). Farmers and ranchers were once de facto first responders and partners with fire response agencies. A Hawai'i Island fire chief (and the son of a Hilo sugarcane processing superintendent) recalled:

When I first started [firefighting] there was still cane agriculture. It was managed very well because that's how they grew crop, it was by fire. So they had fire breaks, knew the language, had equipment. So the first thing we'd do if there was a fire was to call the plantation to see if it was a normal cultivation fire or something else. Access was easy. . . . Where there was cane there wasn't rubbish underneath, nothing for the fire to go through. Everything was harvested and cleaned up and replanted.

Now those fallow sugarcane fields are filled with Guinea grass. One Kaua'i fire chief says he has watched fuel loads of Guinea grass "double, even triple" in some fallow fields. Participants also said they missed being able to draw on farmers' and cowboys' expertise regarding local terrain (including locations of archaeological sites), weather, and fuel types. For example, ranch bulldozer operators partnering with firefighters knew where the lava tubes were in pāhoehoe lava fields because they had "been over that ground on horseback [and] knew the ground intimately."

The increase in lightning fires in upland forests was specifically noted by Big Island participants. According to one, fifteen years ago lightning-strike ignitions were unheard of on the island. During his first six years as a federal fire commander on Hawai'i he had heard of two lightning ignitions. "And in the past seven years, it is well over ten or eleven. . . . We're having these lightning strikes because it's a change in the different types of fuels that is not only fire-tolerant, but depends on fire to reproduce."

Protection of Cultural and Natural Resources

Although the primary responsibility of Division of Forestry and Wildlife firefighters remains wildlife fires, all fire response agencies must factor protection of natural and cultural resources into their suppres-

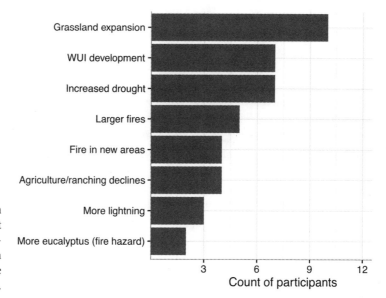

FIGURE 7.7. Changes in Hawaii's environment identified by project participants as having an effect on fire occurrence and suppression.

sion efforts. Fire is generally not a threat to archaeology sites and historic properties. In fact, it can be used as a tool to uncover, clean, and revegetate cultural sites.[18] Interviewees stressed that suppression efforts can do more harm than fire to *heiau* (ritual and religious structures, places of worship), petroglyphs, habitation sites, burials, and other cultural sites. If firefighters do not know where these sites are located, they can damage or destroy them when they are using bulldozers, bucket drops from helicopters, and hoses. A bulldozer can obliterate rock alignments, and the force of water from a bucket drop or hose can easily damage a wall. The former National Park Service IC would tell the helicopter pilots, "Go low and slow. If you're building line, you have to identify these critical areas." A Maui IC said he puts a scout in front of the bulldozer to walk the line and expressed the need to have "eyes on everything" to avoid archeological sites.

The leeward Waiʻanae Coast of Oʻahu (the most flammable region of the island) is a perfect example of a natural and cultural resource hot spot and was mentioned in several interviews. Low-lying coastal areas are ideal for bulldozer operations because they are flat, grassy, and offer easy access. However, they cannot be used in many of these areas because of dense concentrations of archaeological sites. In the uplands, special tactical considerations must be made for threatened and endangered species.[19] Helicopters are the best, and sometimes only, tool for combating fire in the mountains. The ocean is a close source for bucket dips, but agricultural reservoirs and portable diptanks are often used to decrease travel time. One participant said saltwater does not necessarily harm native flora,[20] although it does harm fauna. Knowledge of the location and value of cultural and biological resources is clearly critical, but these resources often pose difficult choices for ICs. One participant with extensive experience in Waiʻanae explained that ICs choreograph an intricate "air show" to avoid critical habitats, such as the exclosures for the endangered *kāhuli* tree snail (*Achatinella* spp. Swainson) on the ridges. "We can keep this fire to 200 acres," he said, "and treat the exclosure as secondary, or we can have the fire grow to 500 acres and prioritize the snails."

The Critical Role of Firefighters' Place-Based Environmental Knowledge

FIGURE 7.8. White ash rings from burnt *mamane* (*Sophora chrysophylla* [Salisb.] Seem.) and *naio* (*Myoporum sandwicense* A.Gray) trees during the 2010 Mauna Kea fire (photo by Jay Hatayama).

Also, as emphasized by many participants, the most important issues for all ICs are team safety, human safety, and public health.[21]

While destructive to native habitats, non-native feral ungulates—horses, cattle, sheep, and goats—provide controversial but critical ecosystem services, including fire pre-suppression. A number of ICs talked about the beneficial role of grazers in reducing fuels. An example repeated by ICs on Hawai'i Island was the Mauna Kea Fire of 2010 (Figure 7.8). Sheep had been removed by order of the federal government to protect the critical habitats of *palila* (Hawaiian honeycreeper, *Loxioides bailleui* Oustalet), mamane trees, and other endemics. According to interviewees, grass fuel loads grew out of control and primed the area for the 1,300 acre fire. As one of them noted, "The forest is still there with the ungulates. But one fire goes through the forest and it's gone. The ungulates have been here since the early 1800s and are now part of the environment."

There is clear evidence that feral ungulates degrade native ecosystems (Mueller-Dombois and Spatz 1975; Murphy et al. 2014; Scowcroft and Conrad 1992), but conservation-based decisions often do not factor in the consequences of their removal for increased fire risk.

Conclusions

In Hawai'i, the local heterogeneity of wildland fire triangle factors (fuels, weather, topography), the spread of pyrophytic savannas and grasslands in the nation's "homogenocene" (Samways 1999) hot spot, the rise of anthropogenic ignitions in wildland-urban interfaces, and increasingly warming, drying conditions present a unique set of challenges for fire suppression and management. In this chapter, based on interviews with fire commanders, we have focused on how fire decision-makers use place-specific environmental knowledge to guide operational responses. To a lesser but equally important extent, we have identified the critical role of local networks in transmitting knowledge. The collaborative dynamics of Hawaii's firefighting community and the knowledge network that has developed across fire agencies and islands are key to survival on multiple levels. Fire agencies overcome infrastructure, equipment, and personnel constraints through creative inter- and intra-agency and island cooperation built largely on personal bonds. In the words of a federal IC who fought fires around the mainland United States and in Europe before his thirteen years in Hawai'i, "I've never seen a more diverse group, different agencies working together and not worrying about what patch they're wearing, worrying about the fire and working together to put the fire out."

Additionally, working together on fires and attending trainings and conferences both support the informal exchange and generational transmission of knowledge and experience between senior commanders and firefighters and strengthens institutional memory.

Indigenous or Native ecological knowledge and resource management systems interpret and respond to ecosystem variability and disturbance (e.g., new crop diseases, climate change) through the complementary use of long-held or revitalized traditional practices (e.g., floral, faunal, air temperature, soil temperature, and moisture indicators in agroecosystems) and conventional science (Tengö and Belfrage 2004). Similarly, the adaptive knowledge of Hawaii's firefighters is dynamic. It draws on native Hawaiian local ecological knowledge, ranching and plantation management practices, technical and analytical training, and new developments in fire science to respond to changing fire regimes. An individual's experience, background, personal and/or family connections, familiarity with place, sensory acumen, and intuition all come into play in decision-making. Several examples were mentioned in the interviews:

- using the Hawaiian moon calendar and consulting with fishers to determine when the wind will die down and be the safest time for a prescribed burn
- spotting lava tube skylights themselves or consulting cowboys familiar with their locations when cutting a fire break through pāhoehoe lava to avoid lava tubes and pōhaku ki'i (pictographs)
- gauging relative humidity when sensors are malfunctioning
- knowing that the wind is likely to reverse direction at certain places during the night
- understanding how fierce winds and rain will differentially impact the fire behavior of native versus non-native grasslands, and when and where being "in the black" really constitutes a safety zone

Like fishers, salt collectors, hula and medicinal plant gatherers, and other cultural practitioners, firefighters are environmental ground-truthers contributing their observations and insights on climate change. One example is the dramatic and relatively recent uptick in lightning fires on Hawai'i Island. New climatology research using a novel storm model, for example, has demonstrated an increase in lightning strikes and the initiation of more wildfires in the United States due to global warming (Romps et al. 2014).

Heuristic devices or rules of thumb employed by Hawaii's firefighters have grown out of collective knowledge and personal perceptual expertise critical for making life or death decisions under immense time pressure. Examples include re-operationalizing strategy when firefighters hear a sound like a freight train (which indicates a fire is about to run), identifying a fuel source by the smell or color of smoke, and calling for backup when a distant fire plume is wider than one's thumb. Klein's (2000) model of naturalistic decision-making by fire ground commanders, emergency room medical personnel, military commanders, and other professionals in high pressure vocations highlights perceptual expertise. For instance, experienced nurses use certain cues to determine if a premature infant has developed a life-threatening infection (e.g., irritability, skin color). Many of these cues are covered in nursing education, but many others are not and may, in fact, be contrary to how sepsis presents in healthy neonates or adults. New nurses tended to miss signs of sepsis, resulting in higher death rates in preterm infants. Septic deaths were reduced significantly after perceptual indicators of infection in preterm neonates were compiled from experienced nurses and incorporated in training material (Klein 2000, 43).

In tactical training, fire commanders learn that decision-making is both analytical and intuitive. One interviewee estimated that the knowledge a commander brings to a fire is 30 percent didactic and 70 percent experience: "You just know from the intuitive cache in your brain what steps to take." He described how this worked for him when he drove around Maui: "I look at an area and think if I got a call right now, what would I do? So I play situational games."

The mental "slide projector" of past fires referred to in a few interviews is not made up of static snapshots but involves what Klein (2000, 45) refers to as a heuristic strategy of mental stimulation involving "the ability to imagine people and objects consciously and to transform those people and objects through several transitions, finally picturing them in a different way than at the start."[22]

The applied significance of these findings suggests that place-based understandings of Hawaii's firefighters can be used to better inform land and resource management practices and fire mitigation programs in the face of climate variability and change. Most of the fire commanders interviewed for the Challenges project referred to climate change implicitly rather than explicitly. Only three respondents used the term *climate change*, and only one used the term *global warming*. Nevertheless, participants did note environmental changes they had recognized over the course of their careers. Observations often related to the intensity, frequency, spread, and seasonality of fires, and their anticipation of more extreme weather and disturbance regimes in the future.

Some participants stressed the need for better development policy and planning. They argued that Hawaii's increasingly incendiary environment be taken into account by restricting new development in wildland-urban interface areas or mandating that planned communities be made firesafe, meaning that they be surrounded by defensible space, fire breaks, and adequate access. Two of the ICs recommended that golf courses surround resorts and residential communities rather than being located at their centers, as is currently most common. And the use of kikuyu grass on golf courses may need to be reconsidered. Although golf courses may seem to planners and developers to offer a green buffer or safe zone, from a firefighting standpoint, kikuyu turf grass can lead to dirty burns and extended fires.

Participants in the Challenges project also urged research and development of new prevention and response procedures based on past Hawaiian and agricultural resource management practices, including more controlled burns, letting fires run their course where assets are not threatened, using ungulates for prescribed grazing to reduce fuels (Elmore et al. 2005; Blackmore and Vitousek 2000), and revegetating with native plants. They also asked for more locally informed public outreach prevention and mitigation programs.[23] For example, the place-based knowledge of Hawaii's seasoned firefighters could be compiled and shared through training courses and guides, as well as through multiagency activities such as training and conferences. This suggestion is especially crucial because Hawaii's firefighting community will become further challenged by the loss of retiring ICs and chiefs with long-term knowledge who, in the words of one interviewee, "know how to put the pieces together in an operational picture—know which way the fire is going, what it's going to be doing, and what is necessary to stop it."

Acknowledgments

This project would not have been possible without the interview participants' time, thoughtfulness, and commitment to fire management in Hawai'i. *Mahalo nui* to all of the fire commanders, chiefs, and educators who were consulted for the Challenges to Rapid Wildfire Containment in Hawai'i Project. Thanks also to Christian Giardina and Trudie Mahoney of the USDA Forest Service for the impetus to undertake the Challenges project. Also, to Heather McMillen for reviewing a draft of the chapter. Funding for this publication was provided by the US Forest Service Region 5 Fire and Aviation Management and the Pacific Southwest Research Station Institute of Pacific Island Forestry (via RJVA #14-JV-1172139-034). Salary support was also provided by the College of Tropical Agriculture and Human Resources at the University of Hawai'i at Mānoa and the Joint Fire Science Program.

Notes

1. Sizeup, a term used universally by firefighters, is a quick initial assessment of environmental conditions in relation to available resources in order to set operational priorities.
2. In the words of one of the participants, "The more opportunities we can get the people together . . . before the fire, after the fire, when there is no fire, and do more training together and do more sit-downs together then we can speed up those relationships. . . . Find the reasons to get together whether, hey, we got a new truck we're blessing [Hawaiian blessing of new fire engines] or we're doing new training on this new truck."
3. In the words of a Hawai'i Island federal IC, "You want to talk about a loss. You know they say [in] government service no one is irreplaceable? That's not true. . . . I'd say about 30 percent of the brain trust of wildland fire response is with those two guys."
4. For consideration of how the alteration of the landscape and cultural practices that followed the introduction and expansion of cattle in the late 1800s were accommodated by the existing Indigenous Polynesian system of natural resource values and management practices, see Maly and Wilcox's (2000) history of range management in Hawai'i, Traditional Polynesian and Western Resource Management: Conflict and Assimilation.
5. A Hawai'i Island participant said, "We have so many mini microclimates. For example, Volcano and the golf course. Volcano Village is 2 miles down the road from Volcano Golf Course. Volcano Village can have 15 inches of rain in a month and the golf course will only have 5. You can have a 60 percent difference in rainfall in a 2 mile area." A Maui Island participant said, "We can be on one ridge and the [relative humidity] could be 20 percent on the windward side, and on the leeward side of the mountain the [relative humidity] could be 8."

6. An example of terrain challenges as described by an Oʻahu participant:

    On Oʻahu our landscape is dramatic. On the Big Island you have more sloping land. Topographically it's a problem. We don't have a lot of roads. We don't have access. We have steep slopes with rolling rocks, burning embers, and debris that can create spot fires. . . . On mild slopes we could scratch a line. But now we have to dig trenches to catch rolling embers and pine cones. Topography sucks on Oʻahu. If it was nice and flat and we didn't have arch[aeology] sites, we could utilize heavy equipment as well. If you have arch sites, you can't bring dozers there. If there are not access roads, you can't pump water out of a truck. So we have to carry backpack pumps and fly blivets. And helicopters are expensive.

7. There was general consensus among participants that instruments that work well for the mainland are often insufficient for Hawaiʻi. A statement that typifies responses is: "You can go online and see what the weather, wind speed, temperature, et cetera was an hour ago. That's all online. But things change so rapidly here. There are so many microenvironments here, that's hard to put online."

    An Oʻahu participant made the point that firefighters should have sufficient "qualitative indicators" in their experiential toolkit to not have to rely fully on "quantitative indicators."

    Discussing the Keetch-Byram Drought Index (KBDI), he described how it is useful for preplanning, but not for suppression:

    You can tell already everything is pretty dry, dead. It's not brown, it's gray, grass is dry, fire is burning strong. . . . There are all these qualitative indicators. Under eucalyptus there is no understory. There are all these twigs and leaves so fire smolders through that stuff. I've seen 18 inch flame heights. That's an indicator that the KBDI is high. [KBDI] is quantitative, while when you are out there you have all of these qualitative indicators. For example Maui, when the KBDI is above 600, they take their fire packs with them everywhere they go. So they use it more for preplanning than suppression.

    From a suppression side, after so many years of experience, hopefully, managers can tell from qualitative indicators than KBDI. That's the slide [projector] mentioned earlier.

8. Nāulu, meaning a "sudden shower" (Pukui and Elbert 1986, 263), can also refer to the sea breeze particular to certain locations, including a nāulu that blows off the coast of Maui influencing Kaʻaholawe Island's weather system. The Hawaiian language includes numerous place-specific terms for elements of the natural environment: winds, cloud formations, rain, ocean currents, and so on.

9. "Burn boss" is standard firefighter terminology for the person who supervises a prescribed burn from start (ignition) to finish (mop-up).

10. We note that lava fire starts are relatively infrequent compared to human ignitions.

11. Cutting line refers to carving a path by hand or bulldozer through exposed soil to block the fire's access to more fuel.

12. There is a prominent temperature inversion in the trade wind that streams over the eastern tropical oceans, including the Hawaiian Islands. Moist air extending above the surface is trapped by dry, warm air above. The inversion forms where the sinking dry, warm air meets the surface flow of cooler maritime air. One participant's observation of a Hawaiian hawk catching thermals high up in the air was a signal for him that the cap was off the inversion layer—in other words, the layer was lifting, a key indicator of atmospheric instability and large fire growth.

13. Curing—the process by which grasses die or become dormant and dry out—is a useful indicator for assessing fire danger. For a good example of a curing guide see the CFA Grassland Curing Guide (CFA, n.d.).

14. The key factor in establishing a safety zone, or "black area," is that it has no reburn potential. The black area also must position firefighters at a safe distance from radiant heat and have adequate egress and ingress routes.

15. Another aspect of the conservation conundrum is that invasive alien species sometimes provide critical habitat for native fauna. This is true of fountain grass and the endangered nēnē. It has been found that the nēnē on the Pōhakuloa

Training Area specifically choose taller fountain grass for nest sites and shorter grasses for food and mobility (US Army Garrison 2010).

16. Kiawe, introduced for cattle fodder, has naturalized culturally as well as botanically. It is a highly valued ethnobotanical resource used primarily as a fuel wood for cooking, and also for a variety of other purposes including making flour from the beans, carving, as a source of honey, and, more recently, making beer. From the perspective of plant users in Hawai'i, there is often not enough kiawe to meet the demand, especially for barbecue or imu (underground oven) fuel. Kiawe is emblematic of several contested or controversial floral and faunal species labeled "alien" and "invasive" that have been targeted for removal by conservation efforts but, at the same time, are regarded as culturally valuable to native Hawaiians and the diversity of ethnic groups that utilize natural-cultural resources. A signature of kiawe and other culturally naturalized species (e.g., waiawī, or yellow strawberry guava [*Psidium cattleianum* Sabine]) is that they often bear a Hawaiian name (Gollin et al. 2004).

17. The wildland-urban interface is the transition zone where human development (structures and infrastructures) meets or overlaps with undeveloped wild areas and/or vegetation, increasing the likelihood of anthropogenic ignitions and fire spread from one zone into the other.

18. Pili has come to play an important role in conservation and restoration projects, replacing invasive species such as fountain grass in island xeriscapes. One of the participants worked with University of Hawai'i researchers on a seven-year study at Pu'ukoholā Heiau burning off the alien buffel grass and reseeding it with native pili grass. Now the area looks like it did in depictions from 1790s. According to the participant who served as the project burn boss, native Hawaiian practitioners visiting the restored site commented that the wind blowing through the pili sounds different and is more pleasing than wind through buffel grass.

19. Three of the five protected Natural Area Reserves Systems (NARS) on O'ahu are in Wai'anae.

20. The firefighting and natural resource management communities do not fully agree that saltwater is harmful to native flora. The use of saltwater (as well as foaming agents) to extinguish wildland fires is a concern for natural resource programs.

21. A forestry IC stated that while the job of the Division Of Forestry and Wildlife may be "to protect the last gardenia on earth," the first order of concern for an IC, "even if it is the last gardenia on the face of this earth," is to never "put my men [in a situation] if I know it's going to be detrimental to them."

22. One of the commanders interviewed for Klein's (2000) study of naturalistic decision-making remarked, "To be a good fire ground commander, you need a rich fantasy life" (14).

23. In their overview of Social Science Findings in the United States, McCaffrey et al. (2015, 23) make the point that situational characteristics such as local ecological conditions are a key consideration for many residents, who have a higher likelihood of adopting fire risk reduction measures they view as appropriate to the local ecological context.

References Cited

Anderson, M. Kat. 2005. *Tending the Wild: Native American Knowledge and the Management of California's Natural Resources.* Berkeley: University of California Press.

Anderson, M. Kat and Michael G. Barbour. 2003. "Native American Practices in National Parks: A Debate: Simulated Indigenous Management: A New Model for Ecological Resoration in National Parks." *Ecological Restoration* 21: 269–277.

Athens, J. Stephen. 2008. "Rattus Exulans and the Catastrophic Disappearance of Hawaii's Native Lowland Forest." *Biological Invasions* 11 (7): 1489–1501.

Beavers, A., R. E. Burgan, F. M. Fujioka, R. Laven, and P. Omi. 1999. *Analysis of Fire Management Concerns at Makua Military Reservation.* Fort Collins, CO: Center for Environmental Management of Military Lands.

Benoit, J., F. M. Fujioka, and D. R. Weise. 2009. "Modeling Fire Behavior on Tropical Islands with High-Resolution Weather Data." In *Proceedings of the Third International Symposium on Fire Economics, Planning, and Policy: Common Problems and Approaches*, edited by A. González-Cabán, 321–330. Gen. Tech Rep. PSW-GTR-227. U.S. Department of Agriculture Forest Service, Pacific Southwest Research Station, Albany, CA.

Berkes, F., J. Colding, and C. Folke, eds. 2002. *Navigating Social-Ecological Systems: Building Resilience for Complexity and Change*. Cambridge: Cambridge University Press.

Berkes, F., and C. Folke, eds. 1998. *Linking Social and Ecological Systems. Management Practices and Social Mechanisms for Building Resilience*. Cambridge: Cambridge University Press.

Bernard, Russell H. 2011. *Research Methods in Anthropology: Qualitative and Quantitative Approaches*. 5th ed. Walnut Creek, CA: AltaMira.

Blackmore M., and P. M. Vitousek. 2000. "Cattle Grazing, Forest Loss, and Fuel Loading in a Dry Forest Ecosystem at Pu'u Wa'awa'a Ranch, Hawaii." *Biotropica* 32: 625–632.

Burney, D. A., R. V. DeCandido, L. P. Burney, F .N. Kostel-Hughes, T. W. Stafford, and H. F. James. 1995. "A Holocene record of climate change, fire ecology and human activity from montane Flat Top Bog, Maui." *Journal of Paleolimnology* 13: 209–217.

Castillo, J., G. Enriques, M. Nakahara, D. R. Weise, L. Ford, R. Moraga, and R. Vihnanek. 2003. "Effects of Cattle Grazing, Glyphosphate, and Prescribed Burning on Fountain Grass Fuel Loading in Hawaii." In *Proceedings of the 23rd Tall Timbers Fire Ecology Conference* [Tallahassee, FL]: *Fire in Grassland and Shrubland Ecosystems*, 230–239.

Chu, P. S., W. Yan, and F. M. Fujioka. 2002. "Fire-Climate Relationships and Long-Lead Seasonal Wildfire Prediction for Hawaii." *International Journal of Wildland Fire* 11: 25–31.

Commissioner of Agriculture and Forestry. 1903. *Territory of Hawaii Report of the Commissioner of Agriculture and Forestry for the Biennial Period Ending December 31st, 1902*. Honolulu: Gazette Print.

CFA (County Fire Authority). N.d. *CFA Grassland Curing Guide*. Accessed January 12, 2016. http://www.cfa.vic.gov.au/fm-files/attachments/Publications/curingguide.pdf.

D'Antonio, C. M., and P. M. Vitousek. 1992. "Biological Invasions by Exotic Grasses, the Grass/Fire Cycle, and Global Change." *Annual Review of Ecology and Systematics* 23: 63–87.

Desmond, M. 2007. *On the Fireline: Living and Dying with Wildland Firefighters*. Chicago: University of Chicago Press.

Dodson, J., and M. Intoh. 1999. "Prehistory and Palaeoecology of Yap, Federated States of Micronesia." *Quaternary International* 59: 17–26.

DOFAW (Division of Forestry and Wildlife). 2010. *Hawai'i Statewide Assessment of Forest Conditions and Trends State of Hawaii*. Honolulu: Hawaii Department of Land and Natural Resources, Division of Forestry and Wildlife.

———. 2015. "Native Ecosystems Protection and Management, Rare Plant Program." Accessed January 12, 2016. http://dlnr.hawaii.gov/ecosystems/rare-plants.

Dolling, K., P. S. Chu, and F. M. Fujioka. 2005. "A Climatological Study of the Keetch/Byram Drought Index and Fire Activity in the Hawaiian Islands." *Agriculture and Forest Meteorology* 133: 17–27.

Ellsworth, L. M., C. M. Litton, A. P. Dale, and T. Miura. 2014. "Invasive Grasses Change Landscape Structure and Fire Behavior in Hawaii." *Applied Vegetation Science* 17: 680–689.

Elmore, Andrew J., Gregory P. Asner, and R. Flint Hughes. 2005. "Satellite Monitoring of Vegetation Phenology and Fire Fuel Conditions in Hawaiian Drylands." *Earth Interactions* 9: 1–21.

Eriksen, C. 2014. *Gender and Wildfire: Landscapes of Uncertainty*. New York: Routledge.

Fowler, Cynthia. 2013. *Ignition Stories: Indigenous Fire Ecology in the Indo-Australian Monsoon Zone*. Durham: Carolina Academic Press.

Giambelluca, Thomas W., Qi Chen, Abby G. Frazier, Jonathan P. Price, Yi-Leng Chen, Pao-Shin Chu, Jon K. Eischeid, and Donna M. Delparte. 2013. Online Rainfall Atlas of Hawaii. *Bulletin of the American Meteorological Society* 94 (3) (March): 313–316.

Gollin, Lisa X., Heather McMillen, and Bruce Wilcox. 2004. "Participant-Observation and Pile Sorting: Methods for Eliciting Local Understandings and Valuations of Plants as a First Step Towards Informed Community Participation in Environment and Health Initiatives in Hawai'i." *Applied Environmental Education and Communication* 3 (4): 259–267.

King, Trevor G. 2004. "Fire on the Land: Livelihoods and Sustainability in Navosa, Fiji." PhD dissertation, Massey University, New Zealand.

Kirch, Patrick V. 1982. "The Impact of the Prehistoric Polynesians on the Hawaiian Ecosystem." *Pacific Science* 36: 1–14.

———. 1986. "Rethinking East Polynesian Prehistory." *Journal of the Polynesian Society* 95: 9–40.

Klein, Gary. 2000. *Sources Of Power: How People Make Decisions*. Cambridge, MA: MIT Press.

Lake, Frank K., and Jonathan W. Long. 2014. "Fire and Tribal Cultural Resources." In *Science Synthesis to Support Socioecological Resilience in the Sierra Nevada and Southern Cascade Range*, edited by J. W. Long, L. Quinn-Davidson, and C. N. Skinner, 173–186. General Technical Report PSW-GTR-247. US Department of Agriculture, Forest Service, Pacific Southwest Research Station, Albany, CA.

LaRosa, A., J. Tunison, A. Ainsworth, J. B. Kauffman, and R. Hughes. 2008. "Fire and Nonnative Invasive Plants in the Hawaiian Islands Bioregion." In *Wildland Fire in Ecosystems: Fire and Non-native Invasive Plants*, edited by K. Zouhar, J. Smith, S. Sutherland, and M. Brooks, 225–241. General Technical Report RMRS-GTR-42. USDA Forest Service, Rocky Mountain Research Station.

Maly, Kepa, and Bruce A. Wilcox. 2000. "A Short History of Cattle and Range Management in Hawai'i." *Rangelands* 22 (5): 21–23.

McCaffrey, Sarah, Eric Toman, Melanie Stidham, and Bruce Shindler. 2015. "Social Science Findings in the United States." In *Wildfire Hazards, Risks, and Disasters,* edited by Douglas Paton and John F. Shroder, 15-34. Waltham, MA: Elsevier.

McEldowney, H. 1979. *Archaeological and Historical Literature Search and Research Design, Lava Flow Control Study, Hilo*. US Army Corps of Engineers. Manuscript N. 050879. Anthropology Department, Bishop Museum, Honolulu.

McWethy, D. B., C. Whitlock, J. M. Wilmshurst, M. S. McGlone, M. Fromont, X. Li, A. Dieffenbacher-Krall, W. O. Hobbs, S. C. Fritz, and E. R. Cook. 2010. "Rapid Landscape Transformation in South Island, New Zealand, Following Initial Polynesian Settlement." *Proceedings of the National Academy of Sciences of the United States of America* 107: 21343–21348.

Mueller-Dombois, Dieter, and Günter Spatz. 1975. "The Influence of Feral Goats on the Lowland Vegetation in Hawaii Volcanoes National Park." *Phytocoenologica* 3: 1–29.

Murphy, Molly J., Faith Inman-Narahari, Rebecca Ostertag, and Creighton M. Litton. 2014. "Invasive Feral Pigs Impact Native Tree Ferns and Woody Seedlings in Hawaiian Forest." *Biological Invasions* 16: 63–71.

NOAA (National Oceanic and Atmospheric Administration). N.d. National Weather Service Storm Prediction Center. Accessed January 12, 2016. http://www.spc.noaa.gov/misc/Critical_Criteria_for_web.doc.

Pukui, Mary Kawena, and Samuel H. Elbert. 1986. *Hawaiian Dictionary*. Honolulu: University Press of Hawaii.

Ray, Lily A., Crystal A. Kolden, and F. Stuart Chapin III. 2012. "A Case for Developing Place-Based Fire Management Strategies from

Traditional Ecological Knowledge." *Ecology and Society* 17 (3): 37. Accessed January 12, 2016. http://dx.doi.org/10.5751/ES-05070-170337.

Romps, David M., Jacob T. Seeley, David Vollaro, and John Molinari. 2014. "Projected Increase in Lightning Strikes in the United States due to Global Warming." *Science* 346(6211): 851–854.

Ryan, K. C., A. T. Jones, C. L. Koerner, and K. M. Lee. 2012. *Wildland Fire in Ecosystems: Effects of Fire on Cultural Resources and Archaeology*. General Technical Report. RMRS-GTR-42-vol. 3. Fort Collins, CO: US Department of Agriculture, Forest Service, Rocky Mountain Research Station.

Samways, Michael J. 1999. "Translocating Fauna To Foreign Lands: Here Comes the Homogenocene." *Journal of Insect Conservation* 3: 65–66.

Schmitt, R. C. 1977. *Historical Statistics of Hawaii*. Honolulu: University Press of Hawaiʻi.

Scowcroft, Paul G., and C. Eugene Conrad. 1992. "Alien and Native Plant Response to Release from Feral Sheep Browsing on Mauna Kea." In *Alien Plant Invasions in Native Ecosystems of Hawaiʻi: Management and Research*, edited by C. P. Stone, C. W. Smith, and J. T. Tunison, 625–665. Honolulu: University of Hawaiʻi, Cooperative National Park Resources Unit.

Sheridan, Daniel P. 2014. "Firefighting: The Odor of Smoke." Accessed January 12, 2016. http://www.fireengineering.com/articles/2014/10/the-odor-of-smoke.html.

Smith, C., and J. Tunison. 1992. "Fire and alien plants in Hawaii: Research and management implications for native ecosystems." In *Alien Plant Invasions in Native Ecosystems of Hawaiʻi: Management and Research*, edited by C. P. Stone, C. W. Smith, and J. T. Tunison, pp. 394–408. Honolulu: University of Hawaiʻi Cooperative National Park Resources Unit.

Tengö, M., and K. Belfrage. 2004. "Local Management Practices for Dealing with Change and Uncertainty: A Cross-scale Comparison of Cases in Sweden and Tanzania." *Ecology and Society* 9 (3): 4. Accessed January 12, 2016. http://www.ecologyandsociety.org/vol9/iss3/art4.

Toman, Eric, Melanie Stidham, Sarah McAfferty, and Bruce Shindler. 2013. *Social Science at the Wildland Urban Interface: A Compendium of Research Results to Create Fire-Adapted Communities*. General Technical Report NRS-111. US Department of Agriculture, Forest Service, Northern Research Station.

Trauernicht, Clay, Lisa X. Gollin, Matthew P. Lucas, and Christian. P. Giardina. 2015. "Challenges to Rapid Wildfire Containment in Hawaii." Technical report.

Trauernicht, Clay, Elizabeth Pickett, Christian P. Giardina, Creighton M. Litton, Susan Cordell, and Andrew Beavers. 2015. "The Contemporary Scale and Context of Wildfire in Hawaii." *Pacific Science* 69: 427–444.

US Army Garrison. 2010. *Implementation Plan Pōhakuloa Training Area Island of Hawaiʻi*. Hawaiʻi Directorate of Public Works, Environmental Division, Pōhakuloa Natural Resources Office. Accessed January 12, 2016. http://www.garrison.hawaii.army.mil/sustainability/Documents/NaturalResources/IP/PTA/PTAIP.pdf.

Veldman, J. W., and F. E. Putz. 2011. "Grass-Dominated Vegetation, not Species-Diverse Natural Savanna, Replaces Degraded Tropical Forests on the Southern Edge of the Amazon Basin." *Biological Conservation* 144: 1419–1429.

Weise, D. R., S. L. Stephens, F. M. Fujioka, T. J. Moody, and J. Benoit. 2010. "Estimation of Fire Danger in Hawai'i Using Limited Weather Data and Simulation 1." *Pacific Science* 64: 199–220.

Welch, James R. 2014. "Xavante Ritual Hunting: Anthropogenic Fire, Reciprocity, and Collective Landscape Management in the Brazilian Cerrado." *Human Ecology* 42 (1): 47–59.

Welch, John R. 2012. "Effects of Fire on Intangible Cultural Resources: Moving Toward a Landscape Approach." In *Wildland Fire in Ecosystems: Effects of Fire on Cultural Resources and Archaeology*, edited by K. C. Ryan, A. T.

Jones, C. L. Koerner, and K. M. Lee, 157–170. General Technical Report RMRS-GTR-42-vol. 3. US Department of Agriculture, Forest Service, Rocky Mountain Research Station, Fort Collins, CO.

Wilmshurst, J. M., T. L. Hunt, C. P. Lipo, and A. J. Anderson. 2011. "High-Precision Radiocarbon Dating Shows Recent and Rapid Initial Human Colonization of East Polynesia." *Proceedings of the National Academy of Sciences of the United States of America* 108: 1815–1820.

Xiang, Wei-Ning. 2013. "Working with Wicked Problems in Socio-Ecological Systems: Awareness, Acceptance, and Adaptation." *Landscape and Urban Planning* 110: 1–4.

CHAPTER 8

Burning Lands

Fire and Livelihoods in the Navosa Hill Region, Fiji Islands

TREVOR KING

INTRODUCING NAVOSA

Anthropogenic fire is a part of the way of life in rural Fiji, and especially in the orographical rain shadow areas of higher islands in the Fiji group, as it is elsewhere in the Southwest Pacific. Fire was an indispensable tool of the Indigenous first peoples who cleared forest for cultivation beginning about 3,000 years ago, and is also an actant within an actor network of both environmental services and degradation. Fire serves various livelihood and development strategies that, in an average climatic year, maintain the productivity, resilience, and sustainability of the land-people nexus, or *vanua*. These livelihood and development strategies network with an unspoken pyrosocial ethos that is tolerant of fire and reflective of local pyrocultural mores and practices under normal climate conditions.

But climates vary. When the El Niño Southern Oscillation (ENSO) shifts toward its El Niño pole, drought and desiccation tend to follow, and normal land management practices, sometimes including uncontrolled and putatively careless burning, can lead to anthropogenic wildfires that ravage the leeward, rain shadow landscapes of the higher islands. The immediate effects include accelerated environmental degradation and increased levels of livelihood vulnerability among local communities. The longer-term effects include ecological changes including lower biodiversity, downstream and distant environmental modification, and the diminishment of local resource capacity.

The subprovince of Navosa can be found in the central-west interior of the island of Viti Levu, which is the highest and largest island in the Fiji group.[1] Navosa occupies a transitional space between the nearly continuously moist, forested uplands of the southeastern windward region (the wet zone) and the seasonally dry leeward plains, hills, and dissected upland ranges of the northwestern part (the dry zone). The name *Navosa* ("the vosa") has disputed origins. Although *vosa* means "word" in standard eastern, Bauan Fijian, my translation of a local participant's interpretation of *vosa* is "vista of [and possibly the sound of] open grassland and bare ridges" (see also Brewster 1920, 19). The name is especially appropriate if one enters Navosa from the northern or eastern forested areas. According to Tanner (1996, 239), "the term '*na vosa*' refers to the bare ridges remaining after a grass fire, a frequent seasonal occurrence in western interior Vitilevu." The region of Navosa is bisected by the Sigatoka River valley, which forms its core, surrounded by low-lying, gentle basin hills in the vicinity of its southern and middle reaches, and steep and rugged hill and mountain slopes around its headwaters, which roughly encircle Navosa. The Sigatoka River traverses almost all of Viti Levu, from the most elevated headwaters around Nadarivatu in the north to its mouth on the southeastern coast.

The predominant vegetation in the dry zone has been grassland or "sedge-fern-grasslands" (Ash 1992, 111),[2] especially on level and rolling terrain, upper hill slopes, and ridges (see also Twyford and Wright 1965, 84–86). An increasing number of non-native shrub and tree species dominate gully floors and are also present on southeast-facing escarpments at medium to higher elevations. This succession of species that are not native to Fiji continues unabated. A current example is the partly deciduous, fire-tolerant tree *masese* (*Gmelina arborea* Roxb. ex Sm.), which has invaded Navosa roadsides where seeds were transported into disturbed soils and is now rapidly spreading across the grasslands.[3] Native forest retains dominance only at the highest elevations and at the higher-moisture eastern margins close to the wet zone.

The interior valleys of Navosa range from about 35 to 270 meters above sea level and are the location of most settlements. Navosa's interior valleys are seasonally dry, similar to the leeward western zone. By contrast, the upland hills and plateaus range from about 550 to more than 1,200 meters above sea level. These zones are nearly continuously moist, like windward areas (King 2004). They are ideal for growing edible aroids and *yaqona* (kava, *Piper methysticum* G.Forst.) but are less suitable for dry-loving yam and cassava crops,[4] which are typically grown in the interior valleys.

The human population of Navosa is primarily composed of Indigenous descendants of the Lapita and more recently arriving peoples, who first arrived in Fiji about 3000 BP and began settling in Navosa around 2000 BP (Field 2004). Distinct increases in charcoal and changes in vegetation have been dated close to these times (Hope et al. 2009; Kumar and Nunn 2003; Southern 1986). Evidence of accelerated erosion in the Sigatoka Valley (the heart of Navosa) dates back to 2000 BP, most likely as a consequence of anthropogenic fire (Dickinson et al. 1998; Kumar et al. 2006) although Indigenous Navosan society has been a sustainable one over the *longue durée* (King 2005).

Local livelihood strategies are diverse, involving a mosaic of crop cultivation and Indigenous arboriculture (Clarke and Thaman 1993; King 2004; Yen 1974) on lands that also support cattle, goat, and horse pastoralism. Gathering, fishing, and hunting—especially in or near creeks, rivers, and forest margins—are prevalent and important during times of environmental or political stress. Root crops predominate. Today cassava (*Manihot esculenta* Crantz) is the main food crop in dry zones. Other common cultigens in the dry zones are yams (*Dioscorea* spp.), tannia (*Xanthosoma sagittifolium* [L.] Schott), plantains and bananas (*Musa* spp.), breadfruit (*Artocarpus altilis* [Parkinson ex F. A. Zorn] Fosberg), sweet potato (*Ipomoea batatas* [L.] Lam.), hibiscus leaves (*Abelmoschus manihot* [L.] Medik.), coconut (*Cocos nucifera* L.), and several other tropical plant and tree crops.

In the higher wet zone areas, the range of cultigens is similar, but the dominant food crop is taro (*Colocasia esculenta* [L.] Schott). Kava is important as a cash crop where edaphic conditions are suitable.

Agroarboriculture (the husbandry of a multistoried and interwoven mixture of food, medicinal, fiber, decorative, and other useful flora) is practiced around villages and settlements as well as in niches in secondary forest areas, where cultivated spaces are typically richly layered and interplanted with many useful plants, shrubs, and trees. These niches are most common in fertile, moderately sloping gullies at medium elevations, but are difficult for an outsider to distinguish using aerial photography or when viewed from afar. From a distance, the agroarboriculture niches often appear homogenous with secondary forest dominated by *Leucaena leucocephala* (Lam.) de Wit and bamboo, together with other native and non-native species, where only the upper stories are visible. Occasionally a tall upper-story fruit tree such as mango or coconut may indicate the location. The understories are densely planted with tannia and other shade-tolerant ground crops. Yam vines (often using *L. leucocephala* trunks for support), kava, papaya, and small trees including juvenile breadfruit occupy the middle stories. Despite being in somewhat vulnerable locations, especially during droughts, fire seldom intrudes into these agroarboriculture niches, which suggests that local people protect and manage the important ecological services and natural resources they provide.

The Indigenous Fijian people of Navosa have lived comparatively traditional lifeways since the time of settlement. An institutionalized, protective polity (France 1968, 1969; Macnaught 1974), together with physical isolation from both the global capitalist world system and the regional political economy, insulates them from modernization. This is despite the various penalizing effects of social and environmental changes attending colonialism (Nicole 2011) and ecological imperialism (Crosby 1986, 1994). The Navosan communities of today have identities based on a mixture of ancestral and political genealogies, cultural icons, and territories where geographical locations such as historic places, settlements, villages, and districts figure highly within larger regional frames of province, division, and nation. Within and between communities, identities interweave intricately with family histories and cultural frames such as local spaces and places, and special maternal family and cross-cousin relationships characterized by distinctive behavioral mores that transcend dominant Fijian social norms. A characteristic cultural feature is *tako/lavo* (father/son) alternate generation descent groups, coinciding with the ethnolinguistic region of Navosa and including part of the provinces of Serua and Namosi (also see Hocart 1931). From the perspective of local culture and heritage, the concept of Navosa articulates in a more extensive and complex way than the geographic Navosa provincial boundary initially drawn under colonial authority. The cultural boundary extends south down the east side of the Sigatoka River as far as the Serua coast, overlapping with the ritual *naga* districts described by Fison (1885).

Navosa is regarded by outsider Fijians as a place where cultural traditions persevere. Group solidarity is more important than individual ambition, especially in the less accessible heartland. Navosa is slowly trending toward articulation with the global capital-oriented economy. Yet the local Navosa economy can still be aptly described as a locally reliant, cultural welfare economy where customary sharing and egalitarianism are the rule, in contrast to an economy based on distant supply and individual gain.

British colonialism and ecological imperialism have had profound effects on the environment; for example, colonial authorities wanted to promote pastoralism through "livestock plantations" (McNeill 2003, 359), and to this end they introduced the invasive and pyric disclimax grass *Pennisetum polystachyon* (L.) Schult. to Fiji in the early twentieth century (see Wright 1920), significantly altering Navosa ecotones (Parham 1955), replacing native and other species (Parry 1987), and exacerbating fire-induced land degradation (Cochrane 1968, 1969; King 2004). An accurate assessment of degradation is difficult, however, not only because of intraregional spatial and topographical differences but also because of a lack of scientific studies and the unavailability of precise data about changes in land cover. The few studies available were often precipitated by disasters such as cyclones and droughts. Unfortunately, they were conducted ad hoc, and there is a lack of continuity in the data. This analysis therefore relies on historical reports and my own data collected through 2013.

Late nineteenth-century and early to mid-twentieth-century reports suggest that the native grass *Miscanthus floridulus* (Labill.) Warb. ex K. Schum. & Lauterb. was dominant at that time. Other species in the landscape included a type of lemon grass (*Cymbopogon refractus* [R.Br.] A. Camus). These species covered substantially greater areas than is the case today (Horne 1881; Twyford and Wright 1965). After 1925, *M. floridulus*, *C. refractus*, and other species began to be replaced by *P. polystachyon*, which is the current vegetation disclimax on medium-fertility soil.[5] Two factors appear to be important in this succession: the soil fertility preference of each species and the frequency of anthropogenic vegetation fires. The tall and dense *M. floridulus* tends to dominate on higher-fertility sites if undisturbed, whereas the medium-sized bunch grass *P. polystachyon* dominates on less fertile sites when aided by fire.[6]

The replacement of *M. floridulus* by *P. polystachyon* indicates a decline in overall soil fertility associated with the former's vulnerability to fire and the latter's fire resilience. *M. floridulus* may die out after two or three burns in quick succession, whereas *P. polystachyon* is well adapted to repeated fire. I concluded that *M. floridulus* is far less supportive of land-degrading frequent-fire regimes than communities of *P. polystachyon*, for which the term *pyric disclimax* is appropriate. Further, *M. floridulus* is a managed species that serves to protect Indigenous cultivation sites and reduce land degradation. This is in contrast to its common portrayal as a weed, especially by pastoralists. *M. floridulus* is also used by local communities for house building, fish fences, and garden stakes (Twyford and Wright 1965) among other things. *P. polystachyon* is also useful for building houses. Like *M. floridulus*, it is harvested for roof thatch and provides soft cushioning between the earth floor and overlaying *Pandanus* mats in traditional Navosan *were* (Fijian *bure*) houses. In fact, *P. polystachyon* is preferred over *M. floridulus* for these purposes, being both a better thatch and a softer cushion. The trend in Navosa villages since about 2008, however, has been to replace smaller thatched houses with larger tin-roofed houses. Thatched houses are now becoming uncommon despite their superior comfort for sleeping and coolness in hot weather. When asked why, community participants replied that traditional building materials were becoming scarce. I will add that the reduced availability of skilled traditional builders, who can now earn higher wages on coastal tourist *bure*-building projects, may also be a factor.

Navosa experiences periodic droughts broadly concomitant with the ENSO (Curtis 2008; Deo 2011; Mataki et al. 2006; MFARD and UNDHA 1994; PCCSP 2014). My initial research on fire and livelihoods (King 2004) began during 1998 when communities complained about the vanua-degrading effects of fire and their consequences, which were accompanying the desiccating effects of a severe drought precipitated by the 1997–1998 El Niño event.[7] My subsequent research during this drought episode revealed the participants' reasons for burning the land, the time of the year when they did this, the amount of uncontrolled burning, the effects of repeated fires, and the specific households who were responsible for excessive burning.[8]

Despite the destruction of large areas of vegetation by uncontrolled fires in severe drought years, it was the normal practice, contiguous with pyrosocial norms, to ignite fires for various well-accepted livelihood reasons (King 2004) in any year. The most important reasons for igniting fires were clearing land for gardens, stimulating new grass for the animals, and harvesting wild yams. Swidden and permanent gardens (each around 50 square meters in area) were predominantly located on often steep hillsides but occasionally on narrow colluvial-alluvial areas at the base of hills. "New" grass refers to the tender, fresh young basal shoots that arise from existing grass clumps after fire, and not new grass plants (Figure 8.1). These tender and palatable shoots provide food for usually free-ranging domestic ungulates during the dry season, when forage is lacking, and especially during a drought period such as that influenced by the 1997–1998 El Niño episode. (New grass is given an extended description here for important reasons that emerge later.) The practice of using fire to stimulate new grass shoots is, in effect, a way of producing seasonal forage for livestock animals retained under loose domestication (fencing, often over rugged terrain, is usually sparse and weak). The practice is ubiquitous in Navosa and other regions of Fiji with a similar biophysical environment.

At the beginning of the dry season, the mature clump grasses, and especially *P. polystachyon*, which dominates the higher hill slopes, have an exterior of tough, unpalatable, dying, and dead stalks and leaves that smother competing vegetation and provide excellent fuel for seasonal grass fires. Throughout the dry season, farmers ignite fires on the downhill margins of these grassland tracts whereby the fires traverse the slopes but often extinguish at the peak of steep ridges; alternatively, the fires can spread relatively uninhibited over rolling lands constrained only by heavily vegetated gullies or steep escarpments. After the fire, bunch grasses quickly regrow: fresh leaves usually emerge in one or two weeks from the basal parts of mature *P. polystachyon* clumps. The recovery of other plant species is either absent or much delayed. There are three main agroecological effects on the grassland landscape after firing: first, fire-sensitive species competing with the clump grasses are constrained or extinguished; second, the clump grasses are stimulated into perennial regrowth; and third, the bared land is highly susceptible to rainstorm-induced sheet erosion—especially risky later in the year near the beginning of the wet season. In November 1998, heavy rains fell on the Navosa landscape, and soil erosion was severe in places where *P. polystachyon* and late burning coincided.

Access to areas where wild yams grow is much easier after the surrounding vegetation (often dense, nearly impenetrable stands of *M. floridulus*) has been opened up through burning. This practice facilitates the task of locating wild yams, as indicated by the shoots that emerge from soil exposed by fire.[9] All of the reasons above were generally accepted by the communities concerned, although local participants considered extensive uncontrolled wildfires in drought years unacceptable outcomes.

Negative proximal effects of excessive wildfire included damage to gardens, fallow areas, potential thatch, potentially valuable timber, wild foods and medicines, fences, and reduced control of livestock. Regarding the latter, despite the usual dry season strategy of burning to reinvigorate grass stocks (involving the stimulation of new basal shoots) as a livestock confinement measure (free-ranging

Figure 8.1. Fresh regrowth from a fire-resistant clump of invasive mission grass (*Pennisetum polystachyon* [L.] Schult.) near Nawairabe, June 1998, during the dry season when mature grass is unpalatable to livestock.

domestic animals are attracted to and locate themselves in areas of new forage), excessive burning from uncontrolled fires increases the number and spatial extent of confinement areas, defeating the original plan to confine livestock to limited areas.

In years with normal precipitation I have observed that fire-induced environmental degradation is less widespread and potentially controllable, although still occasionally destructive, especially as a consequence of burned areas being exposed to sudden, extreme high rainfall events (some associated with cyclone activity). Erosion, land degradation, and downstream sedimentation are regional and national outcomes. The local communities indicated awareness of the first two outcomes, but sedimentation is only recognized as a problem when local streambeds become altered, and fishing or recreation is affected. Thus Navosa communities are alert to and concerned about apparent changes in the resource capacity of their land, but perhaps like communities elsewhere, their concern seldom extends to downstream or wider regional and global effects.

In summary, despite a sustainable *longue durée* history and many good livelihood reasons for igniting fires in normal seasons, Navosa communities consider excess burning in extreme drought seasons to be destructive. Current developmental changes involving sectoral, institutional, and sociocultural fragmentation; modernization; land use changes; and extraction of nonrenewable natural resources (especially native timber) inequitably benefit an educated hierarchy and elites associated with a distant capitalist economy (King 2004). This process involving economic development and modernization appears to be exacerbating the vulnerability to fire associated with land cover desiccation during extended drought periods, with the potential outcomes of lowered resilience and future unsustainability in the vanua and the region.

A Tinderbox Paradise? Pyrosocial Fire and Livelihoods

In 1998 and 1999, my research (King 2004) focused on why fire was used in relation to land use and sustainability in Navosa. The study followed a participatory development (Cancian and Armstead 1992; Chambers 1997; Oakley 1987; Rocheleau 1994) and actor-oriented (Long and Long 1992; Long 2001) approach. A reflection upon that approach more recently led to the observation that fire is a key actant (Akrich and Latour 1992), meaning that it combines both human and nonhuman agency in a pyric network of land and people, herein the vanua. The pyric network is composed of a periodically fuel-laden and fire-prone landscape that exists in concert with a set of anthropogenic, livelihood-oriented ignition events. (As an aside, the particular landscape and the relevant events may also be described as collaborating actants involved in network processes.)

The 1998–1999 research involved three different modes: (a) three to five weeks of in-depth studies in each of two village communities, Nasauvakarua and Nawairabe; (b) a regional survey involving day-long participatory assessments and informal interviews in men's and women's groups in each village or settlement; and (c) research including interviews conducted in Fiji outside of Navosa. The emphasis throughout was on participatory methods, including approximately 133 interviews, of which 68 were in-depth household evaluations (sometimes culminating in post hoc interviews), plus numerous expert interviews. The four-week survey involved eighteen village or settlement communities across the Navosa region.[10] Interviews were conducted informally, consistent with Fijian mores. I used a notebook for initial recording, and notes were later transferred to journals. I also used snowball sampling, where interviewees helped direct the investigative path.

The relationship between fire and land use emerged as a key research question during consultations with local participants concerned about fire levels at that time. I became aware of various norms, mores, and perspectives involving fire use and practices that were integral to the Navosa vanua, its social and cultural lifeways, and livelihoods. While I was immersed in the communities, I began to sort the Indigenous knowledge being shared with me into categories and research criteria. This information was synthesized, and I created a semistructured household questionnaire while lying on the floor inside a Navosan *were* in Nasauvakarua. The questionnaire guided the household evaluations and later provided criteria for the regional survey.

Before describing the study's findings, I will touch on theoretical concerns in an attempt to share some insights into Fijian perspectives and practices around fire. I have decided to use the terms *pyrosocial* (the first use of this term to my knowledge) and *pyrocultural* for social and cultural practices connected to the local ethos toward fire. This ethos was unspoken, indicating an awareness of potential punitive actions by other local persons and communities, and by distant authorities. From the local perspective, inhabitants and communities share a pyrosocial ethos that is usually tolerant of fire, but there are some apparent contradictions. For example, individuals could respond to damage to their livelihoods caused by an uncontrolled local fire and seek retribution if they could identify the culprits.[11] The perspectives of distant authorities, however, were markedly different. Government personnel were generally unsympathetic toward any pyrosocial ethos involving tolerance of fire with some exceptions, including officers associated with prescribed burning. The general government view was that uncontrolled fire, and sometimes fire in general, was a destructive, dangerous, and unsightly risk to the natural and other resources that they are trying to manage. One of these resources is the aesthetics of the landscape, significant for Fiji and its tourist industry. I wonder, though, for whom are they managing these resources? How are they managing them? The answers to these questions are potentially complex and rightfully demand further analysis, but a proper consideration is not attempted here. (Analysis by Kull [2002, 2004] of related issues in Madagascar is apropos.) Nevertheless, some indications emerge later in this chapter in a discussion involving the history of fire in Navosa.

Participants from the eighteen Navosa communities, mainly interviewed in separate female and male groups, all of whom practiced rural livelihood strategies, reported a range of reasons for igniting fires and shared their estimates of land burned by uncontrolled blazes. The separate questionnaire evaluations held in Nasauvakarua and Nawairabe villages involving sixty-eight households (coincidentally, each village had thirty-four households) provided additional and in-depth data. I recorded the groups' and households' reasons for igniting fires and evaluated them through open questioning, the use of visual sorting and quantifying techniques, and triangulation. For example, the visual, open question exercise associated with gathering the reasons for igniting fires involved the use of 10 x 10.5 centimeter cards labeled with reasons in Fijian,[12] which were prepared when each reason was initially proffered. As I spent more and more time in the research site, with more and more communities, the number of reasons increased. The cards were kept in a pack and reused when the same reason was proffered in subsequent interviews or evaluations. For example, in one evaluation the cards were overlaid on the "burning" calendar (Figure 8.2). The efficiency of recording brief translated responses to open questioning was greatly improved by using this technique.

Navosans used seeds from the sacred *vesi* tree (*Intsia bijuga* [Colebr.] Kuntze) as tokens to score their reasons.[13] In another exercise, two differently colored circular cards and a paper pin were used to construct a handmade adjustable "fire wheel" to facilitate interviewees' and participatory groups' estimations of the percentage of land burned by uncontrolled fires. In Figure 8.3 the red portion on the disc represents the amount of land burned due to carelessness or accidents (uncontrolled burning in relation to total land burned). The proportion of the active color (red) over the background color (cream) was indicated by discrete percentage markings on the rim of the 26 centimeter diameter fire wheel (Figure 8.3).[14]

This visual technique greatly aided data collection because most participants were not familiar with percentages, and the expression of fractions using the complex Fijian terms was a relatively slow process for all concerned. These Participatory Rural Appraisal (PRA) techniques were incorporated into both the sixty-eight village household questionnaires and those used with the thirty-five men's,

FIGURE 8.2. A senior research participant scoring reasons for burning with cards placed on a burning calendar and then sorting *vesi* (*I. bijuga*) seeds into heaps below them to show their relative importance.

women's, and combined participatory groups held as part of the regional survey involving all eighteen of the study villages and settlements.

The main question of why Navosans burn land and the importance of different reasons for igniting fires (Figure 8.4) were evaluated during the regional survey in small reflexive, participatory groups of three to twelve people. The Turaga-Ni-Koro (elected headman) organized the groups on the day I visited the community, and all evaluations except one were conducted in that same community on the same day in separate female and male groups. There were seventeen male and seventeen female groups in seventeen communities, plus one mixed group in another community where insufficient numbers were present for two groups. Thus, there were thirty-five groups over eighteen communities.

I asked each group of interviewees and householders the open-ended question "Why is the land burned?" The same question was asked during sixty-eight individual household evaluations in Nasauvakarua and Nawairabe. I recorded their reasons in Fijian and then placed them on the ground in front of the interviewees, who scored the proffered reasons with vesi (*I. bijuga*) seeds. The number of reasons proffered and cards scored and recorded ranged from three to seven for each interview. The scoring process was as follows: I first gave participants ten seeds for each card (or reason) that they had identified (e.g., four cards = forty seeds for each); the participants then sorted the seeds into heaps next to the cards to show the importance of each reason. With the help of an interpreter, the seeds were counted and the data recorded in a notebook. The "ten

FIGURE 8.3. A boy helping one of the older interviewees use a fire wheel to estimate the percentage of land affected by uncontrolled burning.

seeds" approach was taken in an attempt to make data collection more efficient, standardize the method, and minimize any variance that might be introduced if interviewees could choose any number of seeds. In practice, however, the total number of seeds used probably does not matter when the important evaluation involves a visual comparison by the interviewee/s of different amounts (not numbers) of seeds aligned with different reasons represented on cards, recorded as numerical data, and interpreted as percentages. I discuss the research results below.

Why Navosans Burn Land

The first main interview question was "Why is the land burned? The results from

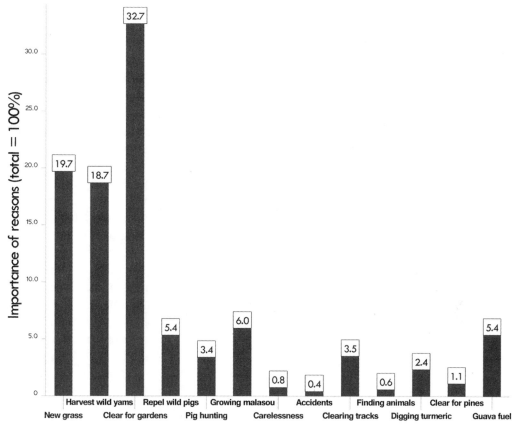

FIGURE 8.4. The reasons identified by Navosans for burning land and their relative importance based on scores from thirty-five evaluations in eighteen communities.

the female and male participant groups are shown in Figure 8.4. The three highest-scoring reasons, consistently mentioned by nearly all groups, were clearing land for gardens, providing new grass for animal forage,[15] and harvesting wild yams (predominantly *Dioscorea nummularia* Lam.).[16] Other reasons were often more localized or had a temporal aspect. Clearing trails (or "tracks" in Fiji) was mentioned in only six of the eighteen villages. Repelling wild pigs was mentioned in seven villages. Clearing land to plant juvenile pine trees was mentioned in two villages. Clearing vegetation for digging turmeric (*Curcuma longa* L.) was mentioned in four villages. The harvesting of dead branches from stands of guava trees (*Psidium guajava* L.) was facilitated by singeing them with a scrub fire, which was mentioned in five villages.[17] These less frequently mentioned reasons were often important in the particular locations concerned. The need to repel wild pigs, for example, was exigent in the dry season for those communities with gardens close to forest margins, such as Nasauvakarua.

Many of the reasons overlapped in purpose. Clearing trails directly helps the movement of both humans and animals (especially horses).[18] Indirectly, clearing trails helps with harvesting of numerous products, including fuelwood, turmeric, and building materials. It is likely that clearing trails is a subsidiary, if implicit, outcome of the three main reasons for burning and therefore may be more important than the numerical data suggest.

Some gender differences in burning practices were apparent. For example, women more often mentioned lighting fires to stimulate growth of the wild green vegetable *malasou* (*Solanum americanum* Mill.). They also reported burning to clear vegetation and facilitate the harvesting of *kari* (turmeric corms) and guava fuelwood more frequently than men. However, they reported burning to hunt pigs less frequently. Malasou grows well after hillside fire and is valued as a relish for cassava and other carbohydrates during droughts when the normally preferred green vegetable species suffer desiccation. Turmeric corms are processed within settlements after harvest in the late dry season and sold at markets or to travelling buyers, as in 1999 when an increase in value of dried turmeric for export to Japan as an organic product occurred. Guava fuelwood is highly regarded for its slow burning but intense heat output and relatively smoke free quality. Only a few native species, now locally rare, are superior as fuelwood.

During individual interviews, Navosans indicated numerous minor or secondary reasons for using fire. A government land use research officer, who was himself a farmer, mentioned that control of soil disease, especially anthracnose (a pest afflicting yams), was important even though local participants did not mention it.[19] Therefore, in the present study, controlling soil disease is assumed to be implicitly involved in clearing for gardens, which was the highest scoring reason for burning. Further research is needed to evaluate this and other details.

In general, carelessness and accidents were reasons that did not score highly, suggesting that igniting fire was largely controlled and intended for specific purposes within communities. As discussed later in this chapter, this finding may also support the notion that pretermission (the act of overlooking as part of pyrosocial norms and the status quo) rather than carelessness (nondeliberate actions contravening sociocultural norms), per se, best describes the causal background here.

Evidence of burning in the landscapes between communities, initiated in proximity to the cooking fires of groups of people traveling between villages, was observed along main valley bottom corridor trails in secluded locations. Lightning tends to be confined to the wet season and is negligible as a cause of ignition in Navosa and elsewhere in Fiji.[20]

Urban, external government officers and other employees of all ethnic backgrounds tended to espouse less informed, even stereotypical reasons for fire ignition, such as hunting wild pigs (as the main and only reason) or careless disposal of lit cigarettes from horseback. These characterizations were generally disparaging of local practices, which are often regarded with barely concealed disdain. For example, a strong disjunction exists between urban officials' and rural persons' perspectives of guava. Government officials consider it a scourge that has ruined pastures and prevented other economic land use, whereas adult villagers regard it as a valuable source of fuel, medicine, and monetary income, including from the local cannery that buys ripe guava. As mentioned above, dead guava branches harvested for fuelwood are valuable as a relatively clean and slow burning cooking fuel with low smoke emissions, and they are especially valued by women. Children, in contrast, have another perspective: they value guava for its enjoyable edible fruits. The nutritive value of guava, although significant from a scientific viewpoint, was not usually mentioned, perhaps because participants categorized it with the medicinal aspect.

When Does Intentional Burning Occur?

The question of when intentional burning occurs was evaluated by asking men's and women's group participants in Nasauvakarua and Nawairabe to construct a participatory burning calendar (Figure 8.2). In Nasauvakarua, intentional burning for most reasons began in June, the start of the dry season, and normally extended to the end of September. An exception was burning for new grass, which was done during dry spells from June through January. In normal years, the wet season arrives around October, but a drought in 1998 had facilitated or demanded the extension of burning for new grass through even later in the season.[21] In the same year, wild yams were harvested much later than usual, continuing until November, apparently in response to high prices caused by drought conditions for tubers at Sigatoka, Nadi, Ba, and other market towns. In Nawairabe, burning typically began in May or June and continued until the end of October. Digging early yams sometimes entailed burning earlier in the year. Also, burning sometimes occurred later for small areas of new grass and clearing gardens.

To summarize, rural community members reported starting fires for various pragmatic livelihood reasons in the early dry season. Carelessness was seldom mentioned. Drought in the late dry season, however, stimulated continued burning for new grass shoots to feed hungry animals because of a dearth of available forage. This practice increases the risk of severe soil loss from sheet erosion if storms arrive at the beginning of the wet season. Travelers were reported to have caused wildfires by allowing cooking fires to escape along secluded access corridors. Many of the reasons for burning mentioned by community members overlapped, with secondary purposes being subsumed by primary purposes. Urban government officials often had a poor understanding of the relationships between livelihoods and fire, although local agricultural experts were aware of the connection between hungry livestock, late burning, and new grass.

Another temporal cause was related to political tensions, mainly affecting land-owning communities that had planted exotic forests with species such as *Pinus caribaea* Morelet var. *hondurensis*. Political tensions were more significant in coastal regions than in heartland Navosa, but jealousies, rivalries, and feelings of injustice could nevertheless make plantations and gardens targets of arson. These tensions occur both within and across community boundaries; some of many possible reasons included: (a) elimination of alternative livelihood sources (e.g., wild turmeric; wild vegetables, fruits, and medicines; fuelwood; grazing pasture) caused by logging or plantation enclosure; (b) contested rights to land; and (c) feelings of injustice from challenges to egalitarian norms. Historically, forests containing exotic species tend to suffer fire damage during drought years (Were 1997), especially during political conflict (as in 1987, when a coup occurred).

Uncontrolled Burning: From Watershed to Soilshed?

Another main question in both the in-depth interviews and group evaluations was "What part of the land that gets burned is because of carelessness or accidents?"; I used the fire wheel (see Figure 8.3) to help gather answers to this question. Each interviewee or group was presented with the fire wheel (set at the 50 percent / 50 percent position) and asked to adjust it so that the red part represented their answer to the question.

The overall average of responses from sixty-eight in-depth interviews of the total area of land burned was 73 percent. In all except one anomalous case, participants showed a high degree of certainty in presenting their answers. The results showed a strong level of consistency with low variance. The standard deviation from the overall mean of the sixty-eight in-depth interviews was 12.6, indicating that local participants had consistent perceptions of the amount of uncontrolled burning.[22] The anomalous case was associated with one senior group participant who was observed to pressure those junior to him to dramatically lower their initially reported figure. The same anomalous figure occurred when that senior participant was interviewed in his household, lowering the mean and increasing the standard deviation of both the group and household assessments. The mean of the thirty-five group evaluations was 71 percent with a standard deviation of 13, but 73 percent with a standard deviation of 11.7 with the anomalous response removed ($n = 34$).

The mean locally reported proportion of land damaged in 1998 by uncontrolled fire (73 percent) was very large that year because grassland fires spread far and wide. The hilly sectors from approximately 35 to 1,000 meters above sea level were dominated by a patchwork of black burnt grassland on slopes, ridges, and plateaus, with browned forest margins and niches, and isolated green pockets of gully vegetation. Young timber plantations and regenerating forests were often destroyed, and even mature exotic plantations were severely damaged by the blazes.

Plantation forests were mainly *P. caribaea* in the dry zone, with some slash pine (*Pinus elliottii* Engelm.) and mahogany (*Swietenia macrophylla* King) in relatively high and moist zones, such as the windward wet zones where wildfire is less common. The *P. caribaea* plantations in dry zones of Navosa were established in the 1980s and 1990s as part of a community forestry development effort led by Fiji's Ministry of Forestry. They were relatively small (from less than 1 to about 5 hectares) and were usually located on steep hillsides prone to fire and difficult to harvest with machinery. These plantations can be controversial because they resist alternate land use, and local chiefs, via their usual twin roles as land and financial managers, tend to have a prerogative over any profits from logging, despite community-oriented and vanua-oriented purposes. It was not unusual for some community members or subclans/clans to claim that they were unjustly served by developments in the vanua, either during past history (e.g., being victimized during the early twentieth-century Native Land Commission transactions) or as a result of recent changes including the development of community forests. Consequently, some community members—particularly those who have rivaled the current leadership, or those who identified or profited more with lands adjoining or outside of the main community—may have felt they had little reason to prevent fire damage to certain community resources and forest plantations, in particular. This complex situation is dynamic and ongoing, and illustrates how land management and political ecology are intertwined in Navosa as elsewhere (Blaikie and Brookfield 1987).

The combination of steep topography and high temperatures in the late dry season dramatically increase fire risk and severely constrain the cutting of firebreaks. Clearing a firebreak 2 fathoms (3.6 meters) wide around cultivation plots on very rough and steep (often 35 degree slopes) hillsides when temperatures are very high is an immensely trying and rather dangerous task. Community members deem firebreaks impossible to construct when they aim to burn larger areas to generate new grass for animals. Even cutting firebreaks around small garden plots is arduous under these conditions. I observed that most cultivators worked barefoot. The hardened soles of their feet were their only foot protection, though usually sufficient.[23] Villages and settlements, generally located downwind on flat land near waterways, did not suffer fire damage, in part because the surrounding grass was kept short by grazing, cutting, and frequent fire, and because wind strength was low (less so at higher elevations) in the dry season.

The rainy season normally begins about October, but in 1998 it was delayed until the middle of November. Initial heavy rainstorms caused severe soil erosion as rain impacted denuded and bared upland grasslands and late-planted cultivation areas. In some places, severe sheet erosion occurred over steep grassland hillsides, with soil loss of 1 centimeter or deeper exposing the roots of bunch grasses and substantially altering waterways. The devastation was plain to see in late 1998, but it is uncertain whether similar amounts of erosion occurred following every drought. What is certain is that gully erosion is advancing relatively rapidly. Many senior participants pointed out historical changes in specific valley landscapes due to erosion. One of these, the historical down-cutting of the Rogorua Creek in the Solikana Valley of Navatusila District, is discussed by King (2012). These edaphic changes appear to be an unaltered, or perhaps accentuated, continuation of prehistoric erosional processes (Parry 1987). Unfortunately, there is a lack of regional data for historical comparison other than general and anecdotal descriptions, particularly from colonial officials and, curiously, also during El Niño drought episodes.[24]

Highland Views: El Niño Soilsheds, Colonial Reaction, and Fire Regulation

The following descriptions are from the botanist Horne (1881), who visited Navosa in 1878, a time described elsewhere as the "the worst El Niño event in the nineteenth century"[25] (Grove and Chappell 2000, 15):

> Unfortunately these fires are still occurring, and only last dry season they laid waste to a great part of the province of Navosa.... (Horne 1881, 134)
>
> By burning the grass off the ground the soil is laid bare, and the surface is cracked and crumbled into dust by the heat of the sun. From the steepness of the ground in some places, the first heavy rain carries off the loose surface soil to a depth of several inches. The frequent recurrence of this has resulted in laying bare the subsoil in some parts of the country. (133)

Horne's survey was requested by Governor Gordon, who a year earlier was overseeing the final stages of the 1876 war in the Colo (hill or mountain) region, which surrounds Navosa in the center of Viti Levu. In late July, the middle of the dry season, he was camped at Fort Nasaucoko on the southern margins of Navosa. Other than details about success in war, Gordon (1901) recorded several comments about fire, most of them derogatory toward those who ignited them (both soldiers and local inhabitants). A few examples follow, dated from mid-July to early August:

> Heavy fires all over the country. I wish those who make them could see what destruction they are causing. (Gordon 1901, 88)
>
> The Goths [sic] had set fire to the reeds close by, spoiling the pretty junction of the rivers. (96)
>
> The number of times one has to cross the stream ... is provoking, especially as the banks are steep and slippery, and most of the crossings in thick bush ... but that is nothing to the vexation which the universal fires, springing up everywhere, cause me. (97)
>
> The only thing that disturbed my enjoyment of the glorious day and delicious air

was the prevalence of smoke everywhere, and fires all over the country. (98)

Later, on August 9 and 12, Gordon's (1901) convictions were expressed in the following strong characterizations:

Bush fires everywhere, far and near; verily this is a land wasted by fire. (110)

Fires everywhere—the whole country burning—land wasted by fire in very truth. (112)

It is not surprising that Gordon (1901, 619) had occasion to "lecture on the iniquity of careless fire-raising" later that year in Nawaka before embarking upon a return trip to Navosa, thus beginning his campaign for fire prevention and the conservation of land.

Despite the likely plaudits for Gordon's fire prevention initiative from his British peers and environmental conservationists, my question is how much, and in what form, did the local communities want such a policy? A lacuna in Gordon's journals is that he made no apparent attempt to ascertain local perspectives on fire and the reasons for ignition, despite being in a very good position to do so and despite a willingness to inquire about other aspects of local society and culture. The local voice about fire is silent in his journals. Also absent is recognition of abnormal El Niño drought conditions.

It is evident that Gordon moved promptly to enact relatively unconditional and strictly worded fire regulations for the dry districts of Fiji without waiting for the completion of Horne's supportive survey. Climate and rainfall at that time are salient considerations. Public records for Navosa rainfall during that period of history are lacking, but rainfall in 1876 was recorded in Bua Province, Vanua Levu, about 150 kilometers northeast of Navosa.[26] That record indicates rainfall in 1876 was below average but not as dry as in May 1877 to September 1878, at the climax of a major El Niño (Holmes 1881, 1887). Horne's Fiji survey was conducted during an El Niño with extreme drought conditions. For this reason, perhaps it is not surprising that Horne's (1881) propositions regarding fire were quite punitive. Similarly, the unconditional framing of fire regulations enacted by Gordon in 1877 and 1878 (Legislative Council 1878; Government of Fiji 1883) was very authoritarian. Much more could be written about Gordon's Scots-British aesthetic and managerial perspective, the role of fire in the Navosa war[27] and in dry zone agriculture, and his experiences in Southern Navosa, which were documented in his memoirs (Gordon 1879, 1901). The coincidence between the occurrence of a major El Niño with drought conditions and Gordon's experience in Navosa undoubtedly inflamed his views on fire and land degradation. The important outcome in this context, however, was his subsequent desire to minimize fire in dry zone Fiji as well as the legal and administrative changes he enacted to effect fire control.

In the following years, various ordinances were introduced attempting to regulate fire in Fiji. In Regulation No. 6 of March 1877 (Government of Fiji 1883), Gordon introduced one-year imprisonment punishments only for Indigenous Fijians who set fires. Later, in June 1878, he commanded that a firebreak "two fathoms wide" (3.6 meters) should be cleared around a place intended for burning, and three days' notice given to the Turaga-Ni-Koro, who would arrange for an inspection and subsequent permission and supervision of the fire (Government of Fiji 1883). In addition, a similar ordinance targeting non-Indigenous Fijians was enacted in 1878, "for the better prevention by accidents by fire" (Legislative Council 1878). The government attempted to enforce fire prevention by hiring fire wardens, and the Turaga-Ni-Koro had the authority to punish offenders with up to six months of imprisonment, later extended to one year. According to senior village interviewees in my 1998–1999 research, these prevention strategies were partially successful in the colonial era, and they blamed the relaxation of enforcement after independence in 1970 for being a retrograde step that hinders fire prevention today.

Nevertheless, these fire regulations and enforcement strategies were perceived by the wider public as part of the former colonial regime. Fijians wanted freedom from the institutional constraints of colonialism. Subsequently, there has been a decline in Fijian government interest in and support for fire regulation. Fire wardens apparently were phased out after independence, and none of the communities I visited in 1998 reported having them. The blanket nature of these regulations was an important issue. They were initially composed around the extreme El Niño drought year of 1878 for the conditions present at that time. In either neutral or wet years—some of which are associated with La Niña—these regulations were for the most part only partly relevant and excessively punitive. Instead of adding qualifications for moister years, a decree in 1969 (Government of Fiji 1985)—just before independence and similarly precipitated by a severe drought—expanded the laws to cover all citizens, not just Indigenous Fijians. Unauthorized burning of vegetation was prohibited every year over a large part of leeward Fiji, including Navosa, in the dry season from 1 July to 31 December. In the Forest Decree of 1992 (Government of Fiji 1992), however, responsibility for regulating fire, including grassland fire, was transferred from provincial authorities and the Fijian Affairs Board to officers of the Forest Department under the conservator of forests and a forestry board. In this statute, the laws were changed and radically simplified to allow fire anywhere outside of forest reserves, nature reserves, and temporary fire hazard areas (FHAs). By means of these FHAs, fire restrictions or bans can be declared anywhere at any time by the conservator of forests. Perhaps such a temporary FHA was in place during my Navosa research in 1998–1999, but ensconced as I was in a landscape of high fire risk, regular ignitions, and pyric disturbance, I was unaware of such a ban. The fire prevention laws in effect were either poorly understood or dismissed as inappropriate by participants in my study. The net result was that fire regulations were commonly ignored and rarely enforced, even during drought years. The only exceptions I observed during the 1998–1999 drought were in villages where notable leaders, including one highly respected hereditary chief, promulgated their own bans against lighting fires.

The 1992 regulation envisaged the appointment of fire rangers (Government of Fiji 1992, Part VII), likely based in provincial centers such as Lautoka and Sigatoka, towns on the coast that are far away from places like Navosa with accentuated grass fire risk. The primary concern of fire rangers is protection of production forests, especially larger commercial plantations of fire-prone *P. caribaea* and higher-altitude *P. elliottii* in the dry zones of Fiji, mainly outside of Navosa. Thus, the rangers' purview is a very poor fit with the primary zones of fire risk associated with land degradation. This change in policy and governance, although arguably advantageous for regulating fire within plantation forests,[28] has been regressive for the dominant areas of seasonally dry grassland in leeward Fiji. Furthermore, plantation forests can be threatened by grassland wildfires originating in areas outside the usual purvey of forest rangers. The previous policy of regulating grassland fire using village fire wardens and Turaga-Ni-Koro imbued with punitive powers had impact. Today, fire rangers have little chance of regulating and enforcing grassland wildfire prevention in remote locations.

The Forest Decree of 1992 is another example of policy and governance that blindly follows the prescriptions of external, neocolonial authority structures without reference to local conditions and effective regulation in the affected zones. In my opinion, the 1992 regulations have been a step backward for wildfire control in Fiji. I would suggest that an appropriate institution for wildfire control, and thus forest protection, should involve and empower local community leadership as an important adjunct to top-down official authority.

Primary enforcement responsibility has now shifted to the local police. This restructuring has undermined fire prevention because police are based in district centers that are geographically remote from the difficult to access village and settlement communities where they are needed. Also, police are few in number, and their relationships with forest and fire rangers in government centers are weak. They are not positioned to enforce fire regulations, unlike the village fire wardens and Turaga-Ni-Koro of the past. In the case of Navosa, the police are based at Navatumali, a government outpost on the southern margins of the subprovince, where seasonal fires are often visible during the day. Accessing and finding those responsible for igniting these fires are prohibitively difficult without the involvement of local community leadership. Again, I would argue that power should be divested to community leaders, especially the Turaga-Ni-Koro, who are knowledgeable about the causes of wildfire and who can facilitate the appointment of community fire wardens. Together with the Turaga-Ni-Koro, these wardens could report infractions to the police. These positions of authority entail substantial risk and responsibility, which should be recognized with appropriate official salaries.

Part of the general problem involved in regulating wildfire is that the most powerful politicians and administrators, regionally based in the moist eastern division capitals (then Levuka, now Suva), historically have shown little active concern for excessive dry zone burning and land degradation. Gordon was one exception, aided by his firsthand experiences there. Perhaps a more general lack of dry zone experience among past administrators, as well as inequalities between regions associated with internal colonialism (Durutalo 1985), influenced the perpetuation of Gordon's regulations, which were drafted immediately after his actual experience of land degrading fire in the dry zone, and consequently allowed them to stand for so long with only minor revision. Perhaps this aspect of colonial imposition has been beneficial, at least in the sense of bringing the problem into national awareness, though others have ignored it. Nevertheless, the question remains as to whether the unconditional and draconian aspects of these regulations were warranted and whether a more locally participatory, livelihood-nuanced, and climate-adaptive set of regulations should be introduced (Ghai and Vivian 1992; Maiava and King 2007; Richards 1985).

Drought, Fire, and Warning in Space and Time

Sequential aerial or satellite imagery could yield reliable data about the actual extent of burning over time, but such information is either very spasmodic, restricted in scope, or generally lacking.[29] My observations made during moister to neutral years, including La Niña years, indicate considerably less uncontrolled burning, together with an absence of reports about excessive burning in Navosa, suggesting that fire-induced land degradation is mainly confined to predominantly dry El Niño climate periods. These have occurred about every four to five years in recent decades, with less frequent severe occurrences

(Mataki et al. 2006). The long-term historical average for El Niño events is every seven years (Parry 1997), with ensuing droughts lasting twelve months on average (Grove and Chappell 2000). Recent advances in the monitoring of ENSO climate patterns have enhanced the ability to predict droughts, whose frequency may increase slightly in the Pacific region in future decades (Cai et al. 2012; Santoso et al. 2013). Again, for the Pacific region, extreme El Niño events are predicted to increase in frequency, but with little change in duration (Cai et al. 2014). For Fiji, the current climate science is less certain, with a low-confidence prediction that the overall frequency of drought occurrences in Fiji will decrease slightly through 2100 (PCCSP 2014). This contrasts with a very high-confidence prediction that El Niños will continue to occur. There is little scientific consensus about changes in the intensity of Fijian droughts, but there is very high confidence that "annual mean temperatures and extremely high daily temperatures will continue to rise" (PCCSP 2014, 94).

Local communities have reported another hazard in the form of torrential rainfall events and flash flooding not associated with cyclones. One example is the flash flooding of the Namada Creek valley in western Navosa in April 2010 (King 2010). Another example is an extreme torrential rainfall event recorded at Tikituru in eastern Navosa on January 8–9, 2009 (PWD n.d., ca. 2010). These unpredicted events are believed to be symptomatic of La Niña episodes in conjunction with global climate change.

Early warning systems that help farmers prepare for drought are now feasible, according to the South Pacific Applied Geoscience Commission (Kaloumaira 2002), and their utility for reducing climatic hazard for agricultural and other communities presents clear community, regional, and national benefits. Unfortunately, so far there has been little effort to plan and systematize a comprehensive warning system for Fijian communities and farming groups located beyond the purview of the Fiji Meteorological Service's routine weather forecasts. The involvement of Fiji radio stations in disseminating long-range warnings would be very advantageous in Fiji's rural areas, where print media is nearly nonexistent. Television reception is also restricted and even where available is rarely watched.

Since the Forest Decree of 1992, fire regulations have either not been commonly known or have been ignored in practice, as previously discussed. Research is needed on what the members of rural communities know about the regulations and what they perceive to be the best ways to regulate fire in their vanua. I suggest that steps should be taken to increase fire-risk monitoring during El Niño climate periods and to plan for the development of prevention strategies centered on village and settlement communities.

What are the Effects of Repeated Burning?

During thirty-two group interviews (sixteen female and sixteen male) in the Navosa communities, I asked participants, "How does repeated burning affect the land?" (The participant groups were the same ones that responded to the questions discussed above, except for two communities for which data were lacking.) Participants gave 115 total responses across all groups, which were recorded in a notebook. The 25 different types of responses were grouped into 8 semantic categories. To aid with interpretation, the raw number of responses in each category was converted to a percentage using the total number of responses (115) as the denominator.

More than a quarter of responses indicated that repeated burning causes "soil erosion" (27.8 percent). Another quarter (25.2 percent) complained of decreasing soil fertility, say-ing fire "makes land infertile," especially on hillsides. Over half of all responses (53.0 percent) identified land degradation as an effect of repeated burning. Another two categories ("dries up land" and "harder to grow crops, less cropland") were indicated in 22.6 percent of responses and specifically addressed drying of the land and its negative effects for crop cultivation. Whereas damage to the general environment ("vanua, trees, wild yams, and wildlife") was indicated in 16.5 percent of responses, damage to waterways and fisheries ("river damaged, less prawns and eels") raised little concern, being mentioned in only two (1.7 percent) responses. The comparatively low ranking of damage to waterways and fisheries may indicate a sediment-resilient fishery or possibly a methodological and translation bias derived from asking about "land" rather than "land and water," which would have been more strongly linked to the overall context of questioning.[30] Another explanation may derive from a tendency for participants to report proximal rather than distal explanations (King 2012); in this case, overt changes to the land as the result of recent fire (proximal) rather than gradual impact upon waterways over many years (distal).

Of particular salience was the apparent lack of concern about "difficulty growing grasses" related to pasture deterioration (1.7 percent of responses) and "less food for animals" (4.3 percent of responses). This finding was initially surprising to me because domestic ungulates (cattle, horses, goats, and bullocks) were visibly suffering from a lack of food during the drought. The hungry animals provoked social conflict in the community as they broke through fences—or were let through them, according to speculation by neighbors—to graze on food crops. Further investigation is needed, but some preliminary inferences can be drawn. This finding of apparent "low concern" about pastoral conditions, together with local pyrosocial norms, could be expected because ignitions intended to stimulate tender, fresh new grass shoots on hillsides to aid domestic ungulates in the dry season was one of the main causes of repeated burning. Except when they eat new grass shoots, most ungulates graze in the shadier, cooler, and often more lush gullies in the hot, dry season, not on the hillside grasslands dominated by mature and unpalatable *P. polystachyon*. In 1998–1999, Navosans likely perceived ungulate animals to be both the causes and beneficiaries of repeated burning, rather than being the victims of repeated burning, hence the low importance ratings given for effects on pasture and grazing. In severe drought years this situation is a potential embarrassment to local pyrocultural mores around the sustainability of grazing techniques among local communities, but it is tolerated with the perspective that drought years are unusual.

The finding also reinforces my observation that igniting fires for "new grass for the animals" is associated with the largest area of land burned and is the most environmentally destructive type of burning in Navosa, although I did not collect quantitative data on the relative land areas burned as related to reasons for ignition.[31] Nevertheless, my general observation is unequivocal: igniting fires for "new grass for the animals" is a deliberate strategy to burn the large areas needed to provide fresh forage in the dry season, and it is particularly severe and prevalent during droughts. This is especially the case near communities (such as those in the lower Nasa Creek of Noikoro District) with heavier reliance on pastoral production for their livelihoods.

Regarding the issue of prevention, research is needed on ways to minimize excessive burning in the dry season of El Niño years, especially the practice of burning large areas to provide new grass for the animals.

Burning Lands

FIGURE 8.5. A study participant using cards to identify households responsible for damage to the environment ("instigators") as well as households seen as protecting the environment ("custodians").

Who Burns Carelessly?

Who is responsible for uncontrolled burning? This question was investigated during the sixty-eight in-depth studies in Nasauvakarua and Nawairabe using a different type of card-sort procedure followed by individual questioning (Figure 8.5).

In this exercise, thirty-four small cards (7 x 7.5 centimeters) labeled with the household names were laid in front of interviewees.[32] Navosa houses each have a one-word identifying name associated with their head. I asked two questions. First, "Who helped the environment?" Second, "Who damaged the environment?" In response to the first question, participants picked up one or more of the cards representing the households. At this point I asked, "Why?" or "What have they done?" for each card. When the interviewee had finished responding, I asked the second question and followed the same procedure as the first. During each evaluation, I recorded the responses and answers in a notebook alongside lettered codes for the identified household. The corner of each card was labeled with a small, discrete identifying code. Every household in both villages was interviewed in this way. The individual responses were collated and tabulated under categories by household and meaning. Responses were converted to percentages of the total number of responses: 336 for the first question, and 305 for the second.

To my initial surprise, participants obliged my inquiries by identifying the households of individuals perceived as culprits responsible for uncontrolled burning and provided reasons why members of those households damaged the environment.[33] The action of burning carelessly appeared in 194 out of 305 responses (63.6 percent). Burning carelessly (or, following an alternative interpretation discussed later, burning in a pretermitted way [a type of intentional overlooking]) was by far the most prevalent reason given for damaging the environment. The next highest response score was 47 out of 305 responses (15.4 percent). The most revealing finding, however, was that two to four households in each village were blamed by others for careless burning at far higher rates than other households. The results indicate that community members were aware of and concerned about excessive burning, that a small proportion of each community was strongly linked to careless burning, and that the individuals concerned were well known within the community for this damaging practice. Conversely, those who tried to prevent careless burning were also well known for their custodial or conservationist roles.

Navosa communities were not homogenous in terms of attitudes and actions involving fire and land degradation during the research period. I believe this finding, based on the attitudes and actions of householders, may be somewhat similar year to year. Nevertheless, I would expect that differences would become more marked during drought years. I also expect that communities would tend to question or limit the assertion of pyrosocial norms; that is, for community leaders to act against the social norms that tend to allow the lighting of fires at these times.

A special difficulty arises when putatively careless burning occurs along main corridors between adjacent communities. I observed many burned areas beside trails in secluded valley bottoms located uphill from the remains of recent cooking fires, some of which were still smoldering. Some corridor travelers were from other Navosa communities, while others were from outside of Navosa altogether. I was unable to interview the persons responsible for these corridor fires. Villagers, accurately or inaccurately,[34] identify disregard and carelessness in these incidents, which occur away from the protective gaze of community observation. On the other hand, these burn sites were almost always near creeks on uneven and slow trails in relatively precipitous, semiforested, and aesthetic terrain that provides ideal shelter and resources for lighting fires and preparing meals on long journeys. In other words, the risk of fire ignition could have been less about the attitude of the fire ignitor and more about social norms and culture in a vulnerable environment. It was difficult, however, to ignore my observation that these burned corridors tended to be in secluded locations where they were less likely to be challenged by local conservationists or custodians. Another conundrum that begs debate is the concept of "carelessness," which was occasionally raised in interviews and identified as a minor reason for burning land. The idea that carelessness is at fault for some fires is questionable given the pyrosocial norms around fire, and the blame placed on locals by outsiders. One cannot be careless if one disregards or does not believe that igniting a fire or leaving it smoldering is a dangerous act. In other words, if people disregard or exhibit pretermission by overlooking their immediate actions as part of their pyrosocial norms, then they should not be accused of carelessness. In other words, the accusation of carelessness may reflect nothing more than derogatory discourse in the absence of direct knowledge about the accused's motivations and intentions. A note of caution must be sounded here because there is always a risk of mistranslation. Subtle differences may exist between what is known and what is expressed, especially when crossing dialect boundaries. Such complaints of carelessness, however, may suggest that personal, cultural, social, and political prejudices or injustices are involved. The accused may have implicit but sound reasons for igniting fires that appear careless to others. For example, in the scenario presented above involving Governor Gordon, I believe it would have been more proper of him to have inquired of local community members why they ignited fires before he condemned their pyrocultural actions as careless and began a draconian campaign with blanket restrictions to prevent the wildfires that he assumed was a continual (rather than a climate-related and episodic) scourge of the dry-season landscape.

Conclusions

A complex network of relationships surrounds traditional livelihood strategies and the Indigenous management of land and vanua in Navosa. These relationships appear to function with the least conflict and the most resilience during periods of normal precipitation and relatively stable weather, when the Navosa land-society nexus, despite its pyrosocial norms, exhibits resilience and sustainability. In addition, when vanua have faced change—for example, in the nineteenth century when epidemics of introduced disease caused social devastation, or when new grasses were introduced and invaded like wildfire, or when promising new food cultigens appeared—adaptation seems to have been the rule. I conducted my research, however, in an abnormal El Niño severe drought year, when the livelihood system in Navosa was under stress due to desiccation. The combination of steep, broken topography and high temperatures in the late dry season accentuated fire risk and severely constrained the cutting of firebreaks. Wildfires in open grasslands and forest margins were started from livelihood-oriented but poorly controlled intentional burning, as well as the intentional disregard as part of pyrosocial norms of travelers between communities. The desire to stimulate the growth of tender new grass to feed domestic ungulates during the drought greatly extended the fire season, increased the area of land burned, and increased the potential for severe sheet erosion (and indirectly, gully erosion) when the wet season rainstorms arrived.[35] Many local leaders tried to prevent excessive burning, but the actions of a few individuals nevertheless led to damaging wildfires that spread over wide areas and caused land degradation with negative effects for livelihoods.

Government officials often exhibited poor knowledge of fire on vanua lands and, in general, did not understand pyrocultural practices except in superficial and derogatory ways. In 1998–1999, there was virtually no enforcement of the overreaching fire regulations initially drafted over a century ago and inconsistently modified several times since then. Consequently, in 1998–1999, stakeholders generally did not have clear understandings of their most recent but obscure laws. As of 1999—and at present, to my knowledge—no concerted attempts had been made to negotiate a better system of regionally adaptive and ENSO-sensitive fire regulations with local vanua or communities. From my observations between 1997 and 2013, little attempt has been made to intervene at the local level and regulate fire ignitions and spread when El Niño conditions are imminent and the risk of fire threatens. The science of climate prediction has improved greatly since colonial times, and some evidence from future modelling of the Pacific climate predicts that severe El Niños will become more frequent as global warming progresses.

Fire regulations in Fiji should be revised in line with a locally formulated and enforced model of fire and environmental management that is adaptive with the interaction of local livelihoods, land use, and climate, and incorporates pyrocultural mores as part of local lifeways. In the short term, greater effort should be made to inform and warn local village and settlement leaders of the need for fire prevention and control during drought-threatening El Niño events. In addition, government resources should be allocated at these times for prevention strategies and enforcement. Over the longer term, participatory research is needed on strategies and mechanisms to minimize the practice of burning for new grass to feed domestic ungulate animals in the dry season.

Notes

1. Viti Levu has a land area of 10,388 square kilometers. Its highest peak is about 1,323 meters above sea level (Derrick 1965; Ward 1965).
2. Viti Levu has few endemic grasses, many of which tolerate shade, indicating a history of forest-dominated land cover. The prehistoric record, however, does indicate arid climate periods when grasslands may have flourished in drier zones (Ash 1992; Brookfield and Overton 1988; Nunn et al. 2001; Southern 1986; Twyford and Wright 1965).
3. Anecdotally, a visiting informant with whom I was travelling joked that "one day Na-vosa [here meaning "a vista of grassland ridges"] will be instead called Na-masese!" In effect, masese is afforesting, or perhaps reforesting (see previous note), the savannas, and perhaps countering to some extent a previous invasion of introduced grasses (discussed later). After facilitating the initial introduction, humanity has had little control over the ecological outcomes in many invasions in this region known for its relatively low native biodiversity.
4. A key issue is the absence of waterlogging, which when present causes plant hypoxia or anoxia and in combination with high rainfall may facilitate diseases (e.g., yam anthracnose).
5. Community elders reported that *P. polystachyon* spread very fast when it arrived in Navosa in the early to mid-twentieth century. *P. polystachyon* is not the dominant species on the very low fertility talasiga (sunburned land). This sometimes denuded, fern-dominated land type is common farther west on Viti Levu and on smaller islands such as Lakeba (Latham 1983), and is often mentioned in the literature on fire and land degradation (Cochrane 1969). Only a small percentage of Navosa is properly talasiga, and I have refrained from common use of this term, preferring "grassland" instead. Another Fijian term, dravuisiga, more accurately characterizes grassland landscapes with "less seriously altered soils" (Twyford and Wright 1965, 103), in contrast to their description of talasiga landscapes as seriously altered. Whether talasiga is the result of naturally impoverished soil or the consequence of long-term repeated anthropogenic fire is uncertain (Ash 1992; Latham 1983; Twyford and Wright 1965), although some specific instances are believed to have a pyric cause (Twyford and Wright 1965).
6. The question of what happens to this gradient over time if there were no grass fires, as far as I am aware, has not been the subject of empirical or longitudinal investigations. Cochrane's (1968, 1969) studies in the Ba closed area have come the closest. The absence of areas free of fire complicates field observations.
7. Not only is the land degraded, but so is the vanua (Batibasaqa et al. 1999; Ravuvu 1983).
8. In Navosa, the drought began in late 1997 and continued through the December to March

period, which is normally the wet season, and severely affected the rice and sugar crops in Fijian coastal regions. The drought continued through the following 1998 dry season until mid-November 1998, when heavy rains began to fall. Although food relief was needed in the largely monocropped coastal areas, inland Navosa communities maintained food security through the drought (despite cash crops being badly affected) and only rarely resorted to emergency foods such as the giant taro (*Alocasia macrorrhizos* [L.] G.Don.). The difference in sustainability could be explained by the difference in resilience and availability of subsistence crops (King 2005).

9. The longer harvesting is delayed beyond the "shoot emergence" stage, the more unpalatable the tubers will become.

10. Respected gatekeepers with local kin connections and knowledge facilitated my entry into the villages. Official consent came later. In retrospect, this unofficial introduction was invaluable in establishing the trust that made many aspects of the research possible.

11. In one 1998 incident, a community member who was able to identify the person or household responsible for burning her cassava garden complained to the Bose Ni Koro (Village Council) who adjudicated her loss with the culprit.

12. Many local dialects exist. The standardized Bauan dialect of eastern Fiji differs from the western Fijian group of dialects and communalects. I preferred the use of local dialects where possible, but interpreters often used standardized Fijian terms. Different local interpreters were employed at Nasauvakarua and Nawairabe.

13. The timber of the endangered vesi tree (*I. bijuga*) is used for yaqona (kava) bowls and other cultural artifacts laden with symbolic value.

14. This visual evaluation technique follows the general pattern of locally innovative Participatory Rural Appraisal (PRA) methods (Chambers 1997). An important aspect of many visual methods is that participants, alone or in groups, can visually compare the importance of so-called objective categories using local measures (e.g., large seeds, cards, adjustable charts) before a ranking or score is reported and then recorded by the researchers. An advantage of these techniques is that they often include their own preliminary verification, belying the apparent informality of the method. In my opinion, this built-in verification process together with the participatory design input increases the empirical relevance of the results.

15. The "new" grass, as previously described, refers to the regrowth shoots of bunch grasses, most specifically *P. polystachyon*, known locally as mauniba, garasiniba, laulau ni manivosi, mission or mongoose-tail grass.

16. Wild yams often grow within rather impenetrable dense stands of *M. floridulus*. A clearing fire at the end of the annual dry season, just before the yam sprouts emerge, enables the harvester to access and locate the new spouts and dig the yams shortly afterward. The thickness of the sprouts indicates their age and size. Harvesters take only older and larger tubers. Ideally, the small head (sprouting end) is cut off and replanted in the same place, but elders complained that youth often failed to do this when they harvested the yams to be taken to market for sale.

17. Stands of guava are burned in the dry season, leaving the dead and bare, but intact, trunks and branches more easily removed for use as fuelwood as needed. The burned guava trees are "fuel on the hoof" that obviates the need for storage and the risk of personal loss to kerekere (borrowing) by fellow community members within the village. The bases of the guava shrubs sprout again later, making the resource sustainable on a three-year or longer cycle.

18. Horses are hurt by guava, lantana (*Lantana camara* L.), and other abrasive, medium-height vegetation that can scratch their hides, leading to skin infections and other health problems.

19. Local cultivators did not have technical knowledge of anthracnose, but they indirectly linked the effects of fire with disease characteristics. The infectious spores of the yam anthracnose fungus *Colletotrichum gloeosporioides* (Penz.) are typically transported to the vine from soil surface debris at the base of the plant via splashing during rain (Penet et al. 2014). Hence, it is assumed that the soil surface, the site of most

infection (Ripoche et al. 2008), becomes disinfected against yam anthracnose during burning.
20. The rate of ignitions caused by lightning was 1 percent in the relatively dry pine plantations farther west on Viti Levu during 1995–1997 (Were 1997).
21. The paradoxical and somewhat tragic demand for fresh grass shoots of *P. polystachyon* as forage for domestic ungulates is greatest during droughts, which is also the period of greatest wildfire risk. Late burning was more common in 1998 because of the perceived necessity to feed starving animals during the drought. In response to enquiry, local community members complained about "too many animals" despite an upsurge in contributions of cattle for feasting events as one way of alleviating livestock food insecurity and grazing pressure on the landscape.
22. I briefly considered the possibility of recording a historical timeline comparing annual variations in the extent of uncontrolled burning, but decided against it in favor of research rigor around other more current sustainability issues. Nevertheless, participatory research in Navosa evaluating local history with timelines comparing the degree of uncontrolled burning with fluctuations in local climates, ENSO, and other historical factors would make a desirable contribution to the knowledge of fire and land management.
23. During a severe drought, the extreme temperatures of exposed ground will cause discomfort. As Holmes observed in 1881, "For the first and only time in Fiji I noticed that the natives suffered much from sore feet, caused by travelling over paths made intensely hot by the sun. Their feet are generally tough as cowhide . . . but at this season foot bandages were in general use among them for protection" (Holmes 1881, 232).
24. I am tempted to assume that this is coincidence, but perhaps not. Perhaps the war of 1876 was initiated in response to predictions of drought?
25. The quote was made with reference to famine circumstances in India at the time.
26 Meteorological records are poor for Navosa. In recent decades, reliability and coverage have declined, and some recording stations have closed.
27. Burned slopes expose both attackers and hill fort defenses. Evidence also suggests that igniting fire was part of the celebration of victory (Gordon 1901).
28. The main fire risk could come from areas of different land use outside of or surrounding plantations, where the effectiveness of forest rangers is limited.
29. Aerial photography was commissioned at approximately decadal intervals after major cyclone damage from the late 1940s until the end of the twentieth century. Recent economic and political instability led to a decline in aerial photography. During recent research, I found that one major historical series of aerial photography negatives was damaged in storage at the Lands Department in Suva and was no longer usable. Satellite imagery of inland Fiji has so far been spatially uneven.
30. The question can be interpreted differently if the translator used the word vanua, meaning the mosaic of all things about a place and its Indigenous people, rather than qele (soil, land) or vei ka bula (all the living things). In this case, the context of the questioning was in reference to vei ka bula, which should stretch the meaning to encompass land and water.
31. This question is more complex than it seems, especially when considering temporal factors and interactions. The frequencies of ignitions or wildfires and the areal extents will vary across wet, neutral, or dry periods. The task of comparing controlled burns with wildfires may also be complex. For example, when a community member sets fires for new grass, intending to refresh large areas, onlookers may mistake the intentional fire for a wildfire. On the other hand, when cultivators clear their small hillside gardens with fire (possibly using firebreaks), their intentional fires sometimes become wildfires and burn much larger areas, which may be more likely to occur during severe droughts. Cultivators may intentionally allow their fires to spread uphill beyond their gardens knowing that they will encourage new fodder or ease yam harvests. These scenarios are based on my observations. They would be worthwhile subjects for future research.

32. Coincidentally, both village communities had the same number of households.
33. The Fijian term chosen to represent environment is vei ka bula (see note 30).
34. This view must be tempered by my lack of actual data. It is possible that some of these travelers were local community members who intended these cooking fires to spread uphill for their own reasons.
35. This is particularly concerning in light of the aforementioned recent occurrences of unpredicted and sudden extreme events of torrential rainfall that were not associated with cyclones.

References Cited

Akrich, Madeleine, and Bruno Latour. 1992. "A Summary of a Convenient Vocabulary for the Semiotics of Human and Nonhuman Assemblies." In *Shaping Technology/Building Society: Studies in Sociotechnical Change*, edited by Wiebe E. Bijker and John Law, 259–264. Cambridge: MIT Press.

Ash, Julian. 1992. "Vegetation Ecology of Fiji: Past, Present, and Future Perspectives." *Pacific Science* 46 (2): 111–127.

Batibasaqa, Kalaveti, John Overton, and Peter Horsley. 1999. "Vanua: Land, People and Culture in Fiji." In *Strategies for Sustainable Development: Experiences from the Pacific*, edited by John Overton and Regina Scheyvens, 100–106. Sydney: University of New South Wales Press.

Blaikie, Piers, and Harold Brookfield, eds. 1987. *Land Degradation and Society*. London: Methuen.

Brewster, Adolph. 1920. "The History of Nadrau." *Transactions of the Fijian Society* 1920: 16–19.

Brookfield, Harold, and John Overton. 1988. "How Old Is the Deforestation of Oceania?" In *Changing Tropical Forests: Historical Perspectives on Today's Challenges in Asia, Australasia and Oceania*, edited by John Dargavel, Kay Dixon, and Noel Semple, 89–99. Canberra: Centre for Resource and Environmental Studies, Australian National University.

Cai, Wenju, Matthieu Lengaigne, Simon Borlace, Matthew Collins, Tim Cowan, Michael J. McPhaden, Axel Timmermann, Scott Power, Josephine Brown, Christophe Menkes, Arona Ngari, Emmanuel M. Vincent, and Matthew J. Widlansky. 2012. "More Extreme Swings of the South Pacific Convergence Zone Due to Greenhouse Warming." *Nature* 488 (7411): 365–369.

Cai, Wenju, Simon Borlace, Matthieu Lengaigne, Peter van Rensch, Mat Collins, Gabriel Vecchi, Axel Timmermann, Agus Santoso, Michael J. McPhaden, Lixin Wu, Matthew H. England, Guojian Wang, Eric Guilyardi, and Fei-Fei Jin. 2014. "Increasing Frequency of Extreme El Niño Events Due to Greenhouse Warming." *Nature Climate Change* 4: 111–116.

Cancian, Fraıncesca M., and Cathleen Armstead. 1992. "Participatory Research." In *Encyclopedia of Sociology*, edited by Edgar F. Borgatta and Marie L. Borgatta, 1427–1432. New York: Macmillan.

Chambers, Robert. 1997. *Whose Reality Counts? Putting the First Last*. London: Intermediate Technology.

Clarke, W. C., and R. R. Thaman, eds. 1993. *Agroforestry in the Pacific Islands: Systems for Sustainability*. Tokyo: United Nations University Press.

Cochrane, G. Ross. 1968. "Land Use Problems in the Dry Zone of Viti Levu." In *Proceedings of the Fifth New Zealand Geography Conference*, Auckland, August 21–27, 1967, 111–118, Christchurch: New Zealand Geographical Society, Department of Geography, University of Canterbury.

———. 1969. "Problems of Vegetation Change in Western Viti Levu, Fiji." In *Settlement and Encounter*, edited by Fay Gale and Graham H. Lawton, 115–147. Melbourne: Oxford University Press.

Crosby, Alfred W. 1986. *Ecological Imperialism*. Cambridge: Cambridge University Press.

———. 1994. *Germs, Seeds and Animals: Studies in Ecological History*. Armonk, NY: M.E. Sharpe.

Curtis, Scott. 2008. "The El Niño–Southern Oscillation and Global Precipitation." *Geography Compass* 2 (3): 600–619.

Deo, Ravinesh C. 2011. "On Meteorological Droughts in Tropical Pacific Islands: Time-Series Analysis of Observed Rainfall Using Fiji as a Case Study." *Meteorological Applications* 18 (2): 171–180.

Derrick, Ronald Albert. 1965 [1951]. *The Fiji Islands: A Geographical Handbook*. Revised by C. A. A. Hughes and R. B. Riddell. Suva: Government Press.

Dickinson, William R., David V. Burley, Patrick D. Nunn, Atholl Anderson, Geoffrey Hope, Antoine De Biran, Christine Burke, and Sepeti Matararaba. 1998. "Geomorphic and Archaeological Landscapes of the Sigatoka Dune Site, Viti Levu, Fiji: Interdisciplinary Investigations." *Asian Perspectives* 37 (1): 1–31.

Durutalo, S. 1985. "Internal Colonialism and Unequal Regional Development: The Case of Western Viti Levu, Fiji." Master's thesis, University of the South Pacific.

Field, Julie S. 2004. "Environmental and Climatic Considerations: A Hypothesis for Conflict and the Emergence of Social Complexity in Fijian Prehistory." *Journal of Anthropological Archaeology* 23 (1): 7–99.

Fison, Lorimer. 1885. "The Nanga, or Sacred Stone Enclosure, of Wainimala, Fiji." *Journal of the Anthropological Institute of Great Britain and Ireland* 14: 14–30.

France, Peter. 1968. "The Founding of an Orthodoxy: Sir Arthur Gordon and the Doctrine of the Fijian Way of Life." *Journal of the Polynesian Society* 77 (1): 6–32.

———. 1969. *The Charter of the Land: Custom and Colonization in Fiji*. Melbourne: Oxford University Press.

Ghai, D., and J. M. Vivian. 1992. "Introduction." In *Grassroots Environmental Action: People's Participation in Sustainable Development*, edited by D. Ghai and J. M. Vivian, 1–19. London: Routledge.

Gordon, Arthur. 1879. *Letters and Notes Written During the Disturbances in the Highlands (Known as the "Devil Country") of Viti Levu, Fiji 1876*. Edinburgh: R. and R. Clark.

———. 1901. *Fiji Records of Private and Public Life 1875–1880*. Edinburgh: R. and R. Clark.

Government of Fiji. 1883. *Regulations of the Native Regulation Board 1877-1882*. London: Harrison and Sons.

———. 1985. "Land Conservation and Improvement (Fire Hazard Period) Order." *Laws of Fiji: 1985*. Rev. ed., vol. 8, chap. 141, sec. 7, S3. Suva: Government of Fiji.

———. 1992. *Forest Decree No. 31 of the Laws of Fiji*. Suva: Government of Fiji.

Grove, Richard H., and John Chappell. 2000. "El Niño Chronology and the History of Global Crises During the Little Ice Age." In *El Niño: History and Crisis: Studies from Asia-Pacific Region*, edited by R. H. Grove and J. Chappell, 5–34. Cambridge: White Horse.

Hocart, A. M. 1931. "Alternate Generations in Fiji." *Man* 31: 222–224.

Holmes, R. L. 1881. "The Climate of Fiji: Result of Meteorological Observations Taken at Delanasau, Bua, Vanua Levu, Fiji, 1871–1880." *Quarterly Journal of the Royal Meteorological Society* 7: 222–243.

———. 1887. "The Climate of Fiji: Results of Observations Taken at Delanasau, Bua, During the Five Years Ending December 31, 1885, with a Summary of Results for Ten Years Previous." *Quarterly Journal of the Royal Meteorological Society* 13: 30–36.

Hope, Geoffrey, Janelle Stevenson, and Wendy Southern. 2009. "Vegetation Histories from the Fijian Islands: Alternative Records of Human Impact." In *The Early Prehistory of Fiji*, edited by Geoffrey Clark and Atholl Anderson, 63–86. Canberra: ANU ePress.

Horne, John. 1881. *A Year in Fiji, or an Inquiry into the Botanical, Agricultural, and Economical Resources of the Colony.* London: Edward Stanford.

Kaloumaira, Atunaisa. 2002. "Reducing the Impacts of Environmental Emergencies Through Early Warning and Preparedness: The Case of El Niño Southern Oscillation (ENSO); The Fiji Case Study." In *SOPAC Technical Report 344.* Suva: SOPAC.

King, Trevor. 2004. "A Burning Question? Fire, Livelihoods and Sustainability in the Navosa Region of the Fiji Islands." PhD dissertation, Massey University, Palmerston North.

———. 2005. "The Sustainable Edge: Indigenous Livelihoods, Resilience and Created Fragility in Navosa". In *Development on the Edge: Proceedings of the Fourth Biennial Conference of the Aotearoa New Zealand International Development Studies Network* (*DevNet*), edited by K. Jackson, N. Lewis, S. Adams, and M. Morten, 167–169. Auckland: Centre for Development Studies, University of Auckland.

———. 2010. "Wauosi Community Adaptation to Climate Change Project: Concept Proposal." Unpublished.

———. 2012. "Fluctuation in *Colocasia* Cultivation and Landesque Capital in Navosa, Viti Levu, Fiji." In *Irrigated Taro* (Colocasia esculenta) *in the Indo-Pacific: Biological, Social and Historical Perspectives*, edited by M. Spriggs, D. Addison, and P. J. Matthews, 155–165. Osaka: National Museum of Ethnology.

Kull, Christian A. 2002. "Madagascar Aflame: Landscape Burning as Peasant Protest, Resistance, or a Resource Management Tool?" *Political Geography* 21 (7): 927–953.

———. 2004. *Isle of Fire: The Political Ecology of Landscape Burning in Madagascar.* Chicago: University of Chicago Press.

Kumar, Roselyn, and Patrick D. Nunn. 2003. "Inland and Coastal Lapita Settlement on Vitilevu Island, Fiji: New Data." *Domodomo* (scholarly journal of the Fiji Museum) 16 (1): 15–20.

Kumar, Roselyn, Patrick D. Nunn, Julie S. Field, and Antoine de Biran. 2006. "Human Responses to Climate Change Around AD 1300: A Case Study of the Sigatoka Valley, Viti Levu Island, Fiji." *Quaternary International* 151 (1): 133–143.

Latham, M. 1983. "Origin of the Talasiga Formation." In *The Eastern Islands of Fiji* (*Iles Fidji Orientales*): *A Study of the Natural Environment, Its Use and Man's Influence on Its Evolution*, edited by M. Latham and H. C. Brookfield, 129–142. Paris: ORSTOM.

Legislative Council (Fiji). 1878. "Ordinance No. VIII, 1878. For the Better Prevention of Accidents by Fire." In *Ordinances of the Colony of Fiji: 1875–1905.* Levuka: Colony of Fiji.

Long, Norman. 2001. *Development Sociology: Actor Perspectives.* London: Routledge.

Long, Norman, and Ann Long, eds. 1992. *Battlefields of Knowledge: The Interlocking of Theory and Practice in Social Research and Development.* London: Routledge.

Macnaught, T. J. 1974. "Chiefly Civil Servants? Ambiguity in District Administration and the Preservation of a Fijian Way of Life, 1896–1940." *Journal of Pacific History* 9: 3–20.

Maiava, Susan, and Trevor King. 2007. "Pacific Indigenous Development and Post-Intentional Realities." In *Exploring Post-Development: Theory and Practice, Problems and Perspectives,* edited by Aram Ziai, 83–98. London: Routledge.

Mataki, Melchior, Kanayathu C. Koshy, and Murari Lal. 2006. "Baseline Climatology of Viti Levu (Fiji) and Current Climatic Trends." *Pacific Science* 60 (1): 49–68.

McNeill, J. R. 2003. "Pacific Ecology and British Imperialism, 1770–1970." In *British Imperial Strategies in the Pacific, 1750–1900*, edited by Jane Samson, 349–364. Aldershot: Ashgate.

MFARD (Ministry of Fijian Affairs and Rural Development) and UNDHA (United Nations

Department of Humanitarian Affairs). 1994. "Assessment of Drought Problems in Fiji." *Water Resources Journal* 183: 94–105.

Nicole, Robert. 2011. *Disturbing History: Resistance in Early Colonial Fiji*. Honolulu: University of Hawai'i Press.

Nunn, P. D., R. R. Thaman, L. Duffy, S. Finikaso, N. Ram, and M. Swamy. 2001. "Age of a Charcoal Band in Fluvial Sediments, Keiyasi, Sigatoka Valley, Fiji: Possible Indicator of A Severe Drought Throughout the Southwest Pacific 4500–5000 Years Ago." *South Pacific Journal of Natural and Applied Sciences* 19 (1): 5–10.

Oakley, Peter. 1987. "State or Process, Means or End? The Concept of Participation in Rural Development." *RRDC* [Reading Rural Development Communications] *Bulletin* 21 (March): 3–9.

Parham, J. W. 1955. *The Grasses of Fiji*. Suva: Fiji Department of Agriculture.

Parry, John T. 1987. "The Sigatoka Valley: Pathway into Prehistory." Suva: Fiji Museum.

———. 1997. *The North Coast of Viti Levu, Ba to Ra: Air Photo Archaeology and Ethnohistory*. Suva: Fiji Museum.

PCCSP (Pacific Climate Change Program). 2014. "Fiji Islands." In *Climate Variability, Extremes and Change in the Western Tropical Pacific: New Science and Updated Country Reports 2014*, 93–112. Melbourne: Pacific Climate Change Program.

Penet, Laurent, Sébastien Guyader, Dalila Pétro, Michèle Salles, and François Bussière. 2014. "Direct Splash Dispersal Prevails Over Indirect and Subsequent Spread During Rains in *Colletotrichum gloeosporioides* Infecting Yams." *PLoS One* 9 (12) (article e115757): 1–15.

PWD (Public Works Department, Government of Fiji). N.d., ca. 2010. "PWD Western Division Rainfall Network Summary." Unpublished report. Public Works Department, Government of Fiji, Lautoka.

Ravuvu, Asesela. 1983. *Vaka i Taukei: The Fijian Way of Life*. Suva: Institute of Pacific Studies, University of the South Pacific.

Richards, P. 1985. *Indigenous Agricultural Revolution: Ecology and Food Production in West Africa*. London: Hutchinson.

Ripoche, A., G. Jacqua, F. Bussière, S. Guyader, and J. Sierra. 2008. "Survival of *Colletotrichum Gloeosporioides* (Causal Agent of Yam Anthracnose) on Yam Residues Decomposing in Soil." *Applied Soil Ecology* 38 (3): 270–278.

Rocheleau, Dianne. 1994. "Participatory Research and the Race to Save the Planet: Questions, Critique, and Lessons from the Field." *Agriculture and Human Values* 11 (2–3): 4–25.

Santoso, Agus, Shayne McGregor, Fei-Fei Jin, Wenju Cai, Matthew H. England, Soon-Il An, Michael J. McPhaden, and Eric Guilyardi. 2013. "Late Twentieth Century Emergence of the El Niño Propagation Asymmetry and Future Projections." *Nature* 504: 126–130.

Southern, Wendy. 1986. "The Late Quaternary Environmental History of Fiji." PhD dissertation, Australian National University.

Tanner, Adrian. 1996. "Colo Navosa: Local History and the Construction of Region in the Western Interior of Vitilevu, Fiji." *Oceania* 66 (3): 230–251.

Twyford, Ian T., and Wright, A. C. S. 1965. *The Soil Resources of the Fiji Islands*. Suva: Government of Fiji.

Ward, R. G. 1965. *Land Use and Population in Fiji*. London: Her Majesty's Stationery Office.

Were, Peni. 1997. "Fiji Pine Limited Fire Management Review 1997." Unpublished report, 1997. Fiji Pine Limited, Lautoka.

Wright, C. H. 1920. "Analyses of Two New Grasses." *Agricultural Journal* (Fiji) 1 (10–12): 183–184.

Yen, Douglas E. 1974. "Arboriculture in the Subsistence of Santa Cruz, Solomon Islands." *Economic Botany* 28 (3): 247–284.

CHAPTER 9

Assessing Causes and Effects of Survival Emissions from Global to Local Scales

Agropastoral Communities in the North Kodi Subdistrict of Sumba Island, Indonesia

CYNTHIA T. FOWLER

ASSESSING THE SIGNIFICANCE OF SURVIVAL EMISSIONS

Humans shape environmental change through specific actions that directly and indirectly affect their immediate surroundings. Human actions and activities are embedded in chains of events that sometimes lead to gas and particle emissions with the potential to change the composition and function of Earth's atmosphere. As concerns about the human causes of environmental changes—such as reductions in forest cover, expansion of nonforested biomes, and global warming—have intensified over the past twenty-five years, the question of who is to blame for environmental degradation has become problematic. The natural sciences can say with some certainty that anthropogenic emissions are increasing; between 1990 and 2005, for example, evidence demonstrates that methane (CH_4) and nitrous oxide (N_2O) emissions rose by 17 percent (Smith et al. 2007). The human sciences, however, provide less certain data about which specific human actions affect environmental change and to what extent (Vayda and Walters 2011).

The main sources of global warming emissions are survival emissions (Costello et al. 2009) from biomass combustion and other nonindustrial sources, industrial emissions from fossil fuel consumption, and naturally occurring greenhouse gases. Like natural processes, human activities produce carbon dioxide (CO_2), carbon monoxide (CO), CH_4, and N_2O—the four most significant greenhouse gases—and an array of other chemicals and particles. Survival emissions result from land use, agriculture, animal husbandry, biomass combustion, and other activities that humans pursue to meet their basic needs. Survival emissions arise from croplands, grasslands, forestlands, and home sites where people work to produce their own food and shelter, and where they access water. In world regions such as Africa, South America, and Asia—where many people are subsistence farmers, pastoralists, and fishers—survival emissions are notable sources of global warming. Through their land use activities alone, humans generate 20 percent of CO_2 and 60 percent of CH_4 globally (FAO 2016b). Agriculture of all types, including subsistence and industrial, produces somewhere between 14 percent and 22 percent of all human-generated global warming emissions (EPA 2017). Almost half of CH_4 emissions come from agriculture activities, including survival and industrial systems (EPA 2014). Worldwide, 20 percent of anthropogenic CH_4 (50 to 100 million tons per year) comes from rice paddies alone, making them the largest agricultural source of CH_4. While all human activities combined cause 40 percent of N_2O emissions globally (EPA 2014), agriculture is the main source of anthropogenic N_2O emissions.

These emissions figures and other numbers and graphics presented in this chapter are informative for assessing worldwide agricultural emissions from nonindustrial as well as industrial operations. If we wish to capture specifically which emissions are directly linked to smallholder survival activities, however, we would need to extract them from global or regional totals. Unfortunately, the goal of calculating the contribution made to climate change by survival emissions is frustrated by a lack of local-level studies and thus a paucity of data about emissions from people who live in rural communities and mostly produce products for their own subsistence. Yet assessing survival emissions among small-scale producers is now necessary because "agriculture" in general is often a central subject of climate change discussions, which, not infrequently, include finger-pointing at smallholders for causing environmental degradation and global warming.

The goal of this chapter is to better understand the production of survival emissions emanating from biomass fires. The ethnographic focus is on Indigenous fire regimes among smallholders in the North Kodi Subdistrict of Southwest Sumba Regency. I contextualize the North Kodi case within broader social groups and assess survival emissions at those social scales. I attempt to assess survival emissions by pairing information available from other sources about emissions at several larger scales with local ethnographic details about North Kodi's Indigenous fire regimes, which I collected. Subsequently, I use various lines of evidence to draw conclusions about survival emissions at the local level. This chapter is intended to be a commentary on scientific reasoning about anthropogenic emissions at multiple scales, as well as a study of how Indigenous fire regimes contribute to global change. To pursue these simultaneous agendas, this chapter contains information about anthropogenic emissions at subnational, national, subregional, regional, and global scales that situate Sumba and help characterize the fire-related actions of Kodi people relative to large-scale emissions from biomass combustion.

North Kodi smallholders practice several forms of agriculture, broadly defined, including domesticated crop cultivation in approximately 1 hectare gardens, husbandry of domesticated animals in pens and pastures, orcharding of economically valuable trees, gathering of semidomesticated and wild products from forests and savannas, and fishing in freshwater reservoirs and ocean waters. Kodi smallholders produce for household consumption and exchange in traditional ritual and barter economies as well as in the market economy.

Although no one word completely captures the diversified economy in North Kodi, I use the term *agropastoralists* in this chapter because it names two, rather than only one, of the main Kodi subsistence strategies. *Agropastoralists* is also an apt term because it identifies two forms of livelihood, or two types of activities, that lead to global warming gas and particle emissions. Among the main national agricultural sectors in Indonesia, emissions come mostly from the following activities, products, and outcomes related to agriculture: rice production (37.6 percent), soil cultivation (24.1 percent), synthetic fertilizers (10.5 percent), and crop residues (3.3

TABLE 9.1. Regional mean carbon emissions from 1997 to 2013 for savanna and grassland fires ranked from most to least carbon emissions

Region	CO grams	CO_2 grams	CH_4 grams
SHAF	828	2,215	255
NHAF	507	1,358	156
SHSA	151	404	46
AUST	137	367	42
SEAS	61	164	19
EQAS	27	73	8
TEAS	22	59	7
NHSA	26	70	8
CEAM	21	55	6
TENA	6	16	2
EURO	3	8	1
BOAS	2	6	1
MIDE	1	2	0
BONA	0	0	0

Source: GFED4 (Global Emissions Fire Database), "Global Fire Emissions Database," Version 4, updated 2014. http://www.globalfiredata.org/index.html.

Abbreviations (in order listed above):

SHAF Southern Hemisphere Africa
NHAF Northern Hemisphere Africa
SHSA Southern Hemisphere South America
AUST Australia
SEAS (Southeast Asia [mainland])
EQAS (Equatorial Southeast Asia [insular Southeast Asia, including Indonesia])
TEAS (Temperate Asia)
NHSA (Northern Hemisphere South America)
CEAM Central America
TENA (Temperate North America)
EURO Europe
BOAS Boreal Asia
MIDE Middle East
BONA Boreal North America

percent) (FAO 2016a). Emissions from animal husbandry in Indonesia are produced in the forms of enteric fermentation (12 percent), manure left on pastures (5.6 percent), manure management (3.9 percent), and manure applied to soils (2.3 percent) (FAO 2016a).

Although biomass burning is not listed in the above figures as a primary emission activity, biomass fire ignition is a human action that cross-cuts agricultural and pastoralist sectors and leads to the emission of global warming gases and particles (Jolly et al. 2015). Agriculturalists and agropastoralists use fire as a subsistence tool to meet a variety of goals. Pastoralists burn surface fuels to provide free browse for grazing herds and to accomplish various other objectives. Agriculturalists burn surface fuels to prepare fields for planting and for myriad additional reasons. Because biomass fires and burn scars are palpable on the ground and visible in satellite imagery, they are the subject of many discussions and publications related to socioecological change. For these reasons, I focus here on biomass fire as a source of survival emissions and anthropogenic climate forcing.

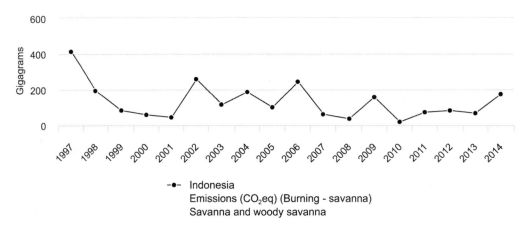

FIGURE 9.1. Emissions of methane and nitrous oxide (CO_2eq) from burning savannas and woody savannas in Indonesia, 1997–2014 (FAO 2016a).

As an indication of the contribution that biomass burning makes toward global warming, consider the following figures. Out of fourteen world regions, equatorial Asia emitted the eighth highest mean carbon emissions (Table 9.1) from agricultural waste fires and the sixth highest from savanna and grassland fires between 1997 and 2013 (GFED4 2014). Globally, Indonesia was the second top emitter of CO_2eq from biomass burning and removal in forests, grasslands, and croplands between 1997 and 2014 (GFED4 2014).[1] CO_2eq emissions from fires in Indonesia's savannas and woody savannas (Figure 9.1) were 173.15 gigagrams in 2014, which is a substantial decrease from 1997, when they were 411.79 gigagrams (FAO 2016a).

As a proportion of the world's regions, Asia contributed 42 percent of CO_2eq from fires in grasslands on organic soils in the period 1997–2014 (Figure 9.2), compared to Africa, which contributed 21.2 percent, followed by Oceania at 0.4 percent, the Americas at 19.5 percent, and Europe at 16.9 percent (FAO 2016b). Between 1997 and 2014, 43 percent of CH_4 and N_2O (CO_2eq) emissions from Indonesia came from burning of its savannas (Figure 9.3). This is relative to 38.2 percent from woody savannas, 9.5 percent from closed shrublands, 7 percent from grasslands, and 2.3 percent from open shrubland (FAO 2016b).

Emissions figures from international development and scientific organizations make evident that not all of humanity is equally to blame for global warming. Differences in anthropogenic emission quantities occur among distinct communities because of geographic, political, economic, and social factors. In terms of geographical differences, "Asians" are producing more CH_4 and N_2O (43.1 percent) for agriculture than people in any other region in the world.[2] By comparison, Americans produce 25.5 percent, Africans 14.9 percent, Europeans 12.4 percent, and Oceanians 4.1 percent (Figure 9.4).

Within Asia, Indonesia often tops the lists of producers of anthropogenic emissions. With 253 million people, Indonesia is the fourth most populous country on Earth, and the total number of agriculturalists is on the rise: from 1980 until 2014 the number rose from 78.6 million to 89.1 million people (FAO 2016a). As the number of agriculturalists is increasing, so too are agricultural emissions. Indonesia emitted 135,346 gigagrams of CO_2eq in 1997 and 165,614 gigagrams in 2014 (Figure 9.5). Worldwide, Indonesia was

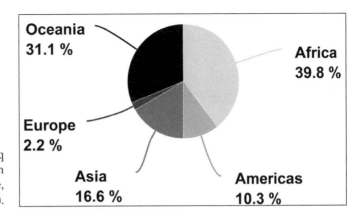

FIGURE 9.2. Emissions of CO_2eq from burning grasslands on organic soils worldwide, 1997–2014 (FAO 2016b).

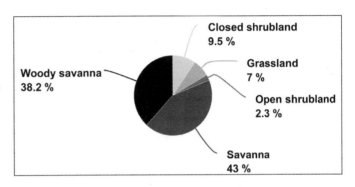

FIGURE 9.3. Percentages of emissions of methane and nitrous oxide (CO_2eq) in Indonesia by land cover, 1997–2014 average (FAO 2016b).

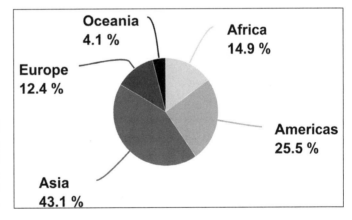

FIGURE 9.4. Emissions of methane and nitrous oxide (CO_2eq) from agricultural activities by region, 1997–2014 average (FAO 2016a).

the sixth largest contributor of CO_2eq from all agricultural activities, emitting an average of 145,692 gigagrams per year between 1997 and 2014 (FAO 2016a) (Figure 9.6).

Assessing the overall contribution of survival emissions of impoverished rural Indigenous agropastoralists in the Global South relative to anthropogenic emissions around the world raises questions about which sectors of global society are most responsible for or most vulnerable to climate change, as well as who should be targeted in projects aiming to reduce human contributions to global warming. What level of global warm-

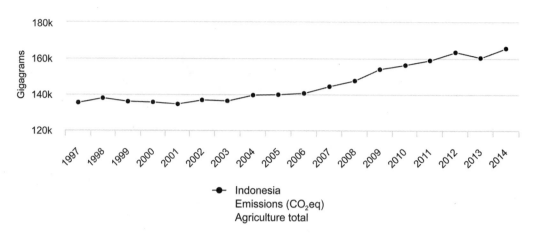

FIGURE 9.5. Emissions of CO_2eq from all agricultural activities in Indonesia, 1997–2014 (FAO 2016a).

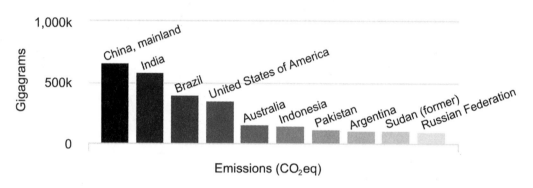

FIGURE 9.6. CO_2eq emissions from all agricultural activities ranked by country, 1997–2014 (FAO 2016a).

ing emissions do agropastoralists in North Kodi produce relative to other types of emissions produced by other categories of people? Do the biomass fires that Kodi people light in the course of performing their agropastoralist activities rival the emissions produced by petroleum-powered vehicles driven in the Global North, or the deep peat fires ignited by resource extractors in wetlands, or the industrial emissions generated in meeting the demands of high-volume consumers in megacities? Identifying specific social groups and determining their activities could help applied professionals decide which climate mitigation programs ought to be prioritized.

My goal here is not to identify who is factually to blame, nor do I recommend where climate-focused NGOs ought to focus their attention. Rather, I draw on multiple scales of emissions data to critique attempts to scapegoat rural Indigenous communities for shared global challenges. Such blame is unwarranted because insufficient empirical evidence is available to support claims that effectively transfer responsibility to people who depend on fire for their livelihoods. My critique in this chapter is specifically aimed at those who argue that Sumbanese burning practices cause environmental degradation because that argument frequently mischaracterizes the fire ecologies of semiarid tropical environments and overgeneralizes fire's effects. Anthropogenic fires do degrade nonpyrophytic ecological communities, but they do not degrade pyrophytic ecosystems, and Sumba's landscapes are diverse. This point launches my analysis of survival emissions on Sumba and my entry into the crucial debate about subsistence fires.

A Debate about Sumba's Anthropogenic Fire Regimes

At one end of the spectrum in the debate over anthropogenic fires on Sumba is the position that landscape burning is an essential practice for managing subsistence regimes, acquiring adequate nutrition, and exercising self-determination. Agropastoralists, agriculturalists, rural folk, forest managers, fire and other types of scientists, and this ethnographer advance this perspective. At the other end of the spectrum is the view of peasant fires as causing deforestation, air pollution, and climate change. Among the camps who often see fire in this way are governmental and nongovernmental organization staffers, the general public in Indonesia and beyond, and some types of scientists. In the case of Indigenous fire ecology on Sumba, some non-agropastoralist, non-Sumbanese people believe that Indigenous peoples' fires are destroying landscapes and causing global warming, whereas Sumbanese agropastoralists believe that fire is the right tool to use for achieving their subsistence goals—as well as a meaningful part of many aspects of their lives. The stakes in debates about subsistence fires are high for Sumbanese agropastoralists whose resource management systems are the subject of much consternation and, at the same time, whose systems are incompletely known to scientists and other stakeholders.

Agropastoralism in North Kodi

Most of Sumba's residents are agropastoralists who speak one or more of the languages or dialects in the Sumba-Hawu group of Central Eastern Malayo Polynesian languages. One of Sumba's ethnolinguistic groups is named Kodi, and its members reside in the Kodi region of the Southwest Sumba Regency. Most of the Kodi region is productive agropastoral space that—at varying temporal and spatial scales—cycles through the land cover types of forests, gardens, and fields. Fire is a key tool for Kodi agropastoralists as they direct their landscapes through these successional cycles.

The Kodi region is subdivided into the three subdistricts of North Kodi, Kodi Bangèdo, and Kodi Bawa. Most residents of the North Kodi Subdistrict are rural agropastoralists who grow food for subsistence and trade. They practice horticulture in croplands and settlements; agroforestry in forests; and husbandry, hunting, and gathering in savannas, shrublands, and forests. They cultivate rice, maize, cassava, peanuts, soybeans, long beans, squash, taro, potatoes, sweet potatoes, yams, greens, chilies, and a wide variety of additional plants. Kodi people foster many types of tree crops in their gardens, fields, and in agroforests, including bananas, coconuts, cashew, cocoa, candlenut, kapok, clove, areca, vanilla, jatropha, mangos, coffee, teak, mahogany, and several others. Kodi also raise pigs, water buffalo, horses, cows, goats, chickens, dogs, and cats.

Spatially, the effects of agropastoralists' activities on the landscape are not uniform since people engage in multiple forms of production in particular land cover types. Kodi's anthropogenic fire regimes vary temporally, spatially, and at different scales from the household to the hamlet, village, district, and the whole island. People vary their productive strategies throughout their lives in response to personal circumstances, social dynamics, financial needs, structural forces, and ecological conditions. Thus their involvement with land cover types varies throughout their life histories.

The annual precipitation cycle—which is shaped by a shorter three- to five-month rainy season (November or December through April or May) and a longer five- to nine-month dry season (May or June through October or November)—also contributes to the variable nature of fire regimes. Average rainfall during the rainy seasons from 1991 to 2001 was 1,449 millimeters (Walker et al. 2005). High rainfall years produce higher precipitation, as in 2002, when western Sumba received 2,367 millimeters of rainfall between January and April.

Population density is another source of variation in fire regimes. Population densities vary from the southern and inland parts of the sector, where water is more abundant, to the drier coastal and northern parts, where water is harder to access. Overall population density in the wetter western sector of Sumba is ninety-seven people per square kilometer (Lansing et al. 2007).

The North Kodi Subdistrict is a less-developed region where people mostly travel by foot to reach their villages and gardens, pastures, forests, freshwater creeks, reservoirs and rivers, schools, marketplaces, and health clinics. Other transportation is available in the form of taxi-vans and motorcycles. An increasing number of people have motorcycles and chainsaws, and a few households have generators. Generally, though, few North Kodi agropastoralists regularly use fossil fuel–burning machines. On the other hand, most North Kodi agropastoralists *do* regularly light biomass fires. Thus I would propose that their main output of emissions comes from biomass burning.

The biomass that agropastoralists in North Kodi burn includes the vegetation grown in their gardens, pastures, and orchards. The fuels in these spaces include crop residues, shrubs, grasses, herbs, and woody materials from trees. The fuels tend to be fast burning, fine fuels; for example, dormant-season grasses or crop residues desiccated from being left out to dry in the sun. Flaming fires in fast burning, fine fuels produce fewer emissions than smoldering fires in heavier fuels or fires with longer residence times (Hartford and Frandsen 1992).

Fire is a pervasive element in Kodi lives and landscapes, and has a prominent role in agropastoral practices. Manipulation of vegetative fuels with fire and other techniques occurs in the course of everyday activities through which people engage with productive spaces. Nearly every day, Kodi people light fires as part of their work to produce plant and animal products to eat and exchange. A typical household in North Kodi has diversified production operations on their roughly 1 hectare garden plots. These are places where Kodi burn weeds, crop residues, woody debris, and other biomass. Some households own their own half-hectare cashew orchards where they may occasionally burn herbaceous and grassy fuels in the understories and woody debris. Nearly all households have ownership in the communal agroforests, savannas, and forests they collectively manage using a variety of techniques including fire and fuel manipulations.

People largely determine the fire ecology of Kodi, but weather events (such as the monsoons, the El Niño Southern Oscillation, the Indian Ocean Dipole) and the semiarid, tropical monsoonal climate also strongly influence fire ecology. In the Köppen-Geiger Climate Classification, Sumba is in the Aw region, meaning it is tropical with dry winters. Southwest Sumba receives most of its rainfall in the summer months of January and February, with dramatically less rainfall during the winter months of September and October. In addition to pervasive anthropogenic fires, it is possible that lightning also ignites fires on Sumba, judging from the presence of lightning deities and myths about lightning in Kodi lore. Two overlapping anthropogenic fire seasons occur in Kodi. One is the *kamuma* (garden) fire regime, which consists of high-frequency, low-intensity, mixed-severity surface fires. The second is the *marada* (grassland, savanna) fire regime, which is lower frequency, higher intensity, and higher severity, and also has a patchy distribution.

Fires in the Sumbanese landscape are typically small. In a study of burned sites in East Sumba, Fisher et al. (2006) found that the median size of fires in 2003–2004 was less than 1 hectare, most were smaller than 5 hectares, and only a few sites had more than 100 hectares burned. These Sumbanese fires are much smaller than those in northern Australia, for example, where the median fire size in Kakadu National Park was 30 to 850 hectares between 1980 and 1994. Since smaller fires generally produce fewer emissions, we might hypothesize that Sumbanese agropastoralists' fires generate fewer emissions than those in northern Australia, but we cannot say for certain due to the absence of emissions data from Sumba.

Kodi Construction of the Fire Environment

North Kodi agropastoralists define the Indigenous fire ecology and construct the fire environment using a combination of mechanical treatments and prescribed burning. Kodi people shape the fire environment by manipulating the composition and arrangement of vegetation, which they do using a variety of tools that include ignition devices (hot coals, burning sticks, matches, lighters), their hands (for pulling weeds, hauling slashed vegetation, creating piles of slash, etc.), and digging and cutting tools, including knives, hoes, machetes, and chainsaws. Kodi agropastoralists sometimes group finer fuels together when they put weeds into piles and align grass into bundles. Other times they distribute coarser fuels into other formations by removing dead wood to use in cooking fires, trimming *komi* (*Schleichera oleosa* [Lour.] Oken) branches to collect lac beetle exudate, and leaving felled tree trunks lying on the ground. They add, remove, and rearrange fuels from the environment as part of the process of crop production, animal husbandry, savanna and grassland management, agroforestry, and forestry.

Kodi agropastoralists use both mechanical treatments and burning to manage fuels in savannas and grasslands, as well as in their gardens. Fuel composition is influenced by a multitude of human actions and activities. Three notable examples of fuel management in savannas and grasslands are harvesting resources, pasturing livestock who selectively graze on the vegetation, and lighting broadcast burns in the dry season. Examples of actions and activities that shape fire environments in gardens are burning off aboveground plant matter two or three times during crop rotation cycles prior to planting and weeding by hand. Mechanical treatments and burning affect fuel-fire cycles because they impact species composition in savanna, grassland, and garden habitats by providing favorable conditions for species that thrive under these conditions, and unfavorable conditions for other species. Numerous species thrive in response to the mechanical treatments and fire regimes that predominate in North Kodi's landscapes. The following taxa of grasses in North Kodi's grasslands, savannas, and gardens respond favorably to fire and construct flammable habitats:

- *Aristida* L.
- *Bothriochloa ischaemum* (L.) Keng
- *Brachiaria* sp. (Trin.) Griseb.
- *Cenchrus brownii* Roem. & Schult.
- *Centotheca lappacea* var. *longilamina* (Ohwi) Bor.

- *Chloris* spp.
- *Chrysopogon subtilis* (Steud.) Miq.
- *Cymbopogon microstachys* (Hook. f.) Soenarko
- *Cynodon dactylon* (L.) Pers.
- *Dactyloctenium aegyptium* (L.) Willd.
- *Digitaria ciliaris* (Retz.) Koel.
- *Digitaria radicosa* (J. Presl) Miq.
- *Eragrostis tenella* (L.) P. Beauv. ex Roem. & Schult.
- *Fimbristylis* spp. Vahl
- *Hackelochloa granularis* (L.) Kuntze
- *Heteropogon contortus* (L.) P. Beauv ex Roem. & Schult.
- *Heteropogon insignis* Thwaites
- *Imperata cylindrica* (L.) Beauv.
- *Ophiuros exaltatus* (L.) Kuntze
- *Paspalidium flavidum* (Retz.) A. Camus
- *Pennisetum macrostachys* (Brongn.) Trin.
- *Pennisetum purpureum* Schumach.
- *Pogonatherum crinitum* (Thundb.) Kunth
- *Rottboellia cochinchinensis* (Lour.) W. D. Clayton
- *Sorghum nitidum* (Vahl) Pers.
- *Sorghum propinquum* (Kunth) Hitchc.
- *Sorghum timorense* (Kunth) Büse
- *Themeda arguens* (L.) Hack.

Numerous shrubs are early colonizers of cleared spaces and thrive in North Kodi's fire environment because they are fire tolerant or fire resistant. The more common shrubs thriving in disturbed spaces are:

- *Eupatorium inulifolium* (Kunth) R. M. King & H. Rob.
- *Calotropis gigantea* (L.) W. T. Aiton
- *Chromolaena odorata* (L.) R. M. King & H. Rob.
- *Crotalaria striata* DC
- *Jatropha gossypifolia* L.
- *Lantana camara* L.
- *Mimosa* spp. L.

These shrubs occur in grasslands and savannas where they are subjected to burning, but their distribution is spatially variable in North Kodi. The shrubs also commonly occur as garden weeds, so they are subjected to both hand weeding and burning. Manipulations through weeding and burning function to encourage the growth and spread of invasive shrubs. These are merely a few among many species growing in North Kodi's fire environment whose fuel loads (i.e., the presence and abundance of vegetation) affect landscape flammability and fire behavior.

Human actions and activities in combination with fire frequencies, intensities, severities, and seasonalities (i.e., the fire regime) affect flora and fauna distributions. Feedback between fuels and fire is a mechanism driving fuel-fire cycles. The fuels and the ways they burn in any given fire influence the species composition and amount of combustible materials that will be available for future burns. The fire environment determines which plant species burn and the amounts of emissions they produce when burned.

The materials that burn during Kodi's kamuma fire season are garden weeds, crop residues, and slash. The fuels created from gardening, weeding, and harvesting are herbaceous, grassy, shrubby, and woody debris. The main crop residues that Kodi farmers burn come from the plants they grow in their gardens. The predominant cultivars that produce residues burned postharvest are maize, rice, cassava, mung bean, sorghum, and millet. The crops that cover the largest number of hectares in the Southwest Sumba Regency where North Kodi is located are rainfed maize and rainfed rice, flooded paddy rice, cassava, and mung beans (Figure 9.7). In Southwest Sumba in 2014, farmers dedicated 30,255 hectares to rainfed maize production; 22,913 hectares to rice (14,568 hectares of rainfed rice and 8,345 hectares of flooded paddy rice); 1,262 hectares to cassava; and

408 hectares to mung beans (Badan Pusat Statistik Kabupaten Sumba Barat Daya 2017). Maize and rice production results in a great bulk of fuels in the form of stalks discarded after cutting maize ears and reaping rice panicles. These stalks are sometimes burned while still standing, but gardeners frequently cut them down or pull them up by the roots and let them dry before burning them.

Scaling up from the agropastoralist community in North Kodi to the broader social contexts of Indonesia, Asia, and worldwide, we find that burning crop residues is one of the consequential activities through which humans contribute to global warming. In Indonesia, the CO_2eq emissions from burning crop residues increased from 762 gigagrams to 919 gigagrams between 1997 and 2014 (Figure 9.8). Between 1997 and 2014, the top producers in Indonesia of CH_4 and N_2O from burning crop residues were rice (63.4 percent or 530 gigagrams) and maize (34.1 percent or 285 gigagrams) (FAO 2016a).

Worldwide, more maize residues are burned than those from any other crop. Wheat, rice, and sugarcane are the other top crop residues that are burned. Indonesia is the sixth largest contributor in the world of CO_2eq from burning maize crop residues. Half (50.3 percent) of total global emissions of CH_4 and N_2O (CO_2eq) from burning crop residues in 1997–2014 came from Asia (Figure 9.9).

Ethnographic evidence tells us that through their small-scale agropastoral operations, North Kodi householders contribute to emissions from burning crop residues at the broader national, regional, and worldwide scales. But we do not know the quantity, proportion, or type of emissions from the relatively small fires in North Kodi. Agropastoralists manage fire and the fire environment for various purposes (Table 9.2). Among the beneficial outcomes of Kodi Indigenous fire ecologies is elimination of unwanted vegetation, but this may also have the unwanted effect of opening spaces where non-native and/or invasive species can grow and potentially become problematic. Major beneficial outcomes of North Kodi agropastoralists' management of fuels and fire through mechanical treatments and prescribed burning are land cover clearing for crop production and encouraging growth of forage, which enables livestock production.

For North Kodi's agropastoralists, having the autonomy to use fire as a tool in their ecological management system is vital for subsistence purposes and is also culturally meaningful in many ways (Fowler 2013). On the other hand, because of cultural politics related to anthropogenic fire at the national and international levels, the fact that North Kodi's residents burn their landscapes increases their vulnerability to further stigmatization as "backward," environmentally destructive swiddeners and increases their risk of dispossession of their freedom to manage and own their traditional territories. I believe that North Kodi agropastoralists ought to be respected and honored for the deeply historical, multigenerational place-based knowledge and practical skills that enable them to survive in their semiarid, tropical, monsoonal landscape.

Linking Levels of Change from the Global Level to the Local Level

We might speculate, based on a reading of broad-level emissions data alongside local-level anthropogenic fire information, that North Kodi agropastoralists enact environmental change locally and beyond. We might also speculate that their collective role is relatively small compared to some other social or economic groups in the world, judging from certain characteristics of their fire culture. Kodi produce emissions mostly through biomass burning; their fires typically burn fast in finer fuels; they manage the fire environment to protect against unintended fire and to prevent escaped burns; their landholdings are small; and their fires are usually contained within their land holdings. To summarize, the actions and activities of these small-scale agropastoralists contribute to global change, principally in the form of emissions from biomass burning, and the impact may be small relative to other groups of contributors.

What, then, are the impacts of global change on Kodi agropastoralists? We can consider two measurable arenas of impact: one is to the island's biophysical conditions and the other is to human health (Table 9.3). Predictions about the effects of climate change on biophysical conditions in Indonesia include:

- an increase in temperature of 2–2.5°C (Met Office 2011);

- a rise in sea levels by .6–1 meter with 10 percent increase in the intensity of one in one-hundred year storm surges (Met Office 2011);

- a decrease in precipitation over Indonesia (CIFOR 2006);

- an increase in precipitation of 0–10 percent (Met Office 2011);

- rising atmospheric CO_2 may produce stronger El Niños (Cobb 2013);

- more severe droughts associated with El Niños (D'Arrigo and Wilson 2008);

- delayed onset of monsoon season (Naylor et al. 2007);

- increased Asian monsoon intensity (D'Arrigo and Wilson 2008);

- decrease in number of cyclone occurrences but increase in severity (Met Office 2011);

- delayed harvests;

- decreased agricultural yields (Naylor et al. 2007);

- greater food insecurity (D'Arrigo and Wilson 2008).

These predictions are highly variable due to researchers' use of differing methodologies, metrics, models, emissions scenarios, and spatial foci. Nationwide, temperatures are predicted to increase by 2–2.5°C (Met Office 2011). Precipitation across Indonesia is expected to either decrease (CIFOR 2006) or increase by up to 10 percent (Met Office 2011). Rising atmospheric CO_2 may produce stronger El Niños (Cobb 2013). Indonesians can expect delays in the onset of the monsoon season (Naylor et al. 2007) as well as increases in the intensity of the monsoons (D'Arrigo and Wilson 2008). El Niño is predicted to lead to more severe droughts (D'Arrigo and Wilson 2008). Cyclones may decrease in number of occurrences but increase in severity (Met Office 2011). Sea levels may rise by .6 to 1 meter, with a 10 percent increase in the intensity of so-called hundred-year storm surges (expected to occur only once every hundred years) (Met Office 2011).

While the implications of national-level climatological changes for the island-level changes of Sumba and the district-level changes of Kodi are unknown, they are likely to influence local fire ecologies. Changes in atmospheric CO_2 could potentially influence the growth of the vegetation that fuels fires. The forecasted changes related to temperature, precipitation, drought, monsoons, and El Niños could also impact fuel loads and fuel moisture, as well as the weather conditions under which fires burn.

Predictions about the effects of climate change on health are not available for Sumba Island or its subdistricts, so we turn to the country-level predictions for Indonesia.

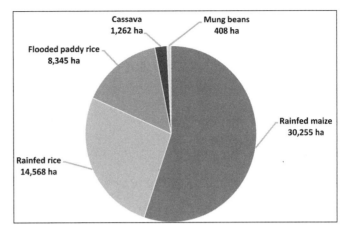

FIGURE 9.7. Land area (hectares) planted in 2014 with the five most widely planted crops in the Southwest Sumba Regency (Badan Pusat Statistik Kabupaten Sumba Barat Daya 2017).

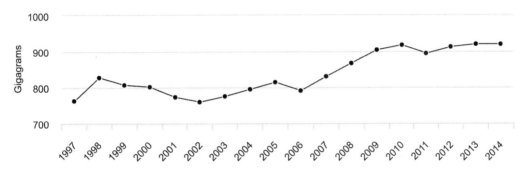

FIGURE 9.8. Emissions of methane and nitrous oxide (CO_2eq) from burning crop residues in Indonesia, 1997–2014 (FAO 2016a).

Climate change may affect the health of Indonesians via its effects on water and food. Greater food insecurities (D'Arrigo and Wilson 2008) are expected to stem from delays in grain harvests and decreased agricultural yields (Naylor et al. 2007). In one forecast, Nelson et al. (2010) predict a decrease in food security of 11 percent by 2050, an increase in the number of children who are malnourished by 2050, and an increase of 19 percent in the number of people of all ages who suffer from malnourishment. The Met Office (2011) provides quite a different outlook in which food security from agricultural products will remain stable for another forty years in Indonesia. Maize harvests in Indonesia are predicted to suffer from decreases in overall yields (Iglesias and Rosenzweig 2009), decreases in yields on irrigated and rainfed fields (Nelson et al. 2009), and/or increases in yields in rainfed gardens (Nelson et al. 2009). Rice production is predicted to fluctuate (Iglesias and Rosenzweig 2009) and/or improve in irrigated fields (Nelson et al. 2009). The predicted effects on marine resources are less ambiguous than predictions for maize and rice. Food from marine resources is expected to decline in Indonesia (Met Office 2011). Indonesia is expected to suffer from a decrease of 23 percent in marine fisheries between 2005 and 2055, which would be one of the largest reductions in marine fisheries in the world (Cheung et al. 2010).[3]

Based on these predictions for Indonesia, we might speculate about the potential effects of climate change on the health of Ko-

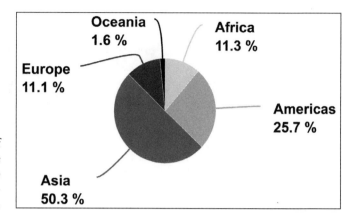

FIGURE 9.9. Emissions of methane and nitrous oxide (CO$_2$eq) from burning crop residues in world regions, 1997–2014 (FAO 2016a).

TABLE 9.2. Detrimental and beneficial effects of North Kodi's anthropogenic fire regimes

Detrimental Effects of Fire	Beneficial Effects of Fire
Opens growing spaces for non-native and/or invasive species.	Temporarily eliminates unwanted vegetation.
Produces emissions of gases and particles.	Clears gardens for crop production.
Accidental and escaped fires sometimes cause injury and damage property.	Promotes forage for livestock production.
Increases vulnerability in an already marginalized Indigenous community.	Is culturally meaningful.

di's agropastoralists. Future food yields are uncertain in the Kodi region. Maize and rice, grown in rainfed gardens, are staples in the North Kodi diet. Marine resources are valuable sources of nutrients and calories for North Kodi agropastoralists as well as for people living in the coastal towns of Bondo Kodi and Pero. Increased yields in any or all of these resources would benefit Kodi people. Decreased yields would further threaten food security in a place where adults as well as children already suffer from undernutrition, and where food and water security is already vulnerable.

Predictions about the effects of climate change on water resources are mixed. Although one prediction shows an increase in water scarcity for Indonesians (Béguin et al. 2011), another outlook foresees moderate to high levels of water security (Vörösmarty et al. 2010), and yet another forecast is for no changes in existing water stress (Met Office 2011). Each of the predicted scenarios has different implications for human health. The health of many of Sumba's residents is already stressed by the challenges of accessing the water they require for drinking, cooking, and bathing, and may be further compromised if water scarcity increases or water security decreases. If climate changes do not affect the availability of potable water, then the current challenges to health may remain.

Over the coming decades in southern Indonesia, as temperatures increase by .1 to .3°C per decade, precipitation will decrease by zero to 25 percent (CIFOR 2006). The southern portion of the country where Sumba is located is expected to become hotter and drier. These changes in climate have the potential to change the environment and impact the availability of food and water, which would have consequences for human health.

TABLE 9.3. Implications of climate change for human health in Indonesia

Water	Food	Maize	Rice	Marine Resources
Increase in water scarcity (Béguin et al. 2011)	Increase in food insecurity (D'Arrigo and Wilson 2008)	A 3.83 percent decrease in maize yields until 2020; 7.52 percent decrease after 2050; 10.60 percent decrease after 2080 (Iglesias and Rosenzweig 2009)	A 1.02 percent decrease in rice yields to 2020; a 1.53 percent increase in rice yields after 2050; and .50 percent decrease after 2080 (Iglesias and Rosenzweig 2009)	Decrease of 23 percent in marine fisheries, one of the largest reductions in marine fisheries in the world (Cheung et al. 2010)
No changes in the number of people who suffer from water stress (Met Office 2011)	Food security from agricultural products remains stable for another forty years (Met Office 2011)	CO_2 enrichment will decrease yields of irrigated maize by 2050 (Nelson et al. 2009)	CO_2 fertilization will improve yields of rainfed and irrigated rice, soybeans, and ground nuts by 2050 (Nelson et al. 2009)	A 23 percent decrease in ten-year averaged maximum marine catch between 2005 and 2055 (Cheung et al. 2010)
Moderate to high levels of water security throughout Indonesia relative to other countries (Vörösmarty et al. 2010)	Decrease of 11 percent by 2050 in kilocalories available from food (Nelson et al. 2010)	CO_2 enrichment may decrease or increase yields of rainfed maize by 2050 (Nelson et al. 2009)		Decrease of food from marine resources (Met Office 2011)
	Increase of 19 percent in malnourishment among people of all ages with 300,000 children malnourished by 2050 (Nelson et al. 2010)			

While the changing climate will probably impact the lives of Sumbanese agropastoralists, the practice of burning biomass causes greenhouse gas emissions that contribute to the Earth's changing climate.

Sumbanese agropastoralists are not, of course, the only humans whose actions are causing climate change. The question of to what degree the Sumbanese are causing climate change compared to other segments of society remains unresolved because we have no direct measurements of survival emissions on Sumba. We might surmise that since Sumbanese belong to the larger social group of Indonesians and the global community of agropastoralists, they are responsible for some degree of climate forcing through the practice of biomass burning. Extrapolating data about emissions from Indonesia, Asia, or the world to comment on emissions from Sumba are risky exercises since Sumba has a complex set of social and environmental conditions that are not consistently distributed across space or time.

Conclusions

In this chapter, I have attempted to bring together multilevel emissions data and ethnographic information. Yet using data about emissions at larger levels to gain insight into local-level survival emissions is a problematic methodology because of the "wickedness" (Churchman 1967; Xiang 2013; see also Gollin and Trauernicht, chapter 7) of climate change science. The preceding figures for Asia and Indonesia contextualize the North Kodi data. Ethnographic evidence suggests that agropastoralists in North Kodi participate in a larger community of rural folk who generate emissions from biomass combustion in the course of performing their work. But we remain unable to comment on the significance of anthropogenic fires in North Kodi for global warming. Anthropogenic biomass fires in the world's rural communities are frequently associated with food production, and the emissions coming from them are unavoidable consequences of sustenance. However, survival emissions are difficult to measure because they are dispersed, frequent, and small. They are also informal, lit as needed, and not publicized. But as efforts to mitigate climate change aim to reduce emissions from human sources, it will be necessary to know more about the contribution of survival emissions to climate change.

This chapter approached a more complete understanding of the contributions of Indigenous fire ecologies to global warming. To accomplish this, I used the material from scientific publications about emissions data at global, regional, and national levels, combined with ethnographic information that I collected about Sumbanese burning practices. These data have enabled an assessment of the causes and effects of survival emissions in the North Kodi subistrict of Sumba. In the wide-ranging conversations about global change that have been occurring in the late twentieth and early twenty-first centuries, anthropogenic fire is one of the issues that frequently appears in debates about the causes and effects of socioecological change.

In the Sumbanese iteration of the debate over Indigenous fire regimes, controversies over rural peoples' fires are especially problematic because the survival of individual persons and ethnolinguistic collectives are at stake, as are their meaningful landscapes. The integrity of life on Earth is at risk.

In the high-stakes debates over anthropogenic climate change, multiple epistemologies and sets of knowledge about anthropogenic environmental change exist. Pluralistic knowledges about the causes of socioecological change on Sumba circulate in a context of national, regional, and global inequalities in the capacities of communities to cause global warming, their unequal exposure to effects of global change, as well as inequalities in peoples' abilities to adapt to change. Resolving differential forces within this global context has not been the objective of this chapter, which has not resolved *any* of the so-called wicked problems associated with disjunct understandings of socioecological change. Instead, I have attempted to focus attention on a few qualities of the problem of socioecological change by using the term *wicked* as a heuristic tool for provoking discussion about the sciences and technologies of global change. The specific wicked qualities of interest in this chapter were the divergent perceptions of anthropogenic fires and environmental change on Sumba and the difficulty of using broad-level data to understand local-level processes.

While a goal of social scientists who study environmental change is to document the human causes of socioecological change and the effects of human actions and activities on the environment, a goal of governmental and nongovernmental organizations that work on climate change is to mitigate anthropogenic contributions to global warming. Ideally, this chapter would have addressed both goals, and also determined who contributes what emissions.

Suggesting science-based policy, however, is very difficult if not impossible due to the current state of science. A lesson that comes from the preceding attempt to assess the state of the science about survival emissions is that we need more local-level studies of biomass fires in diverse environments. As debates continue about the effects of biomass burning and other human actions and activities on environmental change, scientists should continue to seek new data about human actions and airborne emissions associated with anthropogenic fires.

Notes

1. CO_2 equivalent (CO_2e, CO_2eq) measures the potential of greenhouse gases to contribute to global warming. The Intergovernmental Panel on Climate Change (IPCC) calculates CO_2 equivalent by multiplying mass by global warming potential. The number resulting from the calculation is the amount of global warming the gas will potentially cause. Gases vary in their effects on global warming because of differences in their persistence in the atmosphere and their abilities to reflect heat (EPA 2014).

    The reason for choosing 1997 as the historical beginning of this measurement is because 1997 is when I began doing ethnographic fieldwork on Sumba, which I continued doing in 2014. GFED4 provides data through 2014.

2. The terms Asians, Africans, Americans, Europeans, and Oceanians imply homogeneity among all peoples and institutions who live across vast geographic regions. I only use these terms here because they are the categories in the climate change literature. In my thinking, the erasure of diversity effected by use of these terms reinforces my argument for the need to look at local-level human-environment phenomena in climate science.

3. Cheung et al. (2010) arrived at 23 percent using emissions scenario A1B. The sum loss in ten-year averaged maximum marine catch is 7 percent when the authors factored in the possibility of climate change mitigation.

References Cited

Badan Pusat Statistik Kabupaten Sumba Barat Daya. 2017. "Perkembangan Luas Panen Tanaman Pangan di Kabupaten Sumba Barat Daya, (ha), 2013–2015." Accessed March 8, 2017. https://sumbabaratdayakab.bps.go.id/linkTabelStatis/view/id/79.

Béguin, A., S. Hales, J. Rocklöv, C. Åstöm, V. R. Louis, and R. Sauerborn. 2011. "The Opposing Effects of Climate Change and Socio-Economic Development on the Global Distribution of Malaria." *Global Environmental Change* 21: 1209–1214.

Cheung, William W. L., Vicky W. Y. Lam, Jorge L. Sarmiento, Kelly Kearney, Reg Watson, Dirk Zeller, and Daniel Pauly. 2010. "Large-Scale Redistribution of Maximum Fisheries Catch Potential in the Global Ocean Under Climate Change." *Global Change Biology* 16: 24–35.

Churchman, C. West. 1967. "Wicked Problems." *Management Science* 14 (4): B141–B142.

CIFOR (Center for International Forestry Research). 2006. *Climate Change Projections for Indonesia*. Bogor: CIFOR.

Cobb, Kim. 2013. "Highly Variable El Niño Southern Oscillation Throughout the Holocene." *Science* 339 (6115): 67–70.

Costello, Anthony, Mustafa Abbas, Adriana Allen, Sarah Ball, Sarah Bell, Richard Bellamy, Sharon Friel, Nora Groce, Anne Johnson, Maria Kett, Maria Lee, Caren Levy, Mark Maslin, David McCoy, Bill McGuire, Hugh Montgomery, David Napier, Christina Pagel,- Jinesh Patel, Jose Antonio Puppim de Oliveira, Nanneke Redclift, Hannah Ree, Daniel Rogger, Joanne Scott, Judith Stephenson, John Twigg, Jonathan Wolff, and Craig Patterson. 2009. "Managing the Health Effects of Climate Change." *The Lancet* 373: 1693–1733.

D'Arrigo, Roseanne, and Rob Wilson. 2008. "El Niño and Indian Ocean Influences on Indonesian Drought: Implications for Forecasting Rainfall and Crop Productivity." *International Journal of Climatology* 28 (5): 611–616.

EPA (Environmental Protection Agency). 2014. "Overview of Greenhouse Gases." Accessed March 11, 2017. http://www.epa.gov/climatechange/ghgemissions/gases/ch4.html.

———. 2017. "Greenhouse Gas Emissions by Economic Sector." Accessed March 13, 2017. https://www.epa.gov/ghgemissions/global-greenhouse-gas-emissions-data.

FAO (Food and Agriculture Organization of the United Nations). FAOSTAT. 2016a. "Emissions—Agriculture." Accessed May 22, 2017. http://faostat3.fao.org/browse/G1/*/E.

———. 2016b. "Emissions—Land Use." Accessed May 22, 2017. http://faostat3.fao.org/browse/G1/*/E.

Fisher, Rohan, E. Wilfrida, E. Bobanuba, Agus Rawambaku, Greg J.E. Hill, and Jeremy Russell-Smith. 2006. "Remote Sensing of Fire Regimes in semiarid Nusa Tenggara Timur, Eastern Indonesia: Current Patterns, Future Prospects." *International Journal of Wildland Fire* 15: 307–317.

Fowler, Cynthia. 2013. *Ignition Stories: Indigenous Fire Ecologies in the Indo-Australian Monsoon Zone*. Durham: Carolina Academic Press.

GFED4 (Global Emissions Fire Database). 2014. "Global Fire Emissions Database." Version 4. Updated. http://www.globalfiredata.org/index.html.

Hartford, Roberta A., and William H. Frandsen. 1992. "When It's Hot, It's Hot. Or Maybe It's Not! (Surface Flaming May Not Portend Extensive Soil Heating)." *International Journal of Wildland Fire* 2 (3): 139–144.

Iglesias, A., and C. Rosenzweig. 2009. *Effects of Climate Change on Global Food Production under Special Report on Emissions Scenarios (SRES) Emissions and Socio-Economic Scenarios: Data from a Crop Modeling Study*. New York: NASA Socio-Economic Data and Applications Center.

Jolly, W. Matt, Mark A. Cochrane, Patrick H. Freeborn, Zachary A. Holden, Timothy J. Brown, Grant J. Williamson, and David M. J. S. Bowman. 2015. "Climate-Induced Variations in Global Wildfire Danger from 1979 to 2013." *Nature Communications* 6 (7537): 1–10.

Lansing, J. Stephen, Murray P. Cox, Sean S. Downey, Brandon M. Gabler, Brian Hallmark, Tatiana M. Karafet, Peter Norquest, John W. Schoenfelder, Herawati Sudoyo, Joseph C. Watkins, and Michael F. Hammer. 2007. "Coevolution of Languages and Genes on the Island of Sumba, Eastern Indonesia." *Proceedings of the National Academy of Science* 104 (41): 16022–16026.

Met Office. 2011. *Climate: Observations, Projections, and Impacts, Indonesia*. Devon, UK: Met Office. Accessed March 11, 2017. http://www.metoffice.gov.uk/media/pdf/8/f/Indonesia.pdf.

Naylor, R. L., D. S. Battisti, D. J. Vimont, W. P. Falcon, M. B. Burke. 2007. "Assessing Risks of Climate Variability and Climate Change for Indonesian Rice Agriculture." *Proceedings of the National Academy of Sciences of the United States of America* 104: 7752–7757.

Nelson, Gerald C., Mark W. Rosegrant, J. Koo, Richard D. Robertson, Timothy S. Sulser, Tingju Zhu, Claudia Ringler, Siwa Msangi, Amanda Palazzo, M. Batka, M. Magalhaes, R. Valmonte-Santos, M. Ewing, and D. Lee. 2009. *Climate Change Impact on Agriculture and Costs of Adaptation*. Washington, DC: International Food Policy Research Institute. Accessed March 11, 2017. http://www.fao.org/fileadmin/user_upload/rome2007/docs/Impact_on_Agriculture_and_Costs_of_Adaptation.pdf.

Nelson, Gerald C., Mark W. Rosegrant, Amanda Palazzo, Ian Gray, Christina Ingersoll, Richard D. Robertson, Simla Tokgoz, Tingju Zhu, Timothy B. Sulser, Claudia Ringler, Siwa Msangi, and Liangzhi You. 2010. *Food Security, Farming, and Climate Change to 2050: Scenarios, Results, Policy Options*. Washington, DC: International Food Policy Research Institute. Accessed March 11, 2017. http://www.ifpri.org

/publication/food-security-farming-and-climate-change-2050.

Smith, P., D. Martino, Z. Cai, D. Gwary, H. Janzen, P. Kumar, B. McCarl, S. Ogle, F. O'Mara, C. Rice, B. Scholes, and O. Sirotenko. 2007. "Agriculture." In *Climate Change 2007: Mitigation. Contribution of Working Group III to the Fourth Assessment Report of the Intergovernmental Panel on Climate Change*, edited by B. Metz, O. R. Davidson, P. R. Bosch, R. Dave, and L. A. Meyer. Cambridge: Cambridge University Press. Accessed March 11, 2017. http://www.ipcc.ch/publications_and_data/ar4/wg3/en/ch8.html.

Vayda, Andrew P., and Brad B. Walters. 2011. "Introduction: Pragmatic Methods and Causal-History Explanations." In *Causal Explanation for Social Scientists: A Reader*, edited by Andrew P. Vayda and Brad B. Walters, 1–21. Lanham, MD: AltaMira.

Vörösmarty, Charles J., P. B. McIntyre, Mark O. Gessner, David Dudgeon, A. Prusevich, P. Green, S. Glidden, Stuart E. Bunn, Carolina A. Sullivan, C. Reidy Liermann, and P. M. Davies. 2010. "Global Threats to Human Water Security and River Biodiversity." *Nature* 467 (7315): 555–561.

Walker, Jonathan S., Alexis J. Cahill, and Stuart J. Marsden. 2005. "Factors Influencing Nest-Site Occupancy and Low Reproductive Output in the Critically Endangered Yellow-Crested Cockatoo *Cacatua sulphurea* on Sumba, Indonesia." *Bird Life International* 15 (4): 347–359.

Xiang, Wei-Ning. 2013. "Working with Wicked Problems in Socio-Ecological Systems: Awareness, Acceptance, and Adaptation." *Landscape and Urban Planning* 110: 1–4.

Contributors

ANA CAROLINA SENA BARRADAS
Estação Ecológica Serra Geral do Tocantins
Chico Mendes Institute for Biodiversity
Conservation, Brazil

SILVIA LAINE BORGES
Departamento de Ecologia
Universidade de Brasília

RAMONA J. BUTZ
USDA Forest Service, Pacific Southwest Region

LUDIVINE ELOY
French National Center for Scientific Research
Paul Valéry University, Montpellier III, France

CYNTHIA T. FOWLER
Department of Sociology and Anthropology
Wofford College, Spartanburg, South Carolina

LISA GOLLIN
Department of Anthropology
University of Hawaiʻi at Mānoa

TREVOR KING
Crawford School of Public Policy
Australian National University

JOYCE K. LECOMPTE
Program on the Environment
University of Washington, Seattle

ISABEL B. SCHMIDT
Departamento de Ecologia
Universidade de Brasília

JEREMY RUSSELL-SMITH
Darwin Centre for Bushfire Research
Charles Darwin University, Australia

ANGELA MAY STEWARD
Instituto Amazônico de Agriculturas Familiares
Universidade Federal do Pará, Brazil

CLAY TRAUERNICHT
Department of Natural Resources and
Environmental Management
University of Hawaiʻi at Mānoa

JAMES R. WELCH
Escola Nacional de Saúde Pública
Fundação Oswaldo Cruz, Brazil

Index

Africa. *See* savannas (Africa)
Agricultural Ecosystems Management Program, 111–12
agrobiodiversity: is high in Central Amazonia fields, 112; peat swamp fields as repositories for, 92; smallholders largely responsible for, 121
Amalfi Coast, 13
Amanã Sustainable Development Reserve, 105, 106, 107, 111
Amazon Fund, 109
Australia: initial occupation of, 25; issue of fire's impact on landscape and fauna in, 25–26; recognition and support for Aboriginal fire management in, 28–29, 43; replacement of fine-scale mosaic to boom-and-bust fire patterns cause loss of small mammal fauna in, 28; traditional patch-mosaic burning in, 27

big huckleberries, 13, *129*; effect of fire on, 8, 130–31, 134–35, 145; fire suppression as greatest threat to, 140–41; uncertainties of effect of fire on, 143–44
Big Huckleberry Summit, 128–29, 130, 134
Bracken fern, 71, *71*,
Brazil: deforestation responsible for 70% of carbon emissions of, 104; ribeirinhos in, 106. *See also* Central Amazonia; Jalapão, Brazil; peat swamp forests; savannas (South America); swidden agriculture (South America)
Burrows, Neil D., et al., 27

carbon emissions: contribution of deforestation to, 104; in world by regions, 215–17
Central Amazonia: agrobiodiversity in, 112; agroforestry initiatives in, 107–9; field preparation in, 113–14; fire used because it is considered practical, 120; fire use in cultivation of manioc in, 115–19; forest conservation allowance program in, 109–10; reasons fire is used in, 118–20; sustainable development reserves in, 105, 106, 107, 111; swidden agriculture system in, 112–14
Challenges to Rapid Wildfire Containment in Hawai'i project, 157
Coast Salish: can claim that inability to use fire is threat to their sovereignty, 145–46; as co-managers of resources to which they have rights, 132; power relations with Forest Service less unequal than the past, 135; use of fire by, 38–39, 131
Coivara (reburning), 115–16; *116*
Crow, 13

ecological knowledge, traditional; 14, 16; integration into fire management of, 156; summary of fire-related, 2–3; supposed opposition to scientific knowledge is fallacious, 23–24; suppression of fire leads to loss of, 75. *See also* fire management techniques, traditional
El Niño Southern Oscillation (ENSO): causes drought in Fiji, 14, 183, 186, 200–201; effect on Northwest Coast of, 36
Engikareti, Tanzania, 65–66, *66*, 72, 73
environmental change: anthropogenic fire and, 7; changes in fire regimes and, 7, 10–11, 13; difficulty in determining which specific human actions cause, 213–14; effects in Indonesia of, 224–27; in Hawai'i, 168–71, 172, 175; survival emissions and, 213–14, 216–19, 223–24
Escobar, Arturo, 4

Fiji. *See* Navosa, Fiji
fire cues (Hawai'i): animals, 164–65; relative humidity, 162; smoke, 162–63; sound, 164; terrain topography, 163–65, 177n6; wind, 160–62, *161*
firefighters (Hawai'i), 14, 16; 156–57; environmental changes seen by, 169–71, 172; environmental cues used by, 159–65; place-based knowledge of, 14, 158–68, 174–75; use of native Hawaiian ecological knowledge by, 15, 174. *See also* fire cues (Hawai'i)
fire(s): Fire Regime/Condition Classes (FRCC), 8; -grass cycles, 11; and humans are ecosystem engineers, 1; importance of reconstructing histories of, 10; as "reaction," 1. *See also* fire cues (Hawai'i); firefighters (Hawai'i); fires, anthropogenic; fire management techniques, traditional; fire regimes
fire(s), anthropogenic: biomass burning, 7; common public view as factor of degradation, 30; distinctive regimes of did not develop until relatively late, 40; environmental orthodoxies about not always adequately informed, 42; ethnic shaming from misunderstanding of, 43; have shaped the course of environmental history, 10;

interruption of, 41; management with, 41; negative characterization of, 23, 25; negative effects of exclusion/suppression of, 11, 13; positive and negative impacts of, 8–10; possible significance of small-scale, 145; proportion of relative to lightning-caused, 11; resistance, 17; summaries of knowledge of and management techniques for, 2–3. *See also* fire management techniques, traditional; fire regimes; under specific regions and cultures

fire management techniques, traditional: clearing of firebreaks, 2, 34–35, 67, 69, 70, 90, 95, 196; control of extent, 2, 120, 185; fuel management, 3, 114–15, 221; modulation of frequency, 2, 115, 117, 195; need to integrate into modern fire management, 15–19, 156; organization of participants, 3, 115; protection of areas not appropriate to burn, 2; protection of burn participants, 3, 16; reburning of unconsumed wood, 115; selection of burn season, 3; transformations of, 13–15; use of backfires, 3, 163; use of fire lines, 3; use of weather, 3, 15, 160–62. *See also* ecological knowledge, traditional

Fire Otherwise paradigm, 4–5, 17, 18, 19

fire regimes: can alter vegetation structure and productivity, 63–64; causes of variation in, 219–20; create novel ecosystems, 12; difficulty of distinguishing human and nonhuman in Africa, 29; environmental changes and changes in, 7, 10–11, 13

Firewise campaign, 4–5

Forest Conservation Allowance Program, 109–10, 121–22

Franklin, Jerry, 138

Gammage, Bill, 26–27
Garde, Murray, et al., 28
Garry oak, 38, 39
global warming. *See* environmental change
Gobin, Jason, 143, 145
Gordon, Arthur, 197–98, 200
Gran Sabana. *See* savannas (South America)
grasses (Hawai'i): buffel, 166; cane, 166; Fountain, 166, 169, *170*; Guinea, 166, 171; invasion of can create increased landscape flammability, 11–12; Kikuyu, 167, 175; molasses, 166–67; pili, 167–68, 169, 170, 178n18
guava, 194, 207n17

Harrison, Faye, 4
Hawai'i: burned forests rarely recover in, 165; changing fire regimes of, 12; difficulties of fighting fires in, 154; environmental changes observed by firefighters in, 169–71, 172; fuel types in, 166–69; grasses and their fire characteristics, 166–68, 169, 170, 171, 175, 178n18; human caused fires predominate in, 151; increase in area burned related to decline in agricultural production, 153–54; issues of fighting grass fires in, 165; lightning caused fires increasing in, 171; loss of agricultural infrastructure makes fighting fires more difficult, 171; negative impacts of anthropogenic fire in, 9; protecting cultural and natural resources from fire in, 172; role of feral ungulates in fire in, 173; trees and their fire characteristics in, 167, 178n16; wildfires occur year-round in, 154; wildland fire infrequent prior to human settlement, 152. *See also* fire cues (Hawai'i); grasses (Hawai'i); firefighters (Hawai'i)

Holmes, R. L., 208n23
Horne, John, 197

Indonesia: anthropogenic emissions of, 216–17, 218; predicted effects of climate change on, 224–27; sources of carbon emissions in, 214–15, 223. *See also* Kodi, North, Indonesia

Jalapão, Brazil, 86, *86*, 87–89
Jones, Rhys, 26

Keetch-Byram Dought Index, 177n7
KingGeorge, Warren, 132, 134–35, 143
Klein, Gary, 175
Knox, Margaret, 144
Kodi, North, Indonesia: as agropastoralists, 214; burning of crop residues in, 220; causes of variation in fire regimes of, 219–20; differing views of anthropogenic fires in, 219; fire as pervasive element in, 220, 221; positive and negative aspects of anthropogenic fire in, 10, 226; small fire size in, 221, 224; taxa of grasses and shrubs that respond positively to fire in, 221–22
Kruger National Park, South Africa, 31–32

Leavell, Daniel M., 5
Lima, Deborah, et al., 121

Maasai, 65, 66; dispossession of, 75; elaborate system of traditional fire management of, 67; limitations of fire use cause hardship for, 74; reasons for using fire of, 67, 73
maize, 223, 225
Malawi. *See* Nyika National Park
Mamirauá Institute for Sustainable Development. 107
Mamirauá Sustainable Development Reserve, 105, 106, *106*, 111
manioc, 112; firing of fields for, 115–16, 117–18; is possible to produce without fire, 118–19; preparation of fields for, 113–14, 116–17; social importance of, 121
matsutake mushrooms, 130
Miller, R., 92, 94
Miscanthus flocidulus, 186
Moses, Russell, 8, 134, 142

Index

Mount Baker-Snoqualmie National Forest: effect of Northwest Forest Plan on, 136–39; fire planning complex and costly in, 143; Fire Regime/Condition Classes of, 8, 10; generally positive view of fire in, 134; greater public engagement of, 139–40; huckleberry enhancement not high priority in, 142; positive and negative impacts of anthropogenic fires in, 9. *See also* big huckleberry
Muckleshoot Indian Tribe, 128–29, 132

Nabhan, Gary, 19
National Environmental Policy Act (NEPA), 142–43
National Fire Protection Association (NFPA), 4
Navosa, Fiji: agroarborculture in, 184–85; drought, burning, and land degradation in, 14, 186–87, 189, 195, 200–201, 208n31; environment of, 183–84; fire regulations do not include local perspective on fire, 198–99; flash flooding in, 201; frequency of uncontrolled burning in, 196; gender differences in explanation of reasons for burning, 194; introduction of invasive species to, 185–86; livelihood strategies in, 184–85; need for revised fire regulations in, 206; political tensions as cause of fire in, 195; positive and negative impacts of anthropogenic fires in, 9; prehistoric evidence of fire in, 184; responsibility for careless burning in, 203–4; soil erosion in, 187, 188, 195, 197, 201, 205; three primary reasons for burning, 193; timing of burning in, 195; top-down fire regulations inappropriate for, 199–200; unspoken pyrosocial ethos in, 183, 189–90, 204–5; use of fire to stimulate new grass in, 187–88, 193; views on effects of repeated burning in, 201
Nelson, Gerald C., et al., 225
Northwest Coast: "affluent forager" model of, 36, 39; beargrass savannas maintained through burning in, 39; fire history research in conducted under "wilderness" paradigm, 37; increase in Indigenous burning in late Holocene in, 36–37; plant cultivation with fire in, 38. *See also* big huckleberry; Mount Baker-Snoqualmie National Forest
Northwest Forest Plan, 136–38, 141
Nyika National Park, 67, 68; controlled-burn program in, 69; invasive species in, 70–71; traditional occupants of viewed as detrimental, 75
Nyika-Vwaza Trust (NVT), 69

Padoch, Christine, 123n7
Pascoe, Bruce, 27
peat swamp forests (veradas): cultivation in, 89–92; effects of fire in, 83–85; ditches cut to lower water levels in, 90, 91, 92; fields burned once and cultivated for 10-20 years in, 90, 95; fire use around is particularly sensitive issue to environmental managers, 85; as islands of stability, 84; positive and negative impacts of anthropogenic fires in, 9; reforestation after cultivation of, 94, 97. *See also* Serra Geral do Tocantins Ecological Station (SGTES)
Pemon, 33, 34–35, 43
Pennisetum polystachyon, 11, 185–86, 206n5
Phoka, 69, 75
Pinedo-Vasquez, Miguel, 123n7

quilombolas, 14, 15–16, 17

REDD (reduced emissions from deforestation and forest degradation), 104, 122
Restrepo, Eduardo, 4
Ribeiro Filhu, A. A., et al., 120
rice, 88, 97, 214, 222, 223, 227

savannas (Africa): efforts to maintain traditional fire management in, 30–31, 32; fire regimes can alter vegetation structure and productivity in, 63–64; human ignitions more frequent than lightning in, 30, 64; inherent dynamism of, 63; long-term persistence requires balance between tree mortality and tree/shrub encroachment, 72; origin and spread of, 29; positive and negative impacts of anthropogenic fires in, 9; prescribed burns in, 31–32; reduced fire frequency can lead to degradation of, 74; role of anthropogenic fires poorly documented in, 64; seasonal burning prevents destructive fires and increases plant diversity in, 30. *See also* Maasai; Niyika National Park
savannas (South America), 32; positive impacts of anthropogenic fires in, 9; positive value of controlled burning is overlooked because of negative view of fire in rainforests, 42; small steps made to consult with Indigenous people about value of fire in, 35; use of fire by Indigenous inhabitants in, 33–35. *See also* Central Amazonas; peat swamp forests (veradas)
Savo, Valentina, et al., 13
Scott, A. C., et al., 82
Serra Geral do Tocantins Ecological Station (SGTES), 12–13, 85–87, 89
Stewart, Omer C., 39
Sumba Island. *See* Kodi, North, Indonesia
swidden agriculture (South America): in Central Amazonia, 112–20; crop-fallow rotation cycle in, 95, 97; in drained peat swamp forests, 89–97; experiments with not using fire in, 118–19; fields as agrobiodiversity repositories, 92, 93–94; fire as long-term forest management in, 121; misunderstanding of, 42; fire used in, 90, 91, 94–97, 115–16, 117–18, 119–20; high species diversity in fallow fields of, 92; impact of fire differs between floodplain and upland, 120; is not a major cause of forest loss, 43, 122; precise and varied methods to apply fire in, 119–20; of quilombolas, 14, 15–16, 17. *See also* manioc; peat swamp forests

Index

Tanzania. *See* Engikareti, Tanzania
Tsing, Anna, 130, 131
Tulalip Tribes, 132, 134, 135, 142

uluhe fern, 168

Vale, Thomas, 144
Viti Levu. *See* Navosa, Fiji

Whitlock, Cathy, 144

Xavante, 34, 43

UPPER SAN JUAN LIBRARY DISTRICT
SISSON LIBRARY
BOX 849 (970) 264-2209
PAGOSA SPRINGS, CO 81147-0849